Research Notes in Mathem

Submission of proposals for consideration

Suggestions for publication, in the form of outlines and representative samples, are invited by the Editorial Board for assessment. Intending authors should approach one of the main editors or another member of the Editorial Board, citing the relevant AMS subject classifications. Alternatively, outlines may be sent directly to one of the publisher's offices. Refereeing is by members of the board and other mathematical authorities in the topic concerned, throughout the world.

Preparation of accepted manuscripts

On acceptance of a proposal, the publisher will supply full instructions for the preparation of manuscripts in a form suitable for direct photolithographic reproduction. Specially printed grid sheets are provided and a contribution is offered by the publisher towards the cost of typing. Word processor output, subject to the publisher's approval, is also acceptable.

Illustrations should be prepared by the authors, ready for direct reproduction without further improvement. The use of hand-drawn symbols should be avoided wherever possible, in order to maintain maximum clarity of the text.

The publisher will be pleased to give any guidance necessary during the preparation of a typescript, and will be happy to answer any queries.

Important note

In order to avoid later retyping, intending authors are strongly urged not to begin final preparation of a typescript before receiving the publisher's guidelines and special paper. In this way it is hoped to preserve the uniform appearance of the series.

Advanced Publishing Program
Pitman Publishing Inc
1020 Plain Street
Marshfield, MA 02050, USA
(tel (617) 837 1331)

Advanced Publishing Program
Pitman Publishing Limited
128 Long Acre
London WC2E 9AN, UK
(tel 01-379 7383)

Titles in this series

1 Improperly posed boundary value problems
A Carasso and A P Stone
2 Lie algebras generated by finite dimensional ideals
I N Stewart
3 Bifurcation problems in nonlinear elasticity
R W Dickey
4 Partial differential equations in the complex domain
D L Colton
5 Quasilinear hyperbolic systems and waves
A Jeffrey
6 Solution of boundary value problems by the method of integral operators
D L Colton
7 Taylor expansions and catastrophes
T Poston and I N Stewart
8 Function theoretic methods in differential equations
R P Gilbert and R J Weinacht
9 Differential topology with a view to applications
D R J Chillingworth
10 Characteristic classes of foliations
H V Pittie
11 Stochastic integration and generalized martingales
A U Kussmaul
12 Zeta-functions: An introduction to algebraic geometry
A D Thomas
13 Explicit *a priori* inequalities with applications to boundary value problems
V G Sigillito
14 Nonlinear diffusion
W E Fitzgibbon III and H F Walker
15 Unsolved problems concerning lattice points
J Hammer
16 Edge-colourings of graphs
S Fiorini and R J Wilson
17 Nonlinear analysis and mechanics: Heriot-Watt Symposium Volume I
R J Knops
18 Actions of fine abelian groups
C Kosniowski
19 Closed graph theorems and webbed spaces
M De Wilde
20 Singular perturbation techniques applied to integro-differential equations
H Grabmüller
21 Retarded functional differential equations: A global point of view
S E A Mohammed
22 Multiparameter spectral theory in Hilbert space
B D Sleeman
24 Mathematical modelling techniques
R Aris
25 Singular points of smooth mappings
C G Gibson
26 Nonlinear evolution equations solvable by the spectral transform
F Calogero
27 Nonlinear analysis and mechanics: Heriot-Watt Symposium Volume II
R J Knops
28 Constructive functional analysis
D S Bridges
29 Elongational flows: Aspects of the behaviour of model elasticoviscous fluids
C J S Petrie
30 Nonlinear analysis and mechanics: Heriot-Watt Symposium Volume III
R J Knops
31 Fractional calculus and integral transforms of generalized functions
A C McBride
32 Complex manifold techniques in theoretical physics
D E Lerner and P D Sommers
33 Hilbert's third problem: scissors congruence
C-H Sah
34 Graph theory and combinatorics
R J Wilson
35 The Tricomi equation with applications to the theory of plane transonic flow
A R Manwell
36 Abstract differential equations
S D Zaidman
37 Advances in twistor theory
L P Hughston and R S Ward
38 Operator theory and functional analysis
I Erdelyi
39 Nonlinear analysis and mechanics: Heriot-Watt Symposium Volume IV
R J Knops
40 Singular systems of differential equations
S L Campbell
41 N-dimensional crystallography
R L E Schwarzenberger
42 Nonlinear partial differential equations in physical problems
D Graffi
43 Shifts and periodicity for right invertible operators
D Przeworska-Rolewicz
44 Rings with chain conditions
A W Chatters and C R Hajarnavis
45 Moduli, deformations and classifications of compact complex manifolds
D Sundararaman
46 Nonlinear problems of analysis in geometry and mechanics
M Atteia, D Bancel and I Gumowski
47 Algorithmic methods in optimal control
W A Gruver and E Sachs
48 Abstract Cauchy problems and functional differential equations
F Kappel and W Schappacher
49 Sequence spaces
W H Ruckle
50 Recent contributions to nonlinear partial differential equations
H Berestycki and H Brezis
51 Subnormal operators
J B Conway
52 Wave propagation in viscoelastic media
F Mainardi
53 Nonlinear partial differential equations and their applications: Collège de France Seminar. Volume I
H Brezis and J L Lions
54 Geometry of Coxeter groups
H Hiller
55 Cusps of Gauss mappings
T Banchoff, T Gaffney and C McCrory

56 An approach to algebraic K-theory
A J Berrick
57 Convex analysis and optimization
J-P Aubin and R B Vintner
58 Convex analysis with applications in the differentiation of convex functions
J R Giles
59 Weak and variational methods for moving boundary problems
C M Elliott and J R Ockendon
60 Nonlinear partial differential equations and their applications: Collège de France Seminar. Volume II
H Brezis and J L Lions
61 Singular systems of differential equations II
S L Campbell
62 Rates of convergence in the central limit theorem
Peter Hall
63 Solution of differential equations by means of one-parameter groups
J M Hill
64 Hankel operators on Hilbert space
S C Power
65 Schrödinger-type operators with continuous spectra
M S P Eastham and H Kalf
66 Recent applications of generalized inverses
S L Campbell
67 Riesz and Fredholm theory in Banach algebra
B A Barnes, G J Murphy, M R F Smyth and T T West
68 Evolution equations and their applications
F Kappel and W Schappacher
69 Generalized solutions of Hamilton-Jacobi equations
P L Lions
70 Nonlinear partial differential equations and their applications: Collège de France Seminar. Volume III
H Brezis and J L Lions
71 Spectral theory and wave operators for the Schrödinger equation
A M Berthier
72 Approximation of Hilbert space operators I
D A Herrero
73 Vector valued Nevanlinna Theory
H J W Ziegler
74 Instability, nonexistence and weighted energy methods in fluid dynamics and related theories
B Straughan
75 Local bifurcation and symmetry
A Vanderbauwhede
76 Clifford analysis
F Brackx, R Delanghe and F Sommen
77 Nonlinear equivalence, reduction of PDEs to ODEs and fast convergent numerical methods
E E Rosinger
78 Free boundary problems, theory and applications. Volume I
A Fasano and M Primicerio
79 Free boundary problems, theory and applications. Volume II
A Fasano and M Primicerio
80 Symplectic geometry
A Crumeyrolle and J Grifone
81 An algorithmic analysis of a communication model with retransmission of flawed messages
D M Lucantoni
82 Geometric games and their applications
W H Ruckle
83 Additive groups of rings
S Feigelstock
84 Nonlinear partial differential equations and their applications: Collège de France Seminar. Volume IV
H Brezis and J L Lions
85 Multiplicative functionals on topological algebras
T Husain
86 Hamilton-Jacobi equations in Hilbert spaces
V Barbu and G Da Prato
87 Harmonic maps with symmetry, harmonic morphisms and deformations of metrics
P Baird
88 Similarity solutions of nonlinear partial differential equations
L Dresner
89 Contributions to nonlinear partial differential equations
C Bardos, A Damlamian, J I Díaz and J Hernández
90 Banach and Hilbert spaces of vector-valued functions
J Burbea and P Masani
91 Control and observation of neutral systems
D Salamon
92 Banach bundles, Banach modules and automorphisms of C*-algebras
M J Dupré and R M Gillette
93 Nonlinear partial differential equations and their applications: Collège de France Seminar. Volume V
H Brezis and J L Lions
94 Computer algebra in applied mathematics: an introduction to MACSYMA
R H Rand
95 Advances in nonlinear waves. Volume I
L Debnath
96 FC-groups
M J Tomkinson
97 Topics in relaxation and ellipsoidal methods
M Akgül
98 Analogue of the group algebra for topological semigroups
H Dzinotyiweyi
99 Stochastic functional differential equations
S E A Mohammed
100 Optimal control of variational inequalities
V Barbu
101 Partial differential equations and dynamical systems
W E Fitzgibbon III
102 Approximation of Hilbert space operators. Volume II
C Apostol, L A Fialkow, D A Herrero and D Voiculescu
103 Nondiscrete induction and iterative processes
V Ptak and F-A Potra
104 Analytic functions – growth aspects
O P Juneja and G P Kapoor
105 Theory of Tikhonov regularization for Fredholm equations of the first kind
C W Groetsch

106 Nonlinear partial differential equations and free boundaries. Volume I
J I Díaz

107 Tight and taut immersions of manifolds
T E Cecil and P J Ryan

108 A layering method for viscous, incompressible L_p flows occupying R^n
A Douglis and E B Fabes

109 Nonlinear partial differential equations and their applications: Collège de France Seminar. Volume VI
H Brezis and J L Lions

110 Finite generalized quadrangles
S E Payne and J A Thas

111 Advances in nonlinear waves. Volume II
L Debnath

112 Topics in several complex variables
E Ramírez de Arellano and D Sundararaman

113 Differential equations, flow invariance and applications
N H Pavel

114 Geometrical combinatorics
F C Holroyd and R J Wilson

115 Generators of strongly continuous semigroups
J A van Casteren

116 Growth of algebras and Gelfand–Kirillov dimension
G R Krause and T H Lenagan

117 Theory of bases and cones
P K Kamthan and M Gupta

118 Linear groups and permutations
A R Camina and E A Whelan

119 General Wiener–Hopf factorization methods
F-O Speck

120 Free boundary problems: applications and theory, Volume III
A Bossavit, A Damlamian and M Fremond

121 Free boundary problems: applications and theory, Volume IV
A Bossavit, A Damlamian and M Fremond

122 Nonlinear partial differential equations and their applications: Collège de France Seminar. Volume VII
H Brezis and J. L. Lions

123 Geometric methods in operator algebras
H Araki and E G Effros

124 Infinite dimensional analysis–stochastic processes
S Albeverio

125 Ennio de Giorgi Colloquium
P Krée

126 Almost-periodic functions in abstract spaces
S Zaidman

127 Nonlinear variational problems
A Marino, L Modica, S Spagnolo and M Degiovanni

128 Second-order systems of partial differential equations in the plane
L K Hua, W Lin and C-Q Wu

P K Kamthan & M Gupta

Indian Institute of Technology, Kanpur

Theory of bases and cones

Pitman Advanced Publishing Program

BOSTON · LONDON · MELBOURNE

PITMAN PUBLISHING INC
1020 Plain Street, Marshfield, Massachusetts 02050

PITMAN PUBLISHING LIMITED
128 Long Acre, London WC2E 9AN

Associated Companies
Pitman Publishing Pty Ltd, Melbourne
Pitman Publishing New Zealand Ltd, Wellington
Copp Clark Pitman, Toronto

First published 1985

AMS Subject Classifications: (main) 46A05, 46A35, 46A40
(subsidiary) 46A12, 46A45, 45D05

ISSN 0743-0337

Library of Congress Cataloging in Publication Data

Kamthan, P. K., 1938–
Theory of bases and cones.

Bibliography: p.
Includes index.
1. Locally convex spaces. 2. Bases (Linear topological spaces) 3. Cone. I. Gupta, Manjul, 1951– . II. Title.
QA322.K36 1985 515.7′3 84-25587
ISBN 0-273-08657-X

British Library Cataloguing in Publication Data

Kamthan, P.K.
Theory of bases and cones.—(Research notes in mathematics, ISSN 0743-0337; 117)
1. Linear topological spaces
I. Title II. Gupta, M. III. Series
515.7′3 QA322

ISBN 0-273-08657-X

Reproduced and printed by photolithography
in Great Britain by Biddles Ltd, Guildford

DEDICATED to

the sacred memories of
P.S. Kamthan, Shila Kamthan and S.K. Kamthan
parents and brother of the first author

PRESENTED to

R.C. James
for originality in applications

G. Köthe
for enhancement

C.W. McArthur
for geometric interpretation
of the Schauder Basis Theory

and to

B.N. Gupta and Shanti D. Gupta
parents of the second author
for their constant encouragement
and cooperation

Contents

PREFACE

PART ONE

CHAPTER 1 : ESSENTIALS 3

1.1 Introduction and Notation 3

1.2 Topological Vector Spaces 4

(a) Norm-type functions 4
(b) Linear topology 4
(c) Metrizable TVS 5
(d) Locally bounded, convex and quotient spaces 5
(e) Nets and sequences 5
(f) LF-, SRILF- and webbed spaces 6
(g) Continuity of linear maps 7
(h) Extensions of barrelledness and properties of spaces 7
(i) F-spaces and the Mackey topology 9
(j) Mackey and weak topologies 9
(k) Nuclear spaces 9
(l) Other useful results 10

1.3 Sequence Spaces 10

1.4 Infinite Series 13

1.5 Ordered Topological Vector Spaces 15

(a) Ordered vector spaces 15
(b) OTVS and properties of cones 16
(c) Dual cones and wedges 17
(d) Topological vector lattices 18
(e) Order convergence 18
(f) Positive linear mappings 20
(g) Conic bases 20

PART TWO

CHAPTER 2 : TOPOLOGICAL BASES 23

2.1 Introduction and Motivation 23

Comparison with algebraic bases 24

2.2 Schauder Bases 24

Sufficient conditions 25
Extensions of the continuity theorem 28

2.3	Associated Sequence Spaces and Topologies	30
	Sequence spaces	30
	Polar topologies	31
	$\tilde{\sigma}$ and $\sigma\gamma(\delta,\mu)$-topologies	33
	Compatibility of $\tilde{T}$	34
CHAPTER 3 :	SOME APPLICATIONS OF SCHAUDER BASES	36
3.1	Introduction	36
3.2	The Completion Theorem	36
	Weak Schauder bases	39
3.3	Continuity of Maps	40
	Linear functionals	40
	Linear operators and seminorms	43
CHAPTER 4 :	THE BASIS PROBLEM	46
4.1	Introduction	46
4.2	More Counter-Examples and ω-Separability	48
4.3	The Amemiya-Kōmura Example	50
	Construction of Y	52
	Construction of Z	53
4.4	Enfló's Example	55
	Approximation property	55
	The main example	58
	Intermediary lemmas	58
	Construction of the example	61
	Construction of functionals	62
CHAPTER 5 :	CHARACTERIZATION OF SCHAUDER BASES	66
5.1	The Main Result	66
5.2	Examples and Counter-Examples	68
	Applications and examples	70
	Bočkarev's example	79
	Estimation of $\Sigma_N^{(1)}$	83
	Estimation of $\Sigma_N^{(2)}$	86
CHAPTER 6 :	THE WEAK BASIS THEOREM	87
6.1	Bases in Compatible Topologies	87
	Possible extensions	87
6.2	Failure of the WBT	90
	Failure in non-locally convex spaces	91
6.3	Further Extensions and Remarks	97
	WBT in webbed spaces	98

CHAPTER 7 : BIORTHOGONAL SYSTEMS 103

7.1 Introduction 103

7.2 Different Systems 103

Minimality 104
Maximality 105
Completeness and maximality 107
Completeness and totality 108

7.3 Similar Systems 112

7.4 Quasi-regular Systems 114

CHAPTER 8 : TYPES OF SCHAUDER BASES 117

8.1 Introduction 117

8.2 Unconditional Bases 117

Characterizations 119
Examples of conditional bases 125

8.3 Shrinking and γ-Complete Bases 128

Shrinking bases 128
γ-complete bases 130
Applications to reflexivity 133
Duality relationship 135

CHAPTER 9 : BASIC SEQUENCES IN FRÉCHET SPACES 137

9.1 Introduction 137

Some general results 138

9.2 Construction of Basic Sequences 140

CHAPTER 10 : BASIC SEQUENCES IN F-SPACES 148

10.1 Introduction 148

10.2 Compatibility and Polarity 148

10.3 Extraction of Basic Sequences 150

10.4 The Existence Theorem 153

10.5 Applications 155

10.6 Block Extensions 159

PART THREE

CHAPTER 11 : BIORTHOGONAL SYSTEMS AND CONES 167

11.1 Introduction 167

11.2 Cones and Wedges 168

Cones associated with absolute biorthogonal sequences 172
Biorthogonal systems generating the same cone 174

11.3 Bases in Associated Cones 176

CHAPTER 12 : CONES ASSOCIATED WITH BASES 184
12.1 Introduction 184
12.2 Interior Points and Lattice Structure 184
12.3 Normal Cones 187
12.4 Regular Cones 193
CHAPTER 13 : ORDER STRUCTURE 198
13.1 Introduction 198
13.2 Lattice Structure 198
Weak lattice operations 204
Boundedness property 207
CHAPTER 14 : ORDER TOPOLOGY 209
14.1 Introduction 209
14.2 The T^{or}-Topology 209
Further results on (0)- and T^{or}-convergence 215
14.3 Order Topology in Dual Spaces 218
CHAPTER 15 : CONIC BASES 222
15.1 Introduction 222
15.2 Companion Sequences 222
15.3 Types of Conic Bases and Cones 224
Different conic bases 225
15.4 Bounded Conic Bases 229
Unbounded conic bases 233
REFERENCES 236
ABBREVIATIONS 250
INDEX 252

Preface

The significance of the theory of bases in topological vector spaces, as an essential tool of functional analysis, is recognized in several branches of modern analysis - for instance, in investigating the structure of locally convex spaces (in particular, nuclear spaces), and in sequence space theory, matrix transformations, theory of operators and so on. Although the purpose of basis theory is to represent uniquely an arbitrary element of a space in terms of a given sequence, it is even more important in the study of those infinite dimensional vector spaces in which the precise location of Hamel bases becomes virtually impossible.

Several books on basis theory applied exclusively to Banach spaces have already appeared, but our wider purpose in this monograph is to deal comprehensively with the theory of bases in the general setting of topological vector spaces, thus including, as well as Banach spaces, the many interesting and useful spaces, such as nuclear spaces, which are not Banach. There is now a vast literature available on the Schauder basis theory in topological vector spaces; we include here some of its more illuminating features, and hope to deal with the rest in a subsequent work [133]. The present monograph reflects in part the second phase of a larger research project in functional analysis initiated around 1970, of which the first part is already completed in the form of [132].

This book is intended for those who have had a first course in locally convex spaces including sequences spaces, and is divided into three parts. Part One consists of Chapter 1, which covers the necessary preliminaries in the theory of locally convex spaces, sequence space theory, summation of infinite series and ordered topological spaces. Most of the results are not to be found in the standard texts and are given here without proof. The reader is urged to go through this chapter so as to become familiar with commonly used results and with abbreviations and notation that are used throughout the rest of the work.

Part Two includes Chapters 2 through 10, which comprise elementary as well as advanced results (though not exhaustive) on Schauder basis theory.

Besides problems on existence and characterization of bases, some applications are also included: for instance, applications in the study of the structure of locally convex spaces, deriving conclusions such as the continuity of operators and characterization of continuous linear functionals.

Part Three, which contains Chapter 11 through 15, deals with the geometrical interpretation of the theory, especially in terms of the theory of cones. Indeed, this part goes on to discuss the order structure of the theory of cones generated by biorthogonal sequences and topological bases. However, this part also incorporates many results on conic bases and on the relationship of order topologies with the original topologies, normal cones with unconditional bases and so on.

Limited space at our disposal has not only prevented consideration of many other interesting features, but has also forced us to omit the detail of most of the proofs of the results - a feature which has both advantageous and irritating aspects!

For convenience of cross-reference, all definitions, examples, exercises, propositions, theorems, corollaries and a few significant equations are numbered in sequence within each chapter and section: e.g., Section 4.1 is the first section of Chapter 4 and Theorem 4.1.6 is the sixth numbered item within it. The completion of a proof, other than a corollary, is shown by □.

It is now our pleasant duty to acknowledge our gratitude to those who have supported and helped our project in various ways. Firstly, our sincere thanks immediately rush to Professor Czesław Bessaga in Poland and Professor Gottfried Köthe in Germany who persuaded us to include in Chapters 9 and 10 a detailed study of basic sequences, the former having even contributed an English version of his Polish work on this subject. We also extend our thanks to Professor William H. Ruckle of Clemson University, USA, whose discussions with us during his visit to this department led to some improvements to the original draft. Our thanks are also due to Professor Z. Ciesielski of Instytut Matematyczny, Sopot, Poland and Dr. M.A. Sofi of Aligarh Muslim University of India for his academic suggestions.

We heartily thank Professor S. Sampath, Director of this Institute, whose enthusiastic, prompt and timely support helped in the financing of the typing of the original manuscript.

We also owe much to our family and friends whose continuous encouragement in our difficult hours were a great help in completing the project.

We thank Ms Bridget Buckley, Editor, Pitman, for her excellent cooperation and Ms Terri Moss for preparing an impressive typescript of the final draft.

Finally, we shall be amply rewarded if the present work helps to make this theory accessible to a larger community of mathematicians and encourages its application to further problems in mathematics. We will be most grateful for any suggestions and comments that may ultimately lead to amplification and improvement of the results discussed here.

Kanpur
November, 1984

P K K
M G

Part One

1 Essentials

1.1 INTRODUCTION AND NOTATION

To have access to most of the results of Chapter 2 onward, the reader needs at least a basic knowledge of topological vector spaces, especially of the duality theory of locally convex spaces and the preliminaries to the theory of sequence spaces. Besides, we assume the reader to be familiar with a few results on other related topics in order to appreciate the limit and scope of advanced results treated in subsequent chapters.

In this chapter we touch briefly upon the essential requirements, particularly on terminological excerpts and results not commonly found in the available texts. At the outset, we recall the following frequently used notation:

$\mathbb{K}$ = $\mathbb{R}$, the set of all reals or $\mathbb{C}$, the set of all complex numbers endowed with the usual topology

X = an arbitrary nontrivial vector space over $\mathbb{K}$

X' = algebraic dual of X

$\mathbb{N}$ = $\{1,2,\ldots,n,\ldots\}$

Φ = family of all finite subsets of $\mathbb{N}$ ordered by set-theoretic inclusion

$\#(A)$ = cardinality of the set A

$\langle A \rangle$ = balanced convex hull of $A \subset X$

$\mathrm{con}(A)$ = convex hull of $A \subset X$

$\dim X$ = dimension of X

$\aleph_0$ = $\#(\mathbb{N})$

$\aleph$ = $\#([0,1])$, the power of continuum

$\mathrm{sp}\{A\}$ = the subspace spanned by $A \subset X$

Π = set of all permutations of $\mathbb{N}$

Λ = a directed set

$\{x_\alpha\}$ or $\{x_\alpha : \alpha \in \Lambda\}$ = a net in X

$\{x_n\}$ or $\{x_n : n \geqslant 1\}$ = a sequence in X.

1.2 TOPOLOGICAL VECTOR SPACES

Throughout this book we will be working on *Hausdorff* topological vector spaces, abbreviated hereafter as TVS, and their several varieties. For details, the reader is referred to [17], [18], [93], [140], [146]; [147], [192], [200], [205] and [234]; [235]. Most of such results needed by the reader for the overall appreciation of the whole content of this book are also available in Chapter 1 of [132].

(a) Norm-type functions

A function $p:X \to \mathbb{R}$ is called an F-*seminorm* (resp. F-*norm*) if, for all x,y in X and α in $\mathbb{K}$ with $|\alpha| \leq 1$, $p(x) \geq 0$; $p(x+y) \leq p(x) + p(y)$:$p(\alpha x) \leq p(x)$ and $p(\beta x) \to 0$ as $\beta \to 0$ (and also $p(x) = 0$ only when $x = 0$). If the function p satisfies the first two conditions as well as $p(\alpha x) = |\alpha| p(x)$ for all α in $\mathbb{K}$ and x in X, then p is simply called a *seminorm*.

A function $p:X \times \Lambda \to \mathbb{R}$ is called a *pseudonorm function* (cf. [94]) if, for all x,y in X, α in $\mathbb{K}$ and d in Λ, $p(x,d) \geq 0$; $p(\alpha x,d) = |\alpha| p(x,d)$, also for given d in Λ there exists e in Λ such that $p(x+y,d) \leq p(x,e) + p(y,e)$ and for $d \geq e$, $p(x,d) \geq p(x,e)$. For each fixed d in Λ, the function $p_d(\cdot) \equiv p(\cdot,d)$ is called a *pseudonorm*.

(b) Linear topology

We use the symbol (X,T) to mean an arbitrary TVS, where T is the *Hausdorff linear* (1.) *topology* on X; also, we write D_T or D_X for the family $\{p_d : d \in \Lambda\}$ of all *pseudonorms* or F-*seminorms* generating the topology T such that

$$\cap \{p_d^{-1}(0) : d \in \Lambda\} = \{0\}. \tag{$\star$}$$

Conversely, if a family $D_T = \{p_d : d \in \Lambda\}$ of F-seminorms or pseudonorms on X satisfies ($\star$), then there exists a unique Hausdorff linear topology T on X so that (X,T) becomes a TVS. Further, the letter $\mathcal{B}_T$ or $\mathcal{B}_X$ will mean the family of all balanced and T-closed neighbourhoods at the origin of a TVS (X,T).

For a subset A of a TVS (X,T), we write $\bar{A} \equiv \bar{A}^T$ to mean the T-*closure* of A and use the symbol $[A] \equiv [A]^T$ to mean $\overline{sp}^T\{A\} \equiv \overline{sp\{A\}}^T$, whereas $\Gamma(A)^T$ will mean the *balanced*, T-*closed and convex hull* of A; also $X^* \equiv X_T^* \equiv (X,T)^*$ will stand hereafter for the nontrivial *topological dual* of (X,T) wherever this exists. If (X,T) and (Y,S) are two TVS, we write $(X,T) \simeq (Y,S)$ to mean that the spaces are *topologically isomorphic*.

(c) Metrizable TVS

Whenever a TVS (X,T) is metrizable, it will be assumed henceforth that T is generated by an invariant metric $p(\cdot,\cdot) = \|\cdot - \cdot\|$, where D_T is a non-decreasing sequence $\{p_i\}$ of pseudonorms or F-seminorms and the F-norm $\|\cdot\|$ is given by the following *Fréchet combination*:

$$\|x\| = \sum_{n=1}^{\infty} \frac{1}{2^n} \frac{p_n(x)}{1 + p_n(x)}, \qquad x \in X.$$

An X equipped with an F-norm is called an *F*-space*.

(d) Locally bounded, convex and quotient spaces

If a TVS(X,T) is also locally convex, we will abbreviate this as l.c. TVS and read as locally convex space. In this case, D_T consists of the total family of all T-continuous seminorms. Conversely, if D_T is a total family of seminorms on X, then D_T generates a *Hausdorff locally convex* (l.c.) *topology* such that (X,T) becomes an l.c. TVS or a locally convex space. Given an l.c. TVS (X,T), we use the symbols $\sigma(X,X^*)$, $\tau(X,X^*)$ and $\beta(X,X^*)$ to mean respectively the weak, Mackey and strong topologies on X.

Corresponding to an l.c. TVS (X,T) or even a TVS(X,T), let p_u denote the *Minkowski functional* for u in $\mathcal{B}_T$. Write X_u for the quotient space $X/p_u^{-1}(0)$ equipped with *quotient norm* $\hat{p}_u$ with $\hat{p}_u(x_u) = \inf\{p_u(x+y) : y \in p_u^{-1}(0)\}$, where $x_u = x + p_u^{-1}(0)$; $x \in X$. Also, we generally denote by K_u the *canonical embedding* $K_u : X \to X_u$. Further, the *completion* of a TVS (X,T) and that of $(X_u, \hat{p}_u)$ will be respectively denoted by $(\hat{X},\hat{T})$ and $(\hat{X}_u, \hat{p}_u)$.

A TVS (X,T) is called *locally bounded* (l.b. TVS) if there exists a T-bounded u in $\mathcal{B}_T$. A metrizable TVS (resp. l.c. TVS) (X,T) which is also complete, indeed sequentially complete (abbreviated as *ω-complete*), is called an *F-space* (resp. a *Fréchet space*).

(e) Nets and sequences

If a net $\{x_\alpha\}$ converges to x in a TVS (X,T), we will write the same as $T\text{-}\lim_\alpha x_\alpha = x$ or $x_\alpha \xrightarrow{T} x$; if there is no confusion about the choice of the topology T, we will drop the letter T from the limit notation.

A sequence $\{x_n\}$ in a TVS (X,T) is called (i) *regular* if for some p in D_T and $\varepsilon > 0$, $p(x_n) \geqslant \varepsilon$ for all $n \geqslant 1$; (ii) *irregular* if $\alpha_n x_n \to 0$ in T for every sequence $\{\alpha_n\}$ in $\mathbb{K}$; (iii) *normalized* if $\{x_n\}$ is both bounded and regular; and (iv) *normal* if for some $\{\alpha_n\}$ in $\mathbb{K}$, $\{\alpha_n x_n\}$ is normalized.

(f) LF-, SRILF- and webbed spaces

Let $X = \cup\{X_n : n \geqslant 1\}$, where $\{X_n\} \equiv \{(X_n, T_n)\}$ is a sequence of l.c. TVS. The finest l.c. topology T on X such that the natural injections $I_n : X_n \to X$ are continuous, is called the *inductive limit* (IL) *topology* and we write $X = \xrightarrow{\text{ind}} X_n$ to mean the IL of X_n. If $X_n \subset X_{n+1}$ and $T_n \approx T_{n+1}|X_n$ (the symbol $\approx$ stands for the *equivalence of two topologies* and $T_{n+1}|X_n$ means the *restriction* of T_{n+1} to X_n), then the IL topology T on $X = \xrightarrow{\text{ind}} X_n$ is termed the *strict inductive limit* (SIL) *topology;* in this case X is called the SIL of $\{X_n\}$. The IL (resp. SIL) of Fréchet spaces is usually known as the LF (resp. *strict* LF)-*space* $X = \xrightarrow{\text{ind}} X_n$. The strict LF-space $X = \xrightarrow{\text{ind}} X_n$ is called a *sequentially retractive inductive limit of Fréchet spaces* (SRILF-*space*) provided the T-convergence of a sequence $\{x_k\}$ in X implies that $\{x_k\} \subset X_N$ for some N and $\{x_k\}$ converges in (X_N, T_N).

To appreciate the unambiguity in the definition of SRILF-spaces, let us recall (cf. [58])

Proposition 1.2.1: Let (X,T) be an LF-space with $X = \xrightarrow{\text{ind}} X_m = \xrightarrow{\text{ind}} Y_n$, where $X_m \equiv (X_m, T_m)$ and $Y_n \equiv (Y_n, S_n)$ are Fréchet spaces with $X_m \subset X_{m+1}$ and $Y_n \subset Y_{n+1}$ for $m,n \geqslant 1$. Then for every n there is m so that $X_n \hookrightarrow Y_m$ and $Y_n \hookrightarrow X_m$ (the symbol $\hookrightarrow$ means the *continuous embedding* from X_n into Y_m etc.).

Definition 1.2.2: A *web* in an l.c. TVS (X,T) is a collection $\mathfrak{R} = \{A_{n_1,\ldots,n_k}\}$ of subsets of X, satisfying

$$X = \bigcup_{n_1 \geqslant 1} A_{n_1}; \quad A_{n_1,\ldots,n_{k-1}} = \bigcup_{n_k \geqslant 1} A_{n_1,\ldots,n_k},$$

for all $k > 1$ and $n_1,\ldots,n_{k-1}$ in $\mathbb{N}$. If all the members of the web $\mathfrak{R}$ are balanced and convex (resp. closed), $\mathfrak{R}$ is then called an *absolutely convex* (resp. *closed*) *web*. A web $\mathfrak{R}$ is said to be *of type* $\mathfrak{C}$ if for each fixed $\{n_k\}$, $k \geqslant 1$, there exists $\{\alpha_k\}$ such that for all β_k in $[0,\alpha_k]$ and all x_k in $A_{n_1,\ldots,n_k}$, the series $\sum_{k=1}^{\infty} \beta_k x_k$ converges in (X,T). An absolutely convex web $\mathfrak{R}$ is said to be *strict* if for any sequence $\{n_k\}$, there exist $\mu_k > 0$ $(k \geqslant 1)$ such that, for all x_k in $A_{n_1,\ldots,n_k}$ and all β_k in $[0,\mu_k]$, $\sum_{k=1}^{\infty} \beta_k x_k$ converges in (X,T) and $\sum_{k=k_0}^{\infty} \beta_k x_k$ is contained in $A_{n_1,\ldots,n_{k_0}}$ for all $k_0 \geqslant 1$.

Definition 1.2.3: An l.c. TVS (X,T) is called *webbed* (resp. *strictly webbed*) if X contains a web $\mathfrak{R}$ of type $\mathfrak{C}$ (resp. a strict web).

Following [39] (cf. also [147], p. 68), we have

Proposition 1.2.4: Let (X,T) be strictly webbed with a strict web $\mathfrak{R}$, then for each balanced, convex, bounded and ω-complete set M of X, there exist sequences $\{n_k\}$ and $\{\alpha_k\}$, $\alpha_k > 0$ such that $M \subset \alpha_k A_{n_1,\ldots,n_k}$ for $k \geq 1$.

Proposition 1.2.5: Let X be a Fréchet space and Y an l.c. TVS having a web $\mathfrak{R}$ of type $\mathfrak{C}$. Suppose R is a linear operator, $R:X \to Y$, such that R has a sequentially closed graph and $X = R^{-1}[Y]$. Then (a) there exist n_1 in $\mathbb{N}$ so that $R^{-1}[A_{n_1}]$ is of second category and a semiball B_1 in X with $B_1 \subset R^{-1}[\overline{\langle A_{n_1}\rangle}]$ and (b) if $R^{-1}[A_{n_1,\ldots,n_k}]$ is of second category, there exist n_{k+1} and a semiball B_{k+1} of X such that $R^{-1}[A_{n_1,\ldots,n_{k+1}}]$ is of second category and $B_{k+1} \subset R^{-1}[\overline{\langle A_{n_1,\ldots,n_{k+1}}\rangle}]$.

(g) Continuity of linear maps

We recall (cf. [135])

Proposition 1.2.6: Let $\langle X,Y\rangle$ be a dual system and $\mathfrak{H}$ a family of $\sigma(X,Y)$-continuous linear maps from X into itself such that $\{R(x):R \in \mathfrak{H}\}$ is $\beta(X,Y)$-bounded for each x in X. Then $\mathfrak{H}$ is $\beta(X,Y)$-equicontinuous.

An l.c. TVS (X,T) is called *Pták* or *fully complete* if each subspace M of X* is $\sigma(X^*,X)$-closed, whenever $M \cap A$ is $\sigma(X^*,X)$-closed for any balanced, convex, $\sigma(X^*,X)$-closed and equicontinuous subset A of X*; cf. [93]. Following [192] and [39], we have

Theorem 1.2.7: Let X be barrelled and Y Pták. A linear map $R:X \to Y$ is continuous, whenever the graph of R is closed.

Theorem 1.2.8: A sequentially closed linear operator from an ω-complete bornological space (*ultrabornological space*) into a webbed space is continuous.

(h) Extensions of barrelledness and properties of spaces

We have

Definition 1.2.9: Let (X,T) be an l.c. TVS. Then (X,T) is said (a) to have (i) the *Banach-Steinhaus property* (BSP) if each $\sigma(X^*,X)$-bounded subset of X^* is $\beta(X^*,X)$-bounded (such a space is called a W-*space*); (ii) the ω-*property* if there exists $\{p_n\} \subset D_T$ such that $p_n(x) = 0$ for all $n \geq 1$ implies that $x = 0$; (iii) the S-*property* if $(X^*,\sigma(X^*,X))$ is ω-complete (such a space is called an S-*space*) and (b) to be (iv) σ-*barrelled* if every $\sigma(X^*,X)$-bounded countable subset of X^* is equicontinuous; (v) ω-*barrelled* if every $\sigma(X^*,X)$-convergent sequence in X^* is equicontinuous; (vi) σ-*infrabarrelled* if each $\beta(X^*,X)$-bounded sequence in X^* is equicontinuous; (vii) *Mazur* if $X^* = X^+$; (viii) *semibornological* if $X^* = X^b$; (ix) a P-*space* if the *canonical embedding* $\psi: X \to X^{**}$ is $\sigma(X^{**},X^*)$-sequentially dense and (X,T) is barrelled; and (x) a C-*space* if $\sigma(X^*,X)$-bounded subsets of X^* are $\sigma(X^*,X)$-relatively compact, where X^+ is the *sequential dual* of (X,T) and X^b is a subspace of X', transforming bounded subsets into bounded sets.

Note: If T^+ denotes the finest l.c. topology on an l.c. TVS $X \equiv (X,T)$ such that T- and T^+-convergent sequences are the same, then $X^+ = (X,T^+)^*$; cf. [230].

Following [145], we have

Theorem 1.2.10: For an l.c. TVS (X,T), the following are equivalent:

(i) $(X^*,\sigma(X^*,X))$ is barrelled;

(ii) $\sigma(X^*,X) \approx \beta(X^*,X)$;

(iii) each bounded subset A of X is finite dimensional;

(iv) $X^+ = X^b = X'$.

For the next four results, we follow [1], [42], [237], [87]; and [113]; [230].

Proposition 1.2.11: (i) Every barrelled space is an S- and a W-space and if a space is σ-barrelled and separable, then it is barrelled. (ii) A subspace of codimension $\aleph_0$ of a metrizable barrelled space is barrelled.

Proposition 1.2.12: If (X,T) is Mazur, then X^* is $\beta(X^*,X)$-complete.

Proposition 1.2.13: Each ω-barrelled space is a W-space. If (X,T) is ω-complete, then $(X^*,\tau(X^*,X))$ is ω-barrelled; conversely, if (X,T) is a metrizable l.c. TVS with $(X^*,\tau(X^*,X))$ being ω-barrelled, then (X,T) is complete.

Proposition 1.2.14: If (X,T) is an S-Mackey space such that X is $\beta(X,X^*)$-separable, then it is barrelled.

(i) F-spaces and the Mackey topology

After [209]; [211], we reproduce

Proposition 1.2.15: Let (X,p) be an F-space with topology T_p and $X^* \neq \{0\}$. Let $v_n = \text{con}(\{x \in X: p(x) < 1/n\})$ for $n \geqslant 1$. If $\tau \equiv \tau(X,X^*)$; then $\{v_n\}$ generates τ, $\tau \subset T_p$ and τ is metrizable and the following statements are equivalent:

(i) (X,T_p) is not locally convex;

(ii) $\tau \subsetneqq T_p$;

(iii) (X,τ) is not complete.

We have (cf. [117]).

Proposition 1.2.16: Let (X,p) be a non-locally convex separable F-space with separating dual X^*. Then X has a proper closed weakly dense (PCWD) subspace Y, or, equivalently, a p-closed subspace Y with $(X/Y,\hat{p})$ having a trivial dual.

(j) Mackey and weak topologies

From [71], we quote

Proposition 1.2.17: Let $\langle X,Y\rangle$ be a dual system (i) If $\sigma(X,Y) \subsetneqq \tau(X,Y)$, then there exists a nonconvex linear topology T on X with $\sigma \subsetneqq T \subsetneqq \tau$; and (ii) $\sigma \approx \tau$ if and only if $(\hat{X},\hat{\tau}) = Y'$, where $\sigma \equiv \sigma(X,Y)$, $\tau \equiv \tau(X,Y)$ and $\hat{X}$ denotes the completion of (X,T).

(k) Nuclear spaces

Given two Banach spaces B_1 and B_2, a continuous linear operator $R: B_1 \to B_2$ is said to be *nuclear* or ℓ^1*-nuclear* provided that there exist $\alpha = \{\alpha_n\}$ in ℓ^1, an equicontinuous sequence $\{f_n\} \subset B_1^*$ and a bounded sequence $\{y_n\}$ in B_2 so that

$$Rx \equiv R(x) = \sum_{n=1}^{\infty} \alpha_n f_n(x) y_n, \quad \forall x \in B_1.$$

Let (X,T) be an l.c. TVS. Given u and v in $\mathcal{B}_T$, we use the notation $v < u$ to mean that v *is absorbed by* u and this is equivalent to $p_u \preccurlyeq kp_v$ for some $k > 0$. If $v < u$, there is a natural map $K_u^v : X_v \to X_u$ such that $K_u = K_u^v \circ K_v$. Let us write $\hat{K}_u^v$ for the extension of K_u^v from $\hat{X}_v$ into $\hat{X}_u$; that is, $\hat{K}_u^v : \hat{X}_v \to \hat{X}_u$. An l.c. TVS (X,T) is said to be *nuclear* or ℓ^1*-nuclear* if, for each u in $\mathcal{B}_T$, there exists v in $\mathcal{B}_T$ with $v < u$ such that $\hat{K}_u^v$ is nuclear. Our major references for nuclear maps and spaces are [186] and [205], and for their further extensions the reader is referred to [47] and [136]. We will need

Proposition 1.2.18: An l.c. TVS (X,T) is nuclear if and only if, for each u in $\mathcal{B}_T$, the map $\hat{K}_u : \hat{X} \to \hat{X}_u$ is nuclear. Also, arbitrary product of nuclear spaces is nuclear.

(ℓ) Other useful results

Let (X,T) and (Y,S) be two TVS and $R: X \to Y$ a linear map. The *algebraic adjoint* $R': Y' \to X'$ is defined by $\langle Rx, y' \rangle = \langle x, R'y' \rangle$ and the *restriction* of R' to Y^*, written as $R^* \equiv R'/Y^*$, is called the *adjoint* or *transpose* of R.

Proposition 1.2.19: Let (X,T) and (Y,S) be two l.c. TVS and $R: X \to Y$ a linear map. Then R is $\sigma(X,X^*)$-$\sigma(Y,Y^*)$ continuous if and only if R is $\tau(X,X^*)$-$\tau(Y,Y^*)$ continuous.

Following [166] and [172], we have

Lemma 1.2.20: Let $\{R_\alpha\}$ be a family of linear maps from X into a TVS (Y,T_2) such that $\{R_\alpha(x)\}$ is T_2-bounded in Y for each x in X. Then there exists a weakest linear topology T_1 on X such that $\{R_\alpha\}$ is T_1-T_2 equicontinuous; also, T_1 is generated by $D_{T_1} = \{\bar{p} : p \in D_{T_2}\}$, where $\bar{p}(x) = \sup_\alpha p(R_\alpha x)$, $x \in X$.

Lemma 1.2.21: Let A_n, $n \geqslant 1$ and A be transformations from a set M into a TVS (X,T). Then a subset B of M is transformed into a precompact subset $A[B]$ of (X,T) provided that (i) $A_n[B]$ is precompact for $n \geqslant 1$, and (ii) $A_n(x) \to A(x)$ uniformly in $x \in B$.

1.3 SEQUENCE SPACES

We will have plenty of occasions to consider examples and counter-examples, mostly chosen from the theory of sequence spaces. Accordingly, we follow

[132] (cf. also [146]; [198]) to recall some frequently used definitions, notation and theorems from this theory.

Let ω denote the vector space of all $\mathbb{K}$-valued sequences. Let $e = \{1,1,\ldots\}$ and $e^n = \{0,\ldots,0,1,0,0,\ldots\}$, 1 being at the n-th place, denote the *unity* and n-th *unit vector* of ω. Suppose $\phi = \text{sp}\,\{e^n : n \geq 1\}$. A subspace λ of ω with $\phi \subset \lambda$ will hereafter be called a *sequence space* (abbreviated hereafter as s.s.). If $x \in \lambda$, then the length ℓ of x means the largest integer ℓ such that $x_\ell \neq 0$ and the n-th *section* $x^{(n)}$ of x stands for the sum $\sum_{i=1}^{n} x_i e^i$.

Corresponding to an s.s. λ, let the α-*dual* (= *Köthe dual*) and the β-dual of λ be defined as follows:

$$\lambda^\times \equiv \lambda^\alpha = \{y \in \omega : p_y(x) = \sum_{n=1}^{\infty} |x_n y_n| < \infty\ ,\quad \forall x \in \lambda\}\ ;$$

$$\lambda^\beta = \{y \in \omega : q_y(x) \equiv |\sum_{n=1}^{\infty} x_n y_n| < \infty,\quad \forall x \in \lambda\}.$$

If μ is any subspace of λ^β with $\phi \subset \mu$, then $\langle\lambda,\mu\rangle$ forms a dual system, the corresponding bilinear function $\langle\cdot,\cdot\rangle$ being given by

$$\langle x,y\rangle = \sum_{n=1}^{\infty} x_n y_n\ .$$

Hence one can talk of several polar topologies, σ, τ, β etc., on either λ or μ, where, for instance, the $\sigma(\lambda,\mu)$ topology on λ is generated by $D_\sigma = \{q_y : y \in \mu\}$. If $\mu = \lambda^\times$, there is another natural topology on λ called the *normal topology* $\eta(\lambda,\lambda^\times)$ and this is generated by $\{p_y : y \in \lambda^\times\}$; it is known (cf. [132], p. 136; [146], p. 409) that $\sigma(\lambda,\lambda^\times) \subset \eta(\lambda,\lambda^\times) \subset \tau(\lambda,\lambda^\times)$.

An element x of λ is called *positive* (resp. *strictly positive*), to be denoted by $x > 0$ (resp. $x >> 0$), if $x_n \geq 0$ (resp. $x_n > 0$) for $n \geq 1$; let $\lambda_+ = \{x \in \lambda : x > 0\}$.

Definition 1.3.1: An s.s. λ is called *monotone* (resp. *normal*) if $m_0\lambda$ (resp. $\ell^\infty\lambda$) is contained in λ; for the definition of m_0 and ℓ^∞, see Table 3.5; also λ is said to be *perfect* if $\lambda = \lambda^{\times\times} \equiv (\lambda^\times)^\times$.

We quote from [132] and [146] the following two results.

Theorem 1.3.2: An s.s. λ is perfect if and only if λ is $\sigma(\lambda,\lambda^\times)$-$\omega$-complete (resp. $\eta(\lambda,\lambda^\times)$-complete). Also, if λ is monotone, then $\lambda^\times = \lambda^\beta$.

TABLE 1.3.5

No.	Sequence space (s.s.) λ with its natural norm/F-norm	Topological dual λ^*	$\lambda^\times$	λ^β
1	$\phi; \|\|x\|\| \equiv \|\|x\|\|_\infty = \sup \|x_n\|$	ℓ^1	ω	ω
2	$\omega; \|x\|_\omega = \sum_{n=1}^{\infty} \|x_n\|/2^n(1+\|x_n\|)$	ϕ	ϕ	ϕ
3	$c_0 = \{x\in\omega: x_n\to 0\};\ \|\|x\|\| \equiv \|\|x\|\|_\infty = \sup\|x_n\|$	ℓ^1	ℓ^1	ℓ^1
4	$\ell^p = \{x\in\omega: \sum_{n=1}^{\infty} \|x_n\|^p<\infty\}, 0<p<\infty;$ $\|\|x\|\|_p = (\sum_{n=1}^{\infty} \|x_n\|^p)^{1/p}$ with $1\leq p<\infty$, or $\|x\|_p = \sum_{n=1}^{\infty} \|x_n\|^p,\ 0 < p < 1$	$\ell^q (1\leq p<\infty)$ or $\ell^\infty (0<p<1)$ $p^{-1}+q^{-1}=1$	ℓ^q $(1\leq p<\infty)$ or ℓ^∞ $(0<p<1)$ $p^{-1}+q^{-1}=1$	ℓ^q $(1\leq p<\infty)$ or ℓ^∞ $(0<p<1)$ $p^{-1}+q^{-1}=1$
5	$\ell^\infty = \{x\in\omega: \sup\|x_n\|<\infty\};\ \|\|x\|\|_\infty = \sup\|x_n\|$	$ba(\mathbb{N},\Phi_\infty)$	ℓ^1	ℓ^1
6	$m_0 = \{x\in\omega$: the set $\{x_n\}$ is finite$\}$; $\|\|x\|\| \equiv \|\|x\|\|_\infty = \sup\|x_n\|$	$ba(\mathbb{N},\Phi_\infty)$	ℓ^1	ℓ^1
7	$k = \phi \oplus sp\{e\};\ \|\|x\|\| \equiv \|\|x\|\|_\infty = \sup\|x_n\|$	ℓ^1	ℓ^1	cs
8	$\delta = \{x\in\omega: \|x_n\|^{1/n}\to 0\};\ \|x\|_\delta = \sup\|x_n\|^{1/n}$	d	d	d
9	$d = \{x\in\omega: \limsup \|x_n\|^{1/n}<\infty\}$;-	-	δ	δ
10	$cs = \{x\in\omega: \{\sum_{i=1}^{n} x_n\}$ converges$\}$; $\|\|x\|\|_{cs} = \sup\|\sum_{i=1}^{n} x_i\|$	bv	ℓ^1	bv
11	$bs = \{x\in\omega: \sup\|\sum_{i=1}^{n} x_i\|<\infty\};\ \|\|x\|\|_{bs} = \|\|x\|\|_{cs}$	$ba(\mathbb{N},\Phi)$	ℓ^1	bv_0
12	$bv = \{x \in \omega: \sum_{n=1}^{\infty} \|x_{n+1}-x_n\| < \infty\}$; $\|\|x\|\|_{bv} = \sum_{n=1}^{\infty} \|x_{n+1}-x_n\| + \lim\|x_n\|$	$bs \oplus \mathbb{K}$	ℓ^1	cs
13	$bv_0 = \{x\in bv: x_n\to 0\};\ \|\|x\|\|_{bv_0} = \sum_{n=1}^{\infty} \|x_{n+1}-x_n\|$	bs	ℓ^1	bs

Theorem 1.3.3: Let λ be an s.s. and μ a subspace of λ^β. Suppose $\sigma = \sigma(\lambda,\mu)$, $\eta = \eta(\lambda,\lambda^\times)$ and $\tau = \tau(\lambda,\lambda^\times)$. Then, for any x in λ, $x^{(n)} \to x$ in σ or η. Further, if λ is monotone, then $x^{(n)} \to x$ in τ.

Whenever an s.s. λ is equipped with a linear topology T, we will write λ as (λ,T) and call it the same as an l.c. s.s. or an l.s.s., depending upon whether the T is an l.c. topology or a linear topology. An l.s.s. (λ,T) is called a K-*space* if, for each $i \geq 1$, the functionals $f_i:\lambda \to \mathbb{K}$, $f_i(x) = x_i$ are continuous on (λ,T); also a K-space (λ,T) is called an AK-*space* if $x^{(n)} \to x$ in T for each x in λ.

We have (cf. [64]; [132], p. 59; [197])

Theorem 1.3.4: Let (λ,T) be an l.c.s.s. such that it is also an AK-space. Then $\lambda^* \equiv (\lambda,T)^*$ can be identified with $\lambda_s = \{\{f(e^n)\}: f \in \lambda^*\}$ and so $\lambda^* \subset \lambda^\beta$. In addition, if (λ,T) is barrelled, then $\lambda^* = \lambda^\beta$.

Table 1.3.5: In Table 1.3.5, note that $ba(\mathbb{N},\Phi_\infty)$ denotes the collection of all charges μ on Φ_∞, where a charge μ on Φ_∞ (= the ring of all subsets of $\mathbb{N}$) is a $\mathbb{K}$-valued finitely additive set function with $|\mu(A)| < \infty$, for each A in Φ_∞.

Finally we quote from [132], p. 188, the following

Proposition 1.3.6: For a monotone s.s.λ, $(\lambda^\times,\sigma(\lambda^\times,\lambda))$ is ω-complete.

1.4 INFINITE SERIES

Henceforth we will write an infinite series in a TVS (X,T) as $\sum_{n\geq 1} x_n$ and this equal to x means $T\text{-}\lim \sum_{i=1}^{n} x_i = x$ in X.

Definition 1.4.1: A formal infinite series $\sum_{n\geq 1} x_n$ in a TVS (X,T) is called (i) *absolutely convergent* if $\sum_{n\geq 1} p(x_n) < \infty$, $p \in D_T$; (ii) *bounded multiplier* (b.m-) *convergent* (resp. *subseries convergent*) if $\sum_{n\geq 1} \alpha_n x_n$ converges in (X,T) for every choice of α in ℓ^∞ (resp. m_o); (iii) *unconditionally* (u-) *convergent* if $\sum_{n\geq 1} x_{\pi(n)}$ converges in (X,T) for every π in Π; and (iv) *unordered convergent* to x in (X,T) if $x = T - \lim_{\sigma\in\Phi} \sum_{i\in\sigma} x_i$.

Definition 1.4.2: A series $\sum_{n\geq 1} x_n$ in a TVS (X,T) is said to be *unordered bounded* (resp. *Cauchy*) if $\{S_\sigma : \sigma \in \Phi\}$ is bounded (resp. Cauchy) in (X,T), where $S_\sigma = \sum_{i\in\sigma} x_i$.

We recall from [172] (cf. also [132], p. 145) the following

Proposition 1.4.3: Let (X,p) be a seminormed space. For any σ in Φ, let us consider the subsets $\{x_i : i \in \sigma\}$ and $\{\alpha_i : i \in \sigma\}$ in X and $\mathbb{R}$ respectively. If $k = \sup\{|\alpha_i|; i \in \sigma\}$, then

$$p(\sum_{i\in\sigma} \alpha_i x_i) \leq 2k \sup\{p(\sum_{i\in\sigma_1} x_i) : \sigma_1 \subset \sigma\} .$$

Also, we have (cf. [167]; [132], p. 176)

Proposition 1.4.4: If a series $\sum_{n\geq 1} x_n$ in an l.c. TVS (X,T) is unordered Cauchy, then, for each bounded subset B of ℓ^∞, the set $S(B;\Phi) = \{\sum_{i\in\sigma} b_i x_i : b \in B,\ \sigma \in \Phi\}$ is precompact.

Once again, we follow [132] to quote the following:

Theorem 1.4.5: Let (X,T) be an ω-complete l.c. TVS or l.b. TVS, then the notions (i) through (iv) in Definition 1.4.1, are equivalent. In general, b.m-convergence implies subseries convergence implies u-convergence.

The next result is taken from [132], pp. 170, 172, and [241].

Theorem 1.4.6: Let $\sum_{n\geq 1} x_n$ be a formal series in an l.c. TVS (X,T), then it is T-subseries convergent if and only if it is $\sigma(X,X^*)$-subseries convergent. Further, if $(X,\sigma(X,X^*))$ is ω-complete, then the following statements are equivalent:

(i) $\sum_{n\geq 1} x_n$ is $\sigma(X,X^*)$-u-convergent;

(ii) $\sum_{n\geq 1} x_n$ is T-u-convergent;

(iii) $\sum_{n\geq 1} x_n$ is T-unordered bounded.

We will also need the following result [165]; cf. also [132], p. 158.

Theorem 1.4.7: A series $\sum_{n\geq 1} x_n$ in an ω-complete l.c. TVS (X,T) is u-convergent if and only if $\sum_{n\geq 1} |f(x_n)|$ converges uniformly in $f \in A$ for any equicontinuous subset A of X^*.

Finally, we have [168]

Proposition 1.4.8: In a Fréchet space (X,T), one has the following equivalent relations:

(i) There exists an unordered bounded series in (X,T) such that it is not u-convergent.

(ii) There exist an unordered bounded series $\sum_{n\geq 1} x_n$ in (X,T) and u in $\mathcal{B}_T$ such that $x_n \notin u$, $n \geq 1$.

(iii) X contains a subspace Y with $(Y,T|Y) \simeq (c_o, \|\cdot\|_\infty)$.

1.5 ORDERED TOPOLOGICAL VECTOR SPACES

We follow [99], [140], [175], [184], [205], [229] and [236] for most of the results of this last section; [175], [184] and [236] should be consulted for terms not explained hereafter.

(a) Ordered vector spaces

The pair $(X,\leq)$ hereafter stands for an *ordered vector space* (OVS) over the field $\mathbb{R}$. The letter $K \equiv K_X$ in an X over $\mathbb{R}$ means either a cone or a wedge; each cone K in X induces an order structure $\leq$ on X and, conversely, positive elements in $(X,\leq)$ form a cone.

An *order interval* $[x,y]$ in $(X,\leq)$ is the set $\{z \in X: x \leq z \leq y\}$. A subset A of $(X,\leq)$ is *order bounded* if A is contained in an order interval. A wedge W in $(X,\leq)$ is called *generating* if $X = W-W$, or, equivalently, $X = \mathrm{sp}\{W\}$. A cone is always a wedge. An element e in a cone K is called a *weak order-unit* if, for $0 \neq y$ in K, there is a nonzero z in K with $z \leq e, y$. A convex subset A of a wedge W is called an *extreme subset* if $u,v \in A$ whenever $\alpha u + (1-\alpha)v \in A$ for some α with $0 < \alpha < 1$. For a nonzero element x of W, we write $R(0,x)$ to mean the *ray* $\{\alpha x: \geq 0\}$; if $R(0,x)$ is an extreme subset, $R(0,x)$ is then called an *extreme ray*.

We have ([221])

Proposition 1.5.1: Let a wedge W induce an ordering $\leq$ in X and let A be a convex subset of W. Then A is an extreme subset of W if and only if $[0,x] \subset A$

and $R(0,x) \subset A$ for each x in A. In particular, a nonzero x in W yields an extreme ray $R(0,x)$ if and only if $[0,x] = \{tx : 0 \leq t \leq 1\}$.

Two OVS $(X,\leq)$ and $(Y,\leq)$ are said to be *order isomorphic* if there exists an *order isomorphism* R from X onto Y; that is, R is an algebraic isomorphism and $R[K_X] = K_Y$.

Definition 1.5.2: An OVS $(X,\leq)$ is said (i) to be *order complete* if every directed $(\leq)$ subset A of X that is majorized in X has supremum in X; (ii) to be *Archimedean* if $x \leq 0$ whenever $\alpha x \leq y$ for y in K and all $\alpha \geq 0$; (iii) to have the *decomposition property* (DP) if for all $x,y \geq 0$, $[0,x] + [0,y] = [0,x+y]$; and (iv) to have the *property* (R) if for any sequence $\{y_n\}$ from K, there exist $\alpha_n > 0$ and y in K such that $\alpha_n y_n \leq y$, $n \geq 1$.

(b) OTVS and properties of cones

We will use the notation $(X,\leq,T)$ to mean either an *ordered topological vector space* (OTVS) or an *ordered locally convex space* (O l.c. TVS) according as (X,T) is a TVS or an l.c. TVS respectively. An OTVS is called *locally full* or *locally order-convex* if each u in $\mathcal{B}_T$ contains a v in $\mathcal{B}_T$ such that v is *full* or *order-convex* (that is, if $x,y \in v$ and $x \leq z \leq y$, then $z \in v$).

Definition 1.5.3: Let K be a cone of an OTVS $X \equiv (X,\leq,T)$. An x in K is called a *quasi-interior point* if $\overline{sp\{[0,x]\}}^T = X$. Further, K is said to be (i) *regular*, *ω-regular* (resp. *fully regular*, *fully ω-regular*) if every order bounded (resp. T-bounded) increasing net, sequence in K converges to an element of K; (ii) *minihedral* if for x,y in K, $x \vee y$ (= sup $\{x,y\}$) exists in K; (iii) *normal* if $(X,\leq,T)$ is locally full; (iv) an $\mathcal{S}$*-cone* (resp. a *strict* $\mathcal{S}$*-cone*) for a saturated class $\mathcal{S}$ of T-bounded subsets of X with $X = \cup\{S : S \in \mathcal{S}\}$ if the class $\bar{\mathcal{S}}_K = \{\overline{(S \cap K) - (S \cap K)} : S \in \mathcal{S}\}$ (resp. $\mathcal{S}_K = \{(S \cap K) - (S \cap K) : S \in \mathcal{S}\}$) is a fundamental system for $\mathcal{S}$; (v) a b-*cone* (resp. a *strict* b-*cone*) if it is an $\mathcal{S}$-cone (resp. a strict $\mathcal{S}$-cone), where $\mathcal{S}$ is the class of all bounded subsets of (X,T).

The following results, except the first part, are essentially due to McArthur [167].

Theorem 1.5.4: Let K be a cone in an O l.c. TVS $(X,\leq,T)$. Then K is T-normal if and only if D_T consists of *monotone seminorms*; that is, if $p \in D_T$ and $0 \leq x \leq y$, then $p(x) \leq p(y)$. In addition, if (X,T) is a Fréchet space, then

K is normal if and only if it is weakly normal, and if (X,T) is an F-space, then K is normal if and only if [0,x] is bounded in (X,T) for each x in K.

Theorem 1.5.5: Suppose K is a cone in an O l.c. TVS $(X,\leq,T)$. Then (i) K is weakly ω-regular if and only if it is ω-regular, and if (X,T) is an F-space and K is ω-regular closed, then K is normal; (ii) the supremum (resp. infimum) of each increasing (resp. decreasing) net in a compact subset A of $(X,\leq,T)$, with K being closed, exists and the net converges to it in (X,T).

A sort of converse of (i) above is also true; cf. [204], namely,

Proposition 1.5.6: Let (X,T) be $\sigma(X,X^*)$-ω-complete l.c. TVS ordered by a normal closed cone K. Then K is fully ω-regular.

(c) Dual cones and wedges

Let $(X,\leq)$ be an OVS ordered by a cone K. If $K^* = \{f \in X':f(x) \geq 0$ for all x in K$\}$, then K^* is called the *dual wedge* and the subspace X^{or} of X' with $X^{or} = K^* - K^*$ is termed the *order dual* of $(X,\leq)$. The collection $X^{ob} = \{f \in X':f$ is order bounded$\}$ is referred to as the *order bound dual* of $(X,\leq)$. If X^{or} separates the points of X, then X is called *regularly ordered.*

Proposition 1.5.7: (i) Let $\langle X,Y \rangle$ be a dual system, where X is ordered by the cone K and Y is ordered by K^*. Then K is $\sigma(X,Y)$-normal if and only if K^* generates Y. (ii) If (X,T) is normed and ordered by the cone K such that there exists a constant $\alpha > 0$ so that $\|z\| \leq \alpha$, whenever $x \leq z \leq y$ with $\|x\|, \|y\| \leq 1$, then, for f in X^*, $f = f_1 - f_2$, where $f_1, f_2 \in K^*$ and $\|f_1\| + \|f_2\| \leq \alpha \|f\|$.

Now we have ([167]; cf. also [138] and [168])

Theorem 1.5.8: Let (X,T) be a P-space ordered by a closed cone K. Then K^* is fully ω-regular for the topology $\beta(X^*,X)$ provided that K is generating; if, in addition, (X,T) is also a Banach space, then the following statements are equivalent:

(i) K generates X;

(ii) K^* is $\beta(X^*,X)$-fully ω-regular;

(iii) K^* is $\beta(X^*,X)$-ω-regular;

(iv) K^* is $\beta(X^*,X)$-normal.

(d) Topological vector lattices

We will use the abbreviation VL to mean a *vector lattice* $(X,\leq)$; that is, an OVS $(X,\leq)$ such that $x \vee y$ (= sup $\{x,y\}$) and $x \wedge y$ (= inf $\{x,y\}$) belong to X, whenever x and y are in X. For an OVS $(X,\leq)$, we use the symbols x^+ and x^- to mean $x \vee 0$ and $-x \vee 0$ respectively, where $x \in X$. Also, we write $|x|$ to mean $x \vee (-x)$. A subset M of a VLX is called *solid* if $y \in M$, whenever $x \in M$ and $|y| \leq |x|$. An OTVS $(X,\leq,T)$ which is also a VL with $\mathcal{B}_T$ consisting of solid sets, is called a *topological vector lattice* (TVL) and if T is an l.c. topology, then a TVL $(X,\leq,T)$ is called a *locally convex topological vector lattice* (l.c. TVL). It is well known that an O l.c. TVS $(X,\leq,T)$ which is also a VL is an l.c. TVL if and only if T is generated by a family D of solid seminorms (p in D is called a *solid* or *lattice seminorm* if $p(x) \leq p(y)$, whenever $|x| \leq |y|$ for $x,y \in X$). In a VL$(X,\leq)$ with x and y in X, any operation: $(x,y) \to x \wedge y$, $x \vee y$ or $x \to x^+$, x^-, $|x|$ is termed a *lattice operation;* consequently a VL $(X,\leq,T)$ which is also an l.c. TVS is indeed an l.c. TVL if and only if the corresponding cone K is T-normal and one of the lattice operations is continuous at the origin.

Following [184] and [91] respectively, we have

Proposition 1.5.9: (i) Let $\langle X,Y\rangle$ be a dual system such that X is a VL relative to the cone K of which the dual wedge K* in Y is generating. If the lattice operations in X are $\sigma(X,Y)$-continuous, then each order interval in Y is finite dimensional. (ii) If (X,T) is a barrelled l.c. TVL, then $(X^*,\beta(X^*,X))$ is complete.

Proposition 1.5.10: Let $(X,\leq,T)$ be an l.c. TVL such that $(X^*,\beta(X^*,X))$ is a W-space. Then $(X^*,\beta(X^*,X^{**}))$ ordered by the dual cone K* is an l.c. TVL.

(e) Order convergence

Let x, $\{x_\alpha\} \equiv \{x_\alpha : \alpha \in \Lambda\}$ and $\{y_n\}$ be in an OVS $(X,\leq)$. Then we write (i) $x_\alpha \xrightarrow{(0)} x$ to indicate that $\{x_\alpha\}$ *order converges* or (0)-*converges* to x; that is, there exist nets $\{z_\beta : \beta \in \Lambda_1\}$ and $\{y_\delta : \delta \in \Lambda_2\}$ with $z_\beta \uparrow x$ (= $\{z_\beta\}$ *increases to* x) and $y_\delta \downarrow x$ (= $\{y_\delta\}$ *decreases to* x) and, for each β in Λ_1 and δ in Λ_2, there exists α_o in Λ such that $z_\beta \leq x_\alpha \leq y_\delta$ for all $\alpha \geq \alpha_o$; (ii) $x_\alpha \xrightarrow{(*)} x$ to indicate that $\{x_\alpha\}$ (*)-*converges* to x if every cofinal subnet of the net $\{x_\alpha\}$ has a cofinal subnet which (0)-converges to x; (iii) $y_n \xrightarrow{(r)} x$ to indicate that $\{y_n\}$ is *regular convergent* to x if there exist n_o in $\mathbb{N}$ and y in

K_X (y is called the *regulator of convergence* for $\{y_n\}$) such that $|y_n-x|$ exists for $n \geqslant n_0$ and, for each $\varepsilon > 0$, there exists $n_\varepsilon \geqslant n_0$ so that $|y_n-x| \leqslant \varepsilon y$ for all $n \geqslant n_\varepsilon$; and (iv) $x_\alpha \xrightarrow{0} x$ to indicate that $\{x_\alpha\}$ 0-*converges to* x provided that $\{x_\alpha\}$ is order bounded and $|x_\alpha-x|$ exists for each α, and there exists a net $\{y_\alpha:\alpha \in \Lambda\}$ with $y_\alpha \downarrow 0$ and $|x_\alpha-x| \leqslant y_\alpha$ for all α. The (r)-convergence is said to have the *diagonal property* if $x_{m,n} \xrightarrow{(r)} x_m \xrightarrow{(r)} x$ implies the existence of $\{n_m\}$ with $x_{m,n_m} \xrightarrow{(r)} x$.

Note: It is easily seen that a sequence $\{x_n\}$ (0)-converges to 0 if and only if there exist sequences $\{y_n\}$ and $\{z_n\}$ with $y_n \uparrow 0$, $z_n \downarrow 0$ and $y_n \leqslant x_n \leqslant z_n$ for all $n \geqslant 1$. Also, the notion of 0-convergence of nets is stronger than that of (0)-convergence; however, for sequences, these two notions are the same.

Definition 1.5.11: A subset M of an OVS $(X,\leqslant)$ is *order-closed* if M contains all limits of (0)-convergent nets from M. The *topology* T^{or} defined on X by all order-closed sets in X is called the *order-topology*.

Definition 1.5.12: An Archimedean VL X is said to have the *boundedness property* (BP) if each subset M of X is order bounded whenever $\alpha_n x_n \xrightarrow{0} 0$ for every $\{x_n\} \subset M$ and $\{\alpha_n\} \subset \mathbb{R}$ with $\alpha_n \downarrow 0$.

Carter [19] observes the simple

Proposition 1.5.13: (i) (0)-limits are unique. (ii) If $x_\alpha \xrightarrow{(0)} x$, then any cofinal subnet of $\{x_\alpha\}$ also (0)-converges to x. (iii) $x_\alpha \xrightarrow{(0)} x$ if and only if $x_\alpha - x \xrightarrow{(0)} 0$ and $\lambda x_\alpha \xrightarrow{(0)} \lambda x$ for each λ in $\mathbb{R}$. (iv) $x_\alpha \uparrow x$ (resp. $x_\alpha \downarrow x$) if and only if $x_\alpha \xrightarrow{(0)} x$ and $\{x_\alpha\}$ is increasing (resp. decreasing). (v) If $x_\alpha \uparrow x$ and $y_\alpha \uparrow y$ (resp. $x_\alpha \downarrow x$ and $y_\alpha \downarrow y$), then $x_\alpha + y_\alpha \uparrow x + y$ (resp. $x_\alpha + y_\alpha \downarrow x + y$). (vi) (0)-convergence implies (*)-convergence. (vii) (*)-convergence implies T^{or}-convergence.

Definition 1.5.14: An l.c. TVL $(X,\leqslant,T)$ is said to be T-*order complete* if $(X,\leqslant)$ is order complete and (a) any $\{x_n\}$ in X with $x_n \downarrow 0$, converges to 0 in (X,T); and (b) any $\{x_n\}$ in X with $x_n \uparrow$, $x_n \geqslant 0$ and unbounded with $\leqslant$, satisfies $p(x_n) \to \infty$ for some p in D_T.

After Ceitlin [22], we have

Theorem 1.5.15: Let $(X,\leq,T)$ be a T-order complete l.c. TVL. Then (i) $x_n \xrightarrow{(0)} 0$ if and only if $p(x_n,\ldots,x_{n+m}) \to 0$ as $n \to \infty$ uniformly in m for each p in D_T, where $p(x_1,\ldots,x_k) = p(|x_1| \vee |x_2| \vee \ldots \vee |x_k|)$; and (ii) $(X,\leq,T)$ possesses the BP provided it has the ω-property and, in this case, a subset B of X is order bounded if and only if $p(x_1,\ldots,x_n) \leq M_p$ for any finite number of elements in B.

(f) Positive linear mappings

Let (X,T), (X_1,T_1) and (X_2,T_2) be arbitrary O l.c. TVS. Then we have

Theorem 1.5.16: The normality of K_X implies that each f in X^* is equal to f_1-f_2, where f_1,f_2 are positive members of X^*. If K_{X_2} is normal, a positive operators $R:X_1 \to X_2$ is T_1-T_2 continuous provided any of the following conditions is satisfied: (i) K_{X_1} has nonempty interior; (ii) (X_1,T_1) is bornological and K_{X_1} is a ω-complete strict b-cone; and (iii) (X_1,T_1) is an F-space and K_{X_1} is a complete generating cone.

Note: The last part of the foregoing result is often known as the *Nachbin-Namioka-Schaefer theorem.*

(g) Conic bases

Let $(X,\leq)$ be an OVS ordered by a cone K. A *conic base* (c.b.) for K is a nonempty convex subset B of K such that each x in K, $x \neq 0$ has the unique representation $x = \alpha y$ for $\alpha > 0$ and y in B.

From [184] and [202], we respectively have

Proposition 1.5.17: A subset B of an OVS $(X,\leq)$ ordered by a cone K is a c.b. for K if and only if $B = K \cap f^{-1}(1)$ for a strictly positive f in X'.

Theorem 1.5.18: Let K be a cone in a vector space X. Then a subset B of K is a c.b. for K if and only if the following conditions are true:

(i) B is convex;

(ii) if $0 \neq x \in K$, then $x = \alpha y$, for $\alpha > 0$ and $y \in B$; and

(iii) $(B-B) \cap K = \{0\}$.

Part Two

2 Topological bases

2.1 INTRODUCTION AND MOTIVATION

It is generally impossible to find out the exact form of a Hamel base in any arbitrary infinite dimensional vector space X, although it does exist. Thus, for all practical purposes, especially from the point of view of locating a precise Hamel base in spaces of infinite dimensions, the Hamel basis theory is not as helpful as one might be tempted to expect! To overcome this difficulty, at least in many important spaces of interest which have natural linear topologies, we may introduce another notion of a base, namely,

Definition 2.1.1: A sequence $\{x_n\}$ in a TVS (X,T) is called a *topological base* (t.b.) if, for each x in X, there exists a unique sequence $\{\alpha_n\}$ in $\mathbb{K}$ such that

$$x = \sum_{n\geq 1} \alpha_n x_n \equiv T\text{-}\lim_{n\to\infty} \sum_{i=1}^{n} \alpha_i x_i . \qquad (2.1.2)$$

To each t.b. $\{x_n\}$ in a TVS (X,T), one may associate a sequence $\{f_n\}$ in X' such that $f_n(x) = \alpha_n$, where $\{\alpha_n\}$ is the sequence associated with x as in (2.1.2). Clearly $\{x_n;f_n\}$ is a biorthogonal pair of sequences, i.e., $f_n(x_m) = \delta_{mn}$, the Kronecker delta; moreover, the sequence $\{f_n\}$ is the unique sequence in X' associated with $\{x_n\}$, often called the *sequence associated with coordinate functionals* (s.a.c.f.) corresponding to $\{x_n\}$. In order to emphasize the s.a.c.f. $\{f_n\}$ corresponding to a t.b. $\{x_n\}$, the latter will often be written as $\{x_n;f_n\}$ interchangeably; that is, $\{x_n\} \equiv \{x_n;f_n\}$.

If $\{x_n;f_n\}$ is a t.b. for a TVS (X,T), we may then introduce the *expansion operators* S_n, S_n' and S_n^* as follows: $S_n : X \to X$ and $S_n' : X' \to X'$ where, for x in X and f in X',

$$S_n(x) = \sum_{i=1}^{n} f_i(x)x_i ; \; S_n'(f) = \sum_{i=1}^{n} f(x_i)f_i .$$

S_n' is the algebraic adjoint to S_n. If each f_n is continuous, we will write

$S_n^* \equiv S_n'|X^*$ and call S_n^* the *adjoint operator* to the *expansion operator* S_n $(n \geq 1)$. Also, we write R_n for the *remainder operator*, $R_n : X \to X$ and $R_n(x) = x - S_n(x)$.

Definition 2.1.3: A sequence $\{x_n\}$ in a TVS (X,T) is said (i) to be *ω-linearly independent* provided that the convergence of $\sum_{n \geq 1} \alpha_n x_n$ to 0 for any $\{\alpha_n\}$ in $\mathbb{K}$ implies that $\alpha_n = 0$ for $n \geq 1$; (ii) *to ω-span* a subspace Y of X, if each member y of Y is expressible as $\sum_{n \geq 1} \alpha_n x_n$ for some $\{\alpha_n\}$ in $\mathbb{K}$.

Note: For examples and counter-examples on ω-linearly independent sequences, see Chapter 7.

The proof of the following result is immediate.

Proposition 2.1.4: A sequence $\{x_n\}$ in a TVS (X,T) is a t.b. if and only if it ω-spans X and is ω-linearly independent.

Comparison with algebraic bases

At the outset let us mention the advantage of using the concept of t.b. rather than the Hamel basis notion. Indeed, recall the spaces c_0 and E of all entire functions equipped with the usual compact open topology T_E (see, for instance, [96], [121], [122] and [126]). Both of these spaces have the dimension $\aleph$ and as such we do not know the precise form of Hamel bases for these spaces. But the sequence $\{e^n\}$ is a t.b. for $(c_0, \|\cdot\|_\infty)$. Also, if $\delta_n : \mathbb{C} \to \mathbb{C}$ with $\delta_n(z) = z^n$, $n \in \mathbb{N}$, then, from the classical Taylor series theorem, the sequence $\{\delta_n\}$ is a t.b. for (E, T_E). In fact, the value of studying topological bases rather than algebraic bases lies only in those TVS (X,T) where $\dim X > \aleph_0$. These two notions of bases coincide in spaces having dimension $\leq \aleph_0$: a typical example of this is the space $(\phi, \|\cdot\|_\infty)$, for which $\{e^n\}$ is both a Hamel and a topological base. For TVS (X,T) with $\dim X > \aleph_0$, a t.b. can never be a Hamel base; for illustration, consider $(\delta, |\cdot|_\delta)$, for which $\{e^n\}$ is a t.b. but not a Hamel base (here, for instance, $\{1/n!\}$ cannot be expressed as a finite linear combination of $\{e^n\}$).

2.2 SCHAUDER BASES

To make the concept of a t.b. more useful in the context of locally convex spaces, it is natural to introduce

Definition 2.2.1: A t.b. $\{x_n;f_n\}$ for a TVS (X,T) is said to have the *Schauder property* (S.p.), or is simply called a *Schauder base* (S.b.), if each $f_n \in X^*$.

To see that the definition of an S.b. is really meaningful, we consider the following set of examples.

Example 2.2.2: Let λ and μ be two sequence spaces with $\mu \subset \lambda^\beta$. Then $\{e^n;e^n\}$ is an S.b. for $(\lambda,\sigma(\lambda,\mu))$; cf. Theorem 1.3.3.

For the next two examples, we refer to [201], p. 22 (cf. also [234], p. 87) and [67] (cf. also [214]) respectively.

Example 2.2.3: The sequence $\{x^n\}$ is a t.b. but not an S.b. for $(\phi, \|\cdot\|_\infty)$, the corresponding s.a.c.f. being given by $\{f^n\}$, where $x^i = e^i$ $(i = 1,2)$ and $x^n = e^2 + e^n/n$ $(n \geqslant 3)$; $f^1 = e^1$, $f^2 = e^2 - \sum_{n\geqslant 3} ne^n$ and $f^n = ne^n$ $(n \geqslant 3)$. Indeed, if $a \in \phi$, then $a_i = 0$ for $i \geqslant N+1$ for some N in $\mathbb{N}$. Thus

$$a = \sum_{i=1}^{N} f^i(a)x^i,$$

where we observe that, for b in ϕ, $f^1(b) = b_1$, $f^2(b) = b_2 - \sum_{i=3}^{\ell} ib_i$ and $f^n(b) = nb_n$, ℓ being the length of b. Further, $\phi^* = \ell^1$ and $f^2 \notin \ell^1$.

Example 2.2.4: Consider the sequences $\{x^n\}$ and $\{f^n\}$, where $x^1 = e^1$, $x^n = (-1)^{n+1} e^1 + e^n$ $(n \geqslant 2)$; $f^1 = e^1 + e^2 - e^3 + e^4 - e^5 + \ldots$, and $f^n = e^n$ $(n \geqslant 2)$. Then $\{x^n;f^n\}$ is a t.b. for $(\ell^1,\sigma(\ell^1,c_0))$ and, as $f^1 \notin c_0$, this base is not an S.b. for the space in question.

Sufficient conditions

In the definition of a t.b. $\{x_n;f_n\}$ for a Banach space $(X, \|\cdot\|)$, Schauder [206], who had earlier introduced the concept of this base for Banach spaces, assumed the continuity of each f_n. Later on, Banach ([5], p.111) pointed out that the assumption of Schauder was superfluous and showed that each t.b. in a Banach space is necessarily an S.b. Banach's result was subsequently extended to Fréchet and F-spaces by Newns [177] and Arsove [2] respectively. We will derive Arsove's result from a more general theorem for which we follow [77], [191] and [166].

Theorem 2.2.5: Let (X,T) be a TVS having a t.b. $\{x_n;f_n\}$. For each p in the family $D \equiv D_T$ of pseudonorms generating the topology T, let $\bar{p}(x) = \sup \{p(S_n(x)) : n \geqslant 1\}$. Then $\bar{D} = \{\bar{p} : p \in D\}$ defines a linear Hausdorff topology $\bar{T} \supset T$ such that $\{S_n\}$ is $\bar{T}$-T equicontinuous and that $\bar{T}$ is the coarest linear topology on X having these properties. Also, each f_n is continuous on $(X,\bar{T})$. Finally, if (X,T) is locally convex or complete or metrizable, so is then $(X,\bar{T})$.

Proof: Let $p \in D$. There exists q in D so that $p(x + y) \leqslant q(x) + q(y)$ for each x,y in X and, as $S_n(x) \to x$ in T for each x in X, we conclude that $p(x) \leqslant \bar{q}(x)$, giving $T \subset \bar{T}$. The $\bar{T}$-T equicontinuity of $\{S_n\}$ as well as the coarest part of $\bar{T}$ follow from Lemma 1.2.20. For each n there exists p and hence q in D such that $p(x_n) \neq 0$ and

$$|f_n(x)|p(x_n) \leqslant q(S_n(x)) + q(S_{n-1}(x)) \leqslant 2\bar{q}(x), \quad \forall\, x \in X. \tag{*}$$

Hence each f_n is $\bar{T}$-continuous.

It now suffices to prove the completeness of $(X,\bar{T})$ where (X,T) is given to be complete, the other assertions being more or less straightforward. So, let $\{y_\mu\}$ be a $\bar{T}$-Cauchy net in X. Fix p in D and $\varepsilon > 0$ and let $\{\alpha_n\}$ be the sequence given by

$$\lim f_n(y_\mu) = \alpha_n;\ n = 1,2,\ldots .$$

Also there exists η such that $p(S_n(y_\mu - y_\lambda)) \leqslant \varepsilon/2$ for all $n \geqslant 1$ and $\lambda, \mu \geqslant \eta$. Employing the usual arguments, we easily see that, for any given $\lambda \geqslant \eta$,

$$p(S_n(y_\lambda) - \sum_{i=1}^{n} \alpha_i x_i) \leqslant \varepsilon\ , \quad \forall\, n \geqslant 1. \tag{2.2.6}$$

Inequality (2.2.6) also yields the T-Cauchy character of $\sum_{n \geqslant 1} \alpha_n x_n$ and hence this series converges to x in X. The required result now follows from (2.6).□

Remarks (2.2.7): (a) In Theorem 2.2.5, each member of D is a pseudonorm. We may equally take D to be the family consisting of F-seminorms p and in this case the corresponding function $\bar{p}$ is also an F-seminorm. Indeed, let us observe that if A is bounded in (X,T) and $\alpha \in c_0$, then $\sup \{p(\alpha_j x) : x \in A\} \to 0$ as $j \to \infty$ and hence, in particular, for x in X,

$$\limsup_{j\to\infty} \sup_n p(\alpha_j \sum_{i=1}^{n} f_i(x)x_i) = 0;$$

that is, $\bar{p}(\alpha_j x) \to 0$. The other properties of the F-seminorm character of $\bar{p}$ are obvious.

(b) If (X,T) is metrizable in Theorem 2.2.5, then T is generated by a single F-norm p and the corresponding function $\bar{p}$ is also an F-norm satisfying

$$\bar{p}(S_n(x)) \leqslant \bar{p}(x); \ \forall x \in X, n \in \mathbb{N}. \tag{2.2.8}$$

(c) Assume now (X,T) to be an F-space in Theorem 2.2.5, T being generated by an F-norm p. It follows from the closed graph theorem that T is equivalent to $\bar{T}$ and hence, using (2.2.8), we may assume that p satisfies the condition

$$p(S_n(x)) \leqslant p(x); \ \forall x \in X, n \in \mathbb{N}. \tag{2.2.9}$$

(d) Let (X,T) and p be as in (c) above. If p_m denotes the Minkowski functional of conv(v_m), where $v_m = \{x \in X : p(x) < 1/m\}$, then

$$p_m(S_n(x)) \leqslant p_m(x); \ \forall x \in X; m, n \in \mathbb{N}. \tag{2.2.10}$$

To prove (2.2.10), we follow [211]. Let $x \in X$ and $m \in \mathbb{N}$. Choose $\alpha > 0$ with $p_m(x) < \alpha$ and so

$$\alpha^{-1}x = \sum_{i=1}^{M} k_i y_i; \ k_i \geqslant 0, \ \sum_{i=1}^{M} k_i = 1, \ p(y_i) < \frac{1}{m} \quad (1 \leqslant i \leqslant M).$$

Hence

$$\alpha^{-1} \sum_{j=1}^{n} f_j(x)x_j = \sum_{i=1}^{M} k_i z_i; \ z_i = \sum_{j=1}^{n} f_j(y_i)x_j \quad (1 \leqslant i \leqslant M).$$

By (2.2.9), $z_i \in \{x \in X : p(x) < 1/m\}$ and consequently

$$p_m(\sum_{j=1}^{n} f_j(x)x_j) \leqslant \inf \alpha = p_m(x).$$

(e) Finally, let (X,T) be a Fréchet space in Theorem 2.2.5. Using arguments similar to (c) and (d) above, we can show that $D_T = \{p_1 \leqslant p_2 \leqslant \cdots \leqslant p_n \leqslant \cdots\}$, where each p_n satisfies the condition

$$p_n(x) = \sup_{m \geqslant 1} p_n(S_m(x)), \ \forall x \in X. \tag{2.2.11}$$

From remarks (2.2.7), (c) and (e), we can derive the so called *continuity theorem*, namely,

Theorem 2.2.12: Each t.b. $\{x_n;f_n\}$ of an F-space or a Fréchet space is an S.b.

Remarks: The proof of Theorem 2.2.12 for F-spaces, using the idea of an F-norm, is also given in [209], p. 12.

Mitiagin (cf. [174], p. 88) observed that Theorem 2.2.12 still holds if the spaces (X,T) and $(X,\bar{T})$ satisfy the 'closed graph theorem property'; for instance, if (X,T) is barrelled and $(X,\bar{T})$ is Pták, then each t.b. for (X,T) is an S.b.; cf. [77].

To expect a t.b. $\{x_n;f_n\}$ for an arbitrary TVS (X,T) to be an S.b. we have to begin with those spaces which have nontrivial topological duals; for instance, we need to exclude spaces like $L^p[0,1]$, $0 < p < 1$ (cf. [3]; [146], p. 158). Besides, the restrictions of completeness and metrizability on (X,T) are also essential in order to force an arbitrary t.b. to be an S.b., otherwise the desired conclusion may not be true. Spaces $(\ell^1, \sigma(\ell^1,c_0))$ and ϕ considered earlier serve here as counter-examples. Further, these two conditions are not necessary in Theorem 2.2.12, as shown in the following two examples.

Example 2.2.13: $\{e^n;e^n\}$ is an S.b. for $(c_0,\sigma(c_0,\ell^1))$ which is neither complete (indeed, not even ω-complete) nor metrizable.

Example 2.2.14: $\{e^n;e^n\}$ is an S.b. for $(\phi, \|\cdot\|_\infty)$ which is not complete $(\bar{\phi} = c_0)$.

Extensions of the continuity theorem

Let us now discuss some possible extensions of Theorem 2.2.12 to spaces more general than Fréchet spaces. Theorem 2.2.12 was first extended by Arsove and Edwards [3] to a class of inductive limits of Fréchet spaces. Floret [58] further extended this theorem to LF-spaces which, in particular, include spaces considered by Arsove and Edwards; moreover, his proof is also shorter. In the present subsection we reproduce Floret's theorem; for more results on this topic, the reader is referred to Chapter 6.

Theorem 2.2.15: Every t.b. $\{x_n;f_n\}$ for an SRILF-space (X,T) is an S.b.

Proof: For technical reasons, we let $(X,T) = \inf\limits_{\longrightarrow} X_n$, where $X_n \equiv (X_n, |\cdot|_n)$ is a Banach space for $n \geq 1$ (when $X_n \equiv (X_n, \{p_{n,r} : r \geq 1\})$ is a Fréchet space, we proceed in exactly the same way, but using the seminorms $\{p_{n,r}\}$ in place of $|\cdot|_n$). We divide the proof into several stages.

(i) Let

$$Y_n = \{x \in X_n : S_m(x) \in X_n,\ \forall m \geq 1;\ |S_m(x) - x|_n \to 0\}.$$

By the retractivity of (X,T) and continuity of the injections from X_n into X_{n+1}, we have

$$X = \bigcup_{n \geq 1} Y_n;\ Y_n \subset Y_{n+1},\ n \geq 1.$$

(ii) Define another norm $\|\cdot\|_n$ on Y_n by $\|y\|_n = \sup\{|S_m(y)|_n : m \geq 1\}$, $n \geq 1$. Then the natural embedding $i_n : (Y_n, \|\cdot\|_n) \to (X_n, |\cdot|_n)$ is continuous. Further, each $S_k|Y_n$ is $\|\cdot\|_n - \|\cdot\|_n$ continuous.

(iii) We next show that the identity map $I : \operatorname{ind}\limits_{\longrightarrow} X_n \to \operatorname{ind}\limits_{\longrightarrow} Y_n$ is a topological isomorphism. Indeed, we have to prove the continuity of I, the continuity of I^{-1} being a simple consequence of the definition of inductive limits and the continuity of i_n's (cf. (ii)). In this direction, we first show that $(Y_n, \|\cdot\|_n)$ is a Banach space.

Let $\{y_n\}$ be a Cauchy sequence in $(Y_q, \|\cdot\|_q)$ and $\varepsilon > 0$. There exists y in X_q with $y_n \to y$ in $|\cdot|_q$. Further, we find an N in $\mathbb{N}$ so that

$$\|y_i - y_j\|_q = \sup_{m \geq 1} |S_m(y_i - y_j)|_q \leq \varepsilon\ ,\ \forall i,j \geq N. \tag{*}$$

Hence, for each $n \geq 1$, we find z^n in $Z_n = X_q \cap \operatorname{sp}\{x_1,\ldots,x_n\}$ so that $S_n(y_k) \to z^n$ in $|\cdot|_q$. Therefore, by using (*), we find for $n \geq 1$

$$|S_n(y_m) - z^n|_q \leq \sup_{j,k \geq N} \|y_j - y_k\|_q \leq \varepsilon,\ \forall m \geq N. \tag{**}$$

Now we proceed to show that $S_n(y) = z^n$, $n \geq 1$. Since $y_N \in Y_q$, there exists $n_0 \equiv n_0(N)$ such that $|S_n(y_N) - y_N|_q \leq \varepsilon$ for all $n \geq n_0$. Hence, using (**), we obtain

$$|z^n - y|_q < 3\varepsilon, \ \forall n > n_o.$$

Further, $f_n(z^{m_n})\, x_n = f_n(\lim_k S_{m_n}(y_k))\, x_n = z^n - z^{n-1}$, for $m_n > n$ and $n > 1$, yields

$$y = \lim_{n\to\infty} z^n = \lim_{n\to\infty} \sum_{i=1}^{n} f_i(z^{m_i}) x_i,$$

with respect to $|\cdot|_q$. Consequently, $f_n(y) = f_n(z^{m_n})$ and so $S_n(y) = z^n$, $n > 1$. This gives $y \in Y_q$ and it now follows easily from (**) that $\|y_m - y\|_q \to 0$ as $m \to \infty$.

(iv) In order to make use of De Wilde's closed graph theorem (Theorem 1.2.8) to conclude the continuity of I, let us observe that $\underrightarrow{\text{ind}}\ Y_n$ is webbed ([147], p. 63) and $\underrightarrow{\text{ind}}\ X_n$ is ultrabornological ([147], p. 44).

Finally, the desired result follows from the topological isomorphic character of I and the familiar properties of inductive limits (cf. [192], p. 79). □

2.3 ASSOCIATED SEQUENCE SPACES AND TOPOLOGIES

With each S.b. for a TVS (X,T) there are associated two natural structures, the sequence spaces and the topologies, which we shall discuss separately. Unless specified otherwise, it will be assumed throughout this section that $\{x_n;f_n\}$ is a $\sigma(X,X^*)$-S.b. for an l.c. TVS (X,T).

Sequence spaces

Define

$$\delta \equiv \delta\{x_n;f_n\} \equiv \delta_X = \{\{f_n(x)\} : x \in X\};$$

$$\mu \equiv \mu\{x_n:f_n\} \equiv \mu_{X^*} = \{\{f(x_n)\} : f \in X^*\};$$

$$\lambda \equiv \lambda\{x_n;f_n\} \equiv \lambda_X = \{a \in \omega \ : \ \sum_{n \geq 1} a_n x_n \text{ converges in T}\}.$$

Then $\lambda \subset \delta$ and δ , μ form a dual system with respect to the bilinear form $\langle,\rangle : \delta \times \mu \to \mathbb{K}$ with $\langle\{f_n(x)\}, \{f(x_n)\}\rangle = \langle x,f\rangle$. If we consider the maps R and S, $R : X \to \delta$ and $S : X^* \to \mu$ with $R(x) = \{f_n(x)\}$ and $S(f) = \{f(x_n)\}$, then the following is clearly true:

Proposition 2.3.1: R and S are topological isomorphisms from $(X,\sigma(X,X^*))$ onto $(\delta,\sigma(\delta,\mu))$ and from $(X^*,\sigma(X^*,X))$ onto $(\mu,\sigma(\mu,\delta))$ respectively.

We always have $\mu \subset \delta^\beta$ and $\delta \subset \mu^\beta$. On the other hand, using Proposition 2.3.1, the proof of the next result is readily verified.

Proposition 2.3.2: If $(X,\sigma(X,X^*))$ [resp. $(X^*,\sigma(X^*,X))$] is ω-complete, then $\delta = \mu^\beta$ [resp. $\mu = \delta^\beta$].

Exercise 2.3.3: Construct examples to show that ω-completeness in Proposition 2.3.2 cannot be dropped. [Hint: in one of the cases, consider $(c_o,\sigma(c_o,\ell^1))$.]

The space λ can also be topologized in a natural way. Let $Y = \{x \in X : \sum_{i=1}^{n} f_i(x)x_i \to x \text{ in } T\}$, then Y is isomorphic to λ, say, under the map H with $H(y) = \{f_n(y)\}$. Let us write $T|Y$ as T_Y and D_{T_Y} as D. For p in D, let $p^*(\{f_n(y)\}) = p(y)$. Then $\{p^* : p \in D\}$ generates an l.c. topology T^* on λ such that $(Y,T) \simeq (\lambda,T^*)$ under H; in fact, we have transferred the topology T on Y to the space λ and our discussion can be summarized in the form of

Proposition 2.3.4: The sequence space (λ,T^*) is topologically isomorphic to (Y,T) and, in particular, (λ,T^*) is an AK-space, where T^* is generated by the family $D^* = \{p^* : p \in D\}$ with $p^*(\{f_n(y)\}) = p(y)$.

Polar topologies

Our basic assumption remains unchanged, namely, $\{x_n;f_n\}$ is a $\sigma(X,X^*)$-S.b. for an arbitrary l.c. TVS (X,T). For $A \subset X$ and $B \subset X^*$, let

$$\hat{A} = \{x \in A : S_n(x) \in A, \ \forall n \geqslant 1\}; \ \tilde{B} = B \cup \{\bigcup_{n \geqslant 1} S_n^*[B]\}.$$

In this subsection, we discuss the l.c. topology generated by the polars of the sets $\tilde{B}$ by suitably restricting the sets. Our results here are similar in spirit to those given in [36] where the basic assumption is that $\{x_n;f_n\}$ is an S.b. for (X,T) rather than a weak S.b. for the space.

At the outset, let us introduce

Definition 2.3.5: Let $\mathcal{E}$ (resp. $\mathcal{E}_F$) denote the family of all equicontinuous subsets (resp. finite subsets) of X^*. The topology $\tilde{T}$ (resp. $\tilde{\sigma} \equiv \tilde{\sigma}(X,X^*)$) on

X generated by the polars of $\tilde{B}$ with B in $\mathfrak{E}$ (resp. $\mathfrak{E}_F$) is called the (resp. the *weak*) *locally convex topology associated with the weak Schauder base* $\{x_n;f_n\}$ abbreviated as the a.w.S.b. *topology* (resp. the *weak* a.w.S.b. *topology*).

Note: Since $\{f_n;\psi(x_n)\}$ is a $\sigma(X^*,X)$-S.b. for X^* (ψ is the canonical embedding, $\psi : X \to X^{**}$), we may also define an l.c. *topology* $\hat{\sigma} \equiv \hat{\sigma}(X^*,X)$ on X^*, generated by $\{\hat{q}_x : x \in X\}$ where, for f in X^*, $\hat{q}_x(f) = \sup\{|S_n^*(f)| : n \geqslant 1\}$.

The following is useful in appreciating the structure of either $\tilde{T}$ or $\tilde{\sigma}$.

Proposition 2.3.6: For $A \subset X$ and $B \subset X^*$, $\hat{A} \subset A$ and $B \subset \tilde{B}$. Further $(\tilde{B})^o = (B^o)^\wedge$ and if B is equicontinuous, then B is $\sigma(X^*,X)$-bounded.

Proof: If $x \in (\tilde{B})^o$, then for f in B, $|\langle S_n(x),f\rangle| \leqslant 1$ for all $n \geqslant 1$. Hence $(\tilde{B})^o \subset (B^o)^\wedge$ and the other inclusion follows in the same way. For the last part, it is enough to show that $(\tilde{B})^o$ is absorbing. Indeed, if $x \in X$ then $S_n(x) \in \alpha B^o$ for all $n \geqslant 1$ and some constant $\alpha > 0$. Since αB^o is $\sigma(X,X^*)$-closed and $S_n(x) \to x$ in $\sigma(X,X^*)$, we find that $x \in \alpha B^o$ and so $x \in \alpha(\tilde{B})^o$. □

Proposition 2.3.7: The topology $\tilde{T}$ is generated by $\{\tilde{p}_B : B \in \mathfrak{E}\}$ where $\tilde{p}_B(x) = \sup\{p_B(S_n(x)) : n \geqslant 1\}$ with $p_B(S_n(x)) = \sup\{|f(S_n(x))| : f \in B\}$. In particular, $\tilde{\sigma}$ is generated by $\{\tilde{p}_f : f \in X^*\}$, $\tilde{p}_f(x) = \sup\{|f(S_n(x))| : n \geqslant 1\}$.

Proof: It suffices to prove the first part only. Observe that $\tilde{T}$ is generated by $\{p_{\tilde{B}} : B \in \mathfrak{E}\}$, where for any $\sigma(X^*,X)$-bounded subset A of X^*, $p_A(x) = \sup\{|f(x)| : f \in A\}$. We will prove that $\tilde{p}_B = p_{\tilde{B}}$. For x in X and B in $\mathfrak{E}$.

$$p_{\tilde{B}}(x) = \max\{\sup_{f\in B} |f(x)|, \sup_{g\in B^*} |g(x)|\},$$

where $B = \cup\{S_n^*[B] : n \geqslant 1\}$. By the weak basis character of $\{x_n;f_n\}$, $\sup\{|g(x)| : g \in B^*\} = \sup\{p_B(S_n(x)) : n \geqslant 1\} \geqslant \sup\{|f(x)| : f \in B\}$. Hence $p_{\tilde{B}}(x) = \tilde{p}_B(x)$. □

Proposition 2.3.8: We have $T \subset \tilde{T}$ and $\sigma \equiv \sigma(X,X^*) \subset \tilde{\sigma}$. Further, $\tilde{\sigma}$ is the weakest polar topology on X such that $\sigma \subset \tilde{\sigma}$ and $\{S_n\}$ is $\tilde{\sigma}$ - $\tilde{\sigma}$ equicontinuous.

Proof: Let $x \in X$ with $\tilde{p}_B(x) \leqslant 1$ for B in $\mathfrak{E}$. If $U_B = \{y \in X : p_B(y) \leqslant 1\}$, then

U_B is $\sigma(X,X^*)$-closed and, as $S_n(x) \in U_B$ for $n \geq 1$, we find $p_B(x) \leq 1$. Thus $p_B(x) \leq \tilde{p}_B(x)$.

Next, observe that $\{x_n;f_n\}$ is an S.b. for $(X,\tilde{\sigma})$ and also $\tilde{p}_f(S_n(x)) \leq \tilde{p}_f(x)$ for each $n \geq 1$, x in X and f in X^* and this yields the $\tilde{\sigma}$ - $\tilde{\sigma}$ equicontinuity of $\{S_n\}$. Finally, let σ_1 be another polar topology on X such that $\sigma \subset \sigma_1$, $\{x_n;f_n\}$ is an S.b. for (X,σ_1) and $\{S_n\}$ is σ_1-σ_1 equicontinuous. If $f \in X^*$, there exist p_1 and q_1 in D_{σ_1} (q_1 depends upon p_1) such that $\tilde{p}_f(x) \leq \sup\{p_1(S_n(x)) : n \geq 1\} \leq q_1(x)$. Therefore $\tilde{\sigma} \subset \sigma_1$. □

Exercise 2.3.9: (cf. [36]): Let $\{x_n;f_n\}$ be an S.b. for an l.c. TVS (X,T). Prove that $\tilde{T}$ is the weakest polar topology on X such that $T \subset \tilde{T}$, $\{S_n\}$ is $\tilde{T}$-$\tilde{T}$ equicontinuous and $\{x_n;f_n\}$ is an S.b. for $(X,\tilde{T})$.

$\tilde{\sigma}$ and $\sigma\gamma(\delta,\mu)$-topologies

The topology $\tilde{\sigma}$ can be transferred to δ as the topology $\tilde{\sigma}^*$ generated by the family $\{\tilde{p}_f^* : f \in X^*\}$, cf. Proposition 2.3.4 for notation etc. To see that $\tilde{\sigma}^*$ is the same as the topology $\sigma\gamma(\delta,\mu)$ introduced by Garling [64], let us recall the basic concept of the latter from [132]. Suppose that λ is an arbitrary sequence space and $\lambda^\gamma = \{b \in \omega : r_b(a) \equiv \sup_n |\sum_{i=1}^{n} a_i b_i| < \infty\ ,\ \forall a \in \lambda\}$. For any subspace μ of λ^γ , let $\sigma\gamma(\lambda,\mu)$ *denote the* l.c. *topology* on λ generated by $\{r_b : b \in \mu\}$. We will need the following result implicit in the work of Garling [64], p. 973 and [65], p. 998.

Proposition 2.3.10: For any sequence space λ, the space $(\lambda^\beta, \sigma\gamma(\lambda^\beta,\lambda))$ is complete.

Proof: Let $\{b^\alpha\}$ be any $\delta\gamma(\lambda^\beta,\lambda)$-Cauchy net, $\varepsilon > 0$ and $a \in \lambda$. There exists $\alpha_o \equiv \alpha_o(\varepsilon,a)$ such that

$$\sup_n |\sum_{i=1}^{n} (b_i^\alpha - b_i^\delta)a_i| \leq \varepsilon;\ \forall\ \alpha,\ \ \delta \geq \alpha_o. \qquad (*)$$

The coordinatewise convergence of $\{b^\alpha\}$ and the routine use of (*) yields the required result. □

Returning to the notation δ and μ, we find that the $\sigma\gamma(\delta,\mu)$ topology on δ is precisely the one given by $\{\tilde{p}_f^*: f \in X^*\}$. An interesting feature of the compatibility of $\tilde{\sigma} \equiv \tilde{\sigma}(X,X^*)$ is contained in

<u>Proposition 2.3.11</u>: If $(X,\sigma(X,X^*))$ [resp. $(X^*,\sigma(X^*,X))$] is ω-complete, then $(X,\tilde{\sigma})$ [resp. $(X^*,\hat{\sigma})$] is complete. Further, if (X,T) is an S-space, then $\tilde{\sigma}$ is compatible with $\langle X,X^*\rangle$ and it is the weakest compatible topology on X for which $\{x_n;f_n\}$ is an S.b. such that $\{S_n\}$ is $\tilde{\sigma}$-$\tilde{\sigma}$ equicontinuous.

<u>Proof</u>: The first part follows from Propositions 3.2 and 3.10. In the second case, $(\delta,\sigma\gamma(\delta,\mu))$ is an AK-space (cf. [132], p. 66) and therefore, applying Theorem 1.3.4 and using $\mu = \delta^\beta$, we get $(\delta,\sigma\gamma(\delta,\mu))^* = \mu$. For the last part, apply Proposition 2.3.8. □

<u>Exercise 2.3.12</u>: Let $(X,\sigma(X,X^*))$ be ω-complete and $\{S_n\}$ be σ-σ equicontinuous. Show that $(X,\sigma(X,X^*))$ is topologically isomorphic to a closed subspace of some power of $\mathbb{K}$.

<u>Compatibility of $\tilde{T}$</u>

In the rest of this subsection, we assume that (X,T) is an l.c. TVS containing an S.b. $\{x_n;f_n\}$. Our basic interest lies in finding the nature of the compatibility of $\tilde{T}$ and we follow [36] for all the results of this subsection.

<u>Proposition 2.3.13</u>: Consider an arbitrary member B of $\mathfrak{E}$. Then (i) $\tilde{B}$ is $\hat{\sigma}$-precompact; and (ii) $\tilde{B}$ is $\hat{\sigma}$-relatively compact and $\Gamma(\tilde{B})^\sigma$ is $\sigma \equiv \sigma(X^*,X)$-compact provided that $(X^*,\sigma(X^*,X))$ is ω-complete.

<u>Proof</u>: (i) Define $P_i : X^* \to X^*$, $P_i(f) = f(x_i)f_i$. From a known result ([192], p. 50), $P_i[B]$ is $\sigma(X^*,X)$-precompact for $i \geqslant 1$. Next observe that $P_i[B] = P_i[\tilde{B}]$. Further, the set $P_i[B]$ is $\hat{\sigma}$-precompact; indeed, for each $\varepsilon > 0$ and x in X, we find $u_1,\ldots,u_n$ in $P_i[B]$ so that for $u \equiv P_i(f)$, $f \in B$, in $P_i[B]$, there corresponds $u_{i_o} \equiv P_i(g_{i_o})$, $g_{i_o} \in B$ with $|\langle S_m(x), (f(x_i) - g_{i_o}(x_i))f_i\rangle| < \varepsilon$ for $m \geqslant 1$; that is, $\hat{q}_x(u-u_{i_o}) < \varepsilon$ and we are done. This suggests in particular that $S_n^*[\tilde{B}]$ is also $\hat{\sigma}$-precompact for each $n \geqslant 1$. To complete the proof, we proceed to show that $S_n^*(f) \to f$ in $\hat{\sigma}$ uniformly in $f \in \tilde{B}$ and make use of Lemma 1.2.21 (cf. also [171]). In fact,

$$\sup_{f \in B} \hat{q}_x(S_n^*(f)-f) = p_{\tilde{B}}(x-S_n(x))$$

and we then make use of Exercise 2.3.9.

(ii) For the first part, use the completeness of $(X^*,\hat{\sigma})$ and (i). In the second part, let us first observe that $\Gamma(\tilde{B})^{\hat{\sigma}}$ is $\hat{\sigma}$-compact. Therefore $\Gamma(\tilde{B})^{\hat{\sigma}}$ is $\sigma(X^*,X)$-compact and so it is $\sigma(X^*,X)$-closed. Hence $\Gamma(\tilde{B})^{\sigma} = \Gamma(\tilde{B})^{\hat{\sigma}}$. □

Proposition 2.3.14: If (X,T) is an S-space, then $\sigma(X,X^*) \subset \tilde{T} \subset \tau(X,X^*)$; in particular, if (X,T) is a Mackey S-space, then $T \approx \tilde{T}$.

Proof: This follows directly from Proposition 2.3.13 (ii) and the Mackey-Arens theorem. □

Exercise 2.3.15: If (X,T) is an S-space and also $(X,\sigma(X,X^*))$ is ω-complete, show that $(X,\tilde{T})$ is complete (see [36]).

3 Some applications of Schauder bases

3.1 INTRODUCTION

The theory of Schauder bases has several applications in the structural study of l.c. TVS. Since it is not possible to mention all of them here, we choose only a few applications which depend upon the elementary knowledge gained in Chapter 2 and some basic facts from the theory of l.c. TVS. In fact, we explore conditions on an l.c. TVS to find its completeness, apply this result to prove an important theorem in the theory of sequence spaces and finally investigate conditions to infer continuity of linear maps and seminorms on locally convex spaces having S.b.

3.2 THE COMPLETION THEOREM

As pointed out in Chapter 2, the completeness of an l.c. TVS containing a t.b. $\{x_n;f_n\}$ is neither a necessary nor a sufficient condition to enforce the Schauder character of $\{x_n;f_n\}$. In order to infer the completeness of an l.c. TVS (X,T) containing an S.b. $\{x_n,f_n\}$, we have to have some other restrictions on (X,T). In the first instance, (X,T) has to be barrelled. The sole purpose of this condition is to ensure the T-T equicontinuity of $\{S_n\}$. Unless stated to the contrary, we consider throughout this section that (X,T) is an arbitrary l.c. TVS having an S.b. $\{x_n;f_n\}$. To begin with, let us recall the following result from [202], p. 36.

Proposition 3.2.1: Let (X,T) be barrelled and let $(\hat{X},\hat{T})$ denote the completion of (X,T). If $\hat{f}_n$ is the unique extension of f_n to $\hat{X}$, then $\{x_n;\hat{f}_n\}$ is an S.b. for $(\hat{X},\hat{T})$.

Proof: Write $\hat{D}$ for $D_{\hat{T}}$ and introduce operators $\hat{S}_n:\hat{X} \to X$, $n \geq 1$ by

$$\hat{S}_n(\hat{x}) = \sum_{i=1}^{n} \hat{f}_i(\hat{x})x_i, \quad \hat{x} \in \hat{X}.$$

Choose $\hat{x}$ in $\hat{X}$, p in D and $\varepsilon > 0$ arbitrarily. There exist $\hat{q}$ in $\hat{D}$ and y in X

(and hence $N \equiv N(\varepsilon,y,p)$) so that $p(\hat{S}_n(\hat{x})) \leqslant \hat{q}(\hat{x})$; $n \geqslant 1$, $\hat{r}(\hat{x}-y) \leqslant \varepsilon/3$; $\hat{r} = \max(\hat{p},\hat{q})$ and $p(R_n(y)) \leqslant \varepsilon/3$, $n \geqslant N$. Consequently $\hat{p}(\hat{S}_n(\hat{x})-\hat{x}) \leqslant \varepsilon$ for $n \geqslant N$ and as $\hat{f}_n(x_m) = \delta_{mn}$, the required result is proved. □

Theorem 3.2.2: If (X,T) is ω-complete and barrelled, then (X,T) is complete.

Proofs: We follow [130], [231] and [109] respectively to give three proofs.

(i) Let $\hat{x} \in \hat{X}$. Then from the preceding result $\{\hat{S}_n(\hat{x})\}$ is Cauchy in (X,T) and hence it converges to y in X. Therefore $\hat{x} = y \in X$, giving $\hat{X} = X$. □

(ii) For any balanced convex and equicontinuous subset M of X^*, consider the equicontinuous subset M_1 given by $M + \cup \{S_n^*[M]: n \geqslant 1\}$. By the Grothendieck completion theorem, to every $\varepsilon > 0$ there corresponds a neighbourhood of the origin, say, u in $(X^*,\sigma(X^*,X))$ such that $|\langle\hat{x},\hat{f}\rangle| \leqslant \varepsilon$ for all f in $u \cap M_1$. Since $S_n^*(f) \to f$ in $\sigma(X^*,X)$ uniformly on M and $S_n^*(f) - f \in M_1$ for each f in M, for some N we have

$$\sup \{|\langle\hat{S}_n(\hat{x}) - \hat{x},\hat{f}\rangle| : f \in M\} \leqslant \varepsilon, \ \forall \ n \geqslant N.$$

Consequently $\hat{S}_n(\hat{x}) \to \hat{x}$ in $\hat{T}$ and so $X = \hat{X}$. □

(iii) This proof is independent of the Grothendieck completion theorem and depends upon the following lemma (cf. [125]).

Lemma 3.2.3: Let there be a directed family $\{R_i : i \in \Lambda\}$ of linear maps from a vector space X into an arbitrary l.c. TVS (Y,T) such that $\{R_i(x) : i \in \Lambda\}$ is T-bounded in Y for each x in X. Let S denote the weakest l.c. topology on X such that $\{R_i : i \in \Lambda\}$ is S-T equicontinuous. Further, assume that $\{x_\alpha: \alpha \in \Delta\}$ is an S-Cauchy net in X such that $\{R_i(x_\alpha) : \alpha \in \Delta\}$ converges in (Y,T) for i in Λ. (a) Let $\{R_i(X_\alpha) : i \in \Lambda\}$ be T-Cauchy for each α, then (1) $\{\lim_\alpha R_i(x_\alpha) : i \in \Lambda\}$ is Cauchy in (Y,T) and (2) $y = T\text{-}\lim_i \lim_\alpha R_i(x_\alpha)$ provided that $y = \sigma(Y,Y^*)\text{-}\lim_i \lim_\alpha R_i(x_\alpha)$. (b) If $X = Y$ and $x_\alpha = T\text{-}\lim_i R_i(x_\alpha)$ for each α and $y = \sigma(X,X^*)\text{-}\lim_i \lim_\alpha R_i(x_\alpha)$, then $y = T\text{-}\lim_\alpha x_\alpha$.

Proof: To begin with, let us observe, in view of Lemma 1.2.19, that to every $\varepsilon > 0$ and p in D_T, there exists β in Δ such that

$$p(y_i - R_i(x_\beta)) < \frac{\varepsilon}{3}, \ \forall i \in \Lambda, \qquad (*)$$

where $y_i = T - \lim_\alpha R_i(x_\alpha)$, $i \in \Lambda$.

(a) Case (1) is a direct consequence of (*); on the other hand, if $p \in D_T$ and $\varepsilon > 0$, there exist an equicontinuous subset M of Y* and k such that $p(y_i - y_j) < \sup\{|f(y_i - y_j)| : f \in M\} < \varepsilon$ for $i,j > k$ (cf. (1)). This immediately yields (2).

(b) This follows from (a) part (2) and an inequality of the type (*) with β being replaced by $\alpha > \beta$. □

(iii) (cont'd): We use Lemma 2.3(b) with $\Lambda = \mathbb{N}$ and $R_i = S_i$. Since (X,T) is barrelled, the topology $S \approx T$. Let now $\{x_\alpha : \alpha \in \Delta\}$ be a T-Cauchy net in X. If $\alpha_i = \lim_\alpha f_i(x_\alpha)$, one discovers by using Lemma 3.2.3(a) that $\{\lim_\alpha S_i(x_\alpha): i > 1\}$ is Cauchy in (X,T) and hence it converges to y. Now apply Lemma 3.2.3(b) to conclude that $x_\alpha \to y$. □

Remark: Let us consider the importance of the conditions laid down in Theorem 3.2.2; indeed, the only important restrictions are the barrelled character of (X,T) and the presence of a t.b. $\{x_n; f_n\}$ in (X,T), as illustrated in

Example 3.2.4: The space $(\ell^1, \sigma(\ell^1, c_0))$ has an S.b. $\{e^n, e^n\}$; however, this space is ω-complete but not complete and not barrelled (cf. [132], p. 121).

Example 3.2.5: We follow [143] and [145] for this example. Let A = [0,1] and consider $\omega_0(A) = \{f \in \mathbb{R}^A : f(\alpha) = 0$ for all except countable α's in A$\}$ and $\phi(A) = \{f \in \mathbb{R}^A : f(\alpha) = 0$ for all except finite α's in A$\}$. In the usual manner, $\langle \omega_0(A), \phi(A) \rangle$ forms a dual system. We will show that $(\omega_0(A), \sigma)$ is (i) ω-complete, (ii) barrelled, (iii) inseparable and (iv) incomplete, where $\sigma \equiv \sigma(\omega_0(A), \phi(A))$.

(i) Let $\{f_n\}$ be σ-Cauchy in $\omega_0(A)$. If $B_n = \{x \in A : f_n(x) \neq 0\}$, $n > 1$, let us write B for the union of B_n's. Define $f: A \to \mathbb{R}$ by

$$f(x) = \begin{cases} \lim f_n(x), & x \in B, \\ 0, & x \in A \setminus B. \end{cases}$$

Clearly $f \in \omega_o(A)$ and is the σ-limit of $\{f_n\}$.

(ii) By Theorem 1.2.10, it suffices to prove that each countable linearly independent subset $\{g_n\}$ of $\phi(A)$ is $\sigma(\phi(A), \omega_o(A))$-unbounded. Define e^α in $\phi(A)$ with α in A such that $e^\alpha(\beta) = \delta_{\alpha\beta}$, the generalized Kronecker delta. By the induction process, we can choose a subsequence $\{n_k\}$ with $n_k < n_{k+1}$ $(k \geq 1)$, $n_k \to \infty$, and a corresponding sequence $\{\alpha_{n_k}\}$ from A such that $g_{n_j}(\alpha_{n_k}) = 0$ for $1 \leq j \leq k-1$ and $g_{n_k}(\alpha_{n_k}) \neq 0$. Introduce f in $\omega_o(A)$ so that $f(\alpha_{n_k})g_{n_k}(\alpha_{n_k}) \geq k$ and $f(\alpha) = 0$ for $\alpha \neq \alpha_{n_k}$, $k \geq 1$. Clearly $\{f\}^o$ does not absorb $\{g_n\}$.

(iii) Let $D = \{f_n\}$ be an arbitrary countable dense subset of $(\omega_o(A),\sigma)$. If $A_n = \{\alpha \in A: f_n(\alpha) \neq 0\}$ and B is the union of A_n's, then there exists β in $A \setminus B$. Define f in $\omega_o(A)$ with $f(\alpha) = 1$ for $\alpha = \beta$ and $f(\alpha) = 0$ for $\alpha \neq \beta$. Then

$$D \cap \{f + (2e^\beta)^o\} = \emptyset,$$

and so D is not dense in $(\omega_o(A),\sigma)$.

(iv) Since $\omega_o(A)$ is dense in $(\mathbb{R}^A, \sigma(\mathbb{R}^A, \phi(A)))$, the required conclusion is proved.

Weak Schauder bases

In this subsection, we once again take up the question of finding the completeness of an l.c. TVS (X,T) which we assume throughout to have a $\sigma(X,X^*)$-S.b. $\{x_n;f_n\}$. We follow [130] for most of the results: however, the main theorem which contains a similar weaker result in [231] is

Theorem 3.2.6: Let (X,T) be ω-barrelled with $(X,\sigma(X,X^*))$ being ω-complete. Then (X,T) is complete.

Proof: It is easily seen (e.g. [130], Lemma 2.2) that

$$\hat{X} \subset [X^*,\sigma(X^*,X)]^+.$$

Hence $\langle \hat{x}, S_n^*(f)\rangle \to \langle \hat{x},\hat{f}\rangle$ for every $\hat{x}$ in $\hat{X}$ and f in X^*. However, $\langle \hat{x}, S_n^*(f)\rangle = \langle \hat{S}_n(\hat{x}), f\rangle$ for $n \geq 1$. Therefore $\{\hat{S}_n(\hat{x})\}$ is Cauchy in $(X,\sigma(X,X^*))$. □

On the basis of Proposition 1.2.13, one can derive

Corollary 3.2.7: Suppose (X,T) is a Mackey S-space with $(X,\sigma(X,X^*))$ being ω-complete, then (X,T) is complete.

As an application of Theorem 3.2.6, we obtain (cf. [146], p. 414)

Proposition 3.2.8: If λ is an arbitrary perfect sequence space, then $(\lambda,\tau(\lambda,\lambda^\times))$ is complete.

Proof: By Theorem 1.3.2, the sequence space in question is an S-space and at the same time $(\lambda,\sigma(\lambda,\lambda^\times))$ is ω-complete. Now make use of Example 2.2.2 and Corollary 3.2.7. □

Exercise 3.2.9: If (X,T) is ω-barrelled, prove that $\{x_n;\hat{f}_n\}$ is an S.b. for $(\hat{X},\sigma(\hat{X},X^*))$.

3.3 CONTINUITY OF MAPS

The presence of an S.b. in an l.c. TVS is quite useful in ascertaining the continuity of maps by way of determining their continuity through certain simple sequences converging to zero. We consider two types of probem in this direction: (i) continuity of linear functionals and (ii) continuity of operators and seminorms.

Linear functionals

Our basic result on the continuity of linear functionals is Theorem 3.3.5. We first collect the necessary background and assume that (X,T) is an l.c. TVS containing an S.b. $\{x_n;f_n\}$. To begin with, let $\mathcal{B}^\#$ denote the collection of all subsets B of $X \equiv (X,T)$, which are absorbing, balanced, convex and satisfy the condition: for each x in X, there exists an $n_0 \equiv n_0(x,B)$ in $\mathbb{N}$ such that $R_n(x) \in B$ for $n \geqslant n_0$. We recall (cf. [143], Chapter 2)

Proposition 3.3.1: There is a strongest l.c. topology $T^\#$ on X for which $R_n(x) \to 0$ for each x in X and $T \subset T^+ \subset T^\#$. Further, an f in X' is in $X^\# = (X,T^\#)^*$ if and only if $f(R_n(x)) \to 0$ for each x in X.

Proof: For the first part, observe that $\mathcal{B}^\#$ is a fundamental neighbourhood system at the origin for $T^\#$. For the other part, if $f \in X'$ with $f(R_n(x)) \to 0$, then $D_{T^\#} \cup \{p_f\}$ generates $T^\#$ also. □

The next definition, used in the present form in [144], is essentially given in [230].

Definition 3.3.2: A subset K of X', $X \equiv (X,T)$ is said to be *sums-limited* provided that, for each x in X, $\sup\{|g(R_n(x))|: g \in K\} \to 0$ as $n \to \infty$.

Proposition 3.3.3: (a) If K is sums-limited, then $K \subset X^{\#}$. (b) If K and K_1 are sums-limited, then each of the following sets is sums-limited: (i) αK, α a scalar, (ii) $K + K_1$, (iii) $\bar{K}^{\sigma}$, $\sigma \equiv \sigma(X^{\#},X)$ and (iv) K^{oo}, the polar being taken relative to $\langle X,X'\rangle$. (c) $u^o \subset X^{\#}$ and is sums-limited for each u in $\mathcal{B}^{\#}$; conversely, if K is $\sigma(X^{\#},X)$-bounded and sums limited, then K^o is a 0-neighbourhood of $(X,T^{\#})$.

Proof: Straightforward computation. □

The above proposition immediately leads to the following result and hence its proof is omitted.

Proposition 3.3.4: The topology $T^{\#}$ is an $\mathcal{S}$-topology where $\mathcal{S}$ is the collection of all balanced, convex, sums-limited and $\sigma(X^{\#},X)$-bounded subsets of $X^{\#}$. If $f \in X^{\#}$, then $S_n^*(f) \to f$ in $\sigma(X^{\#},X)$.

Theorem 3.3.5: Let (X,T) be an S-space. Then an f in X' is a member of X* if and only if, for each x in X,

$$\lim_{n\to\infty} f(R_n(x)) = 0. \tag{3.3.6}$$

Note: In proving the above theorem, it is enough to show that $X^* = X^{\#}$ (cf. Proposition 3.3.1). We give two proofs: the first is a slight modification of [144] and is contained in [128], while the other is to be found in [37]. For a similar theorem with stronger hypothesis, one may see [103].

First proof: By Proposition 3.3.1, $\{S_n^*(f)\}$ is $\sigma(X^*,X)$-Cauchy for each f in $X^{\#}$ and so there exists g in X* such that $S_n^*(f) \to g$ in $\sigma(X^*,X)$. But $S_n^*(f) \to f$ in $\sigma(X^{\#},X)$ also. Therefore $X^{\#} \subset X^*$. □

Second proof: Let $f \in X^{\#}$. Then

$$f(x) = \sum_{n\geq 1} f_n(x)\, f(x_n), \quad \forall x \in X.$$

Recall the associated sequence spaces δ and μ, and note that $\mu = \delta^\beta$ by Proposition 2.3.2. Here $\{f(x_n)\} \in \delta^\beta$ and so, for some g in X^*, $f(x_n) = g(x_n)$ for $n \geq 1$; thus

$$f(x) = \sum_{n\geq 1} f_n(x) g(x_n) = g(x), \ \forall\, x \in X.$$

Hence $X^\# \subset X^*$. □

As a consequence of Theorem 3.3.5, we have

Proposition 3.3.7: Let (X,T) be an S-space. Then $X^* = X^+ = X^\#$ and $(X^*,\beta(X^*,X))$ is complete.

Proof: This directly follows from Proposition 1.2.12 and the Theorem 3.3.5. □

Remark: The condition that (X,T) is an S-space in Theorem 3.3.5 and in Proposition 3.3.7 cannot be dropped. For instance, we have

Example 3.3.8: This is the space considered in Example 2.2.14. Here $\phi^* = \ell^1$ and for any member of $\phi' = \omega$, condition (3.3.6) is satisfied. Therefore $\phi^\# = \phi' = \omega$, but $\phi^* = \phi^+ = \ell^1 \subsetneqq \phi^\#$. Finally, note that $(\ell^1,\sigma(\ell^1,\phi))$ is not ω-complete, for the sequence $\{e^{(n)}\}$ is $\sigma(\ell^1,\phi)$-Cauchy but is not convergent in $(\ell^1,\sigma(\ell^1,\phi))$.

Example 3.3.9: This is the space $(\ell^1,\tau(\ell^1,c_o))$ for which $\{e^n,e^n\}$ is an S.b.; cf. [132], p. 119. This is not an S-space, as $(c_o,\sigma(c_o,\ell^1))$ is not ω-complete; cf. [132], p. 118. Here $(\ell^1)^\# = \ell^\infty$, in fact, by Proposition 3.3.1, $\ell^\infty \subset (\ell^1)^\#$ and one easily finds that $(\ell^1)^\# \subset (\ell^1)^\beta = \ell^\infty$. Note that $(\ell^1)^* = c_o$.

Exercise 3.3.10: Prove that if (X,T) is a Mackey S-space, then $T = T^+ = T^\#$ and hence show that there cannot be any strictly finer l.c. topology on X for which $\{x_n;f_n\}$ is an S.b. Also show that $(\phi, \|\cdot\|_1)$ is not an S-space and that $(\ell^1,\tau(\ell^1,c_o))$ is not σ-barrelled. [Hint: for the last part, use Proposition 1.2.11.]

An interesting application of Theorem 3.3.5 is contained in

Proposition 3.3.11: Let (λ,T) be a monotone l.c.s.s, where T is compatible with $\langle\lambda,\lambda^{\times}\rangle$, then $\lambda^* = \lambda^{\times} = \lambda^{\#} = \lambda^{+}$ and $(\lambda^{\times},\beta(\lambda^{\times},\lambda))$ is complete.

Proof: By Theorem 1.3.3, $\{e^n;e^n\}$ is a $\tau(\lambda,\lambda^{\times})$-S.b. and hence a T-S.b. for λ and so we have the topology $T^{\#}$. Observe that $\lambda^* = \lambda^{\times}$ and, as $(\lambda^{\times},\sigma(\lambda^{\times},\lambda))$ is ω-complete (cf. Proposition 1.3.5), we conclude that $\lambda^{\times} = \lambda^* = \lambda^{+} = \lambda^{\#}$ and $(\lambda^{\times},\beta(\lambda^{\times},\lambda))$ is complete. □

Remark: The monotonic character of λ in the above proposition is indispensable to infer $\lambda^{\times} = \lambda^{\#}$. For, we have

Example 3.3.12: For the space $(k,\sigma(k,\ell^1))$, the element $y = \{(-1)^n/n\}$ belongs to $k^{\#}$; however $y \notin k^{\times}$. Hence $k^{\times} \subsetneqq k^{\#}$.

Linear operators and seminorms

The results of this last subsection on the continuity of linear operators and seminorms are taken from [103]; their extensions when the hypothesis of Schauder bases is replaced by Schauder decompositions are to be found in [131] (cf. also [81]). As before, we consider (X,T) to be an arbitrary l.c. TVS containing an S.b. $\{x_n;f_n\}$. Let us begin with

Proposition 3.3.13: Let (X,T) be barrelled. If p is a seminorm on X satisfying the condition:

$$\lim_{n\to\infty} p(S_n(x)) = p(x), \quad \forall x \in X, \tag{3.3.14}$$

then p is continuous. Conversely, if every seminorm p on $X \equiv (X,T)$ satisfying (3.3.14) is T-continuous, then (X,T) is infrabarrelled.

Proof: Consider the sequence $\{p \circ S_n\}$ of continuous seminorms on (X,T). Since p is the pointwise limit of this sequence, p is continuous.

Conversely, let u be an arbitrary bornivorous barrel and let p denote the Minkowski functional of u. Suppose that $p_1(x) = \sup\{p(S_n(x)): n \geq 1\}$ for each x in X. Since $\{p_1(S_n(x))\}$ is increasing, it satisfies (3.3.14) and so

p_1 is continuous. Also, the closed character of u ensures that $p \prec p_1$, therefore p is continuous. □

Theorem 3.3.15: The following statements are equivalent in (X,T):

(i) There exists no l.c. topology $T_1 \supset T$ such that $\{x_n;f_n\}$ is an S.b. for (X,T_1).

(ii) If p is a seminorm on X with

$$p(R_n(x)) \to 0, \quad \forall x \in X, \tag{3.3.16}$$

then p is continuous on (X,T).

(iii) If R is a linear operator from (X,T) into another l.c. TVS (Y,S) such that

$$R(R_n(x)) \to 0, \quad \forall x \in X, \tag{3.3.17}$$

then R is T-S continuous.

Proof: (i) ⇒ (ii). For p as in (ii), let $D_1 = \{p\} \cup D_T$. Let T_1 be the l.c. topology on X generated by D_1. Using (3.3.16), we conclude that $\{x_n;f_n\}$ is an S.b. for (X,T_1) and so $T \approx T_1$.

(ii) ⇒ (iii). For q in D_S, let $Q_q = q \circ R$. Then Q_q satisfies (3.3.16) and this yields the continuity of R.

(iii) ⇒ (i). Suppose T_1 is another l.c. topology on X such that $T \subset T_1$ and $\{x_n;f_n\}$ is an S.b. for (X,T_1). The identity operator $I:(X,T) \to (X,T_1)$ satisfies (3.3.17), and so $T_1 \subset T$, giving $T \approx T_1$. Hence (i) is true. □

Theorem 3.3.18: Let (X,T) be a Mackey space. Then the statements (i) through (iii) of Theorem 3.3.15 are equivalent to

(iv) If f in X' satisfies the relation:

$$f(R_n(x)) \to 0, \ \forall \ x \in X,$$

then $f \in X^*$.

Proof: Since (iii) implies (iv), it is enough to show that (iv) implies (i). Suppose T_1 is the topology considered in the implication (iii) ⇒ (i). Note that $X^* \subset X_1^* \equiv (X,T_1)^*$. By (iv), we therefore conclude that $X_1^* = X^*$. Hence

$T \approx T_1$. □

Remark: Let us consider the following condition on (X,T):

(v) If p is a seminorm on X satisfying:

$$p(S_n(x)) \to p(x), \forall x \in X, \tag{3.3.19}$$

then p is T-continuous.

It is clear that (v) ⇒ (ii) of Theorem 3.3.15. The converse is not true, for we have

Example 3.3.20: This is the space $(\ell^\infty, \tau(\ell^\infty, \ell^1))$ for which $\{e^n; e^n\}$ is an S.b.; cf. Theorem 1.3.3. It is known that this space is not infrabarrelled (cf. [132], p. 111; [152]). Hence by Proposition 3.3.13, (v) is not satisfied. Let now $f \in (\ell^\infty)'$ such that

$$f\left(\sum_{i>n} \alpha_i e^i\right) \to 0, \forall \alpha \in \ell^\infty.$$

Then $f \in (\ell^\infty)^\#$ by Proposition 3.3.1. But $(\ell^\infty)^\# = (\ell^\infty)^* = \ell^1$ and so (iv) of Theorem 3.3.18 is satisfied and so is (ii) of Theorem 3.3.15.

4 The basis problem

4.1 INTRODUCTION

The presence of a Schauder base in a TVS is useful from several points of view. We have already encountered some of its advantages in Chapter 3 and we will find many more applications of the S.b. theory in subsequent chapters. Thus, keeping in mind the vast importance the theory of S.b. enjoys, it becomes natural to enquire as to when a given sequence in a TVS is an S.b. Before we treat this problem, it is equally important to ascertain whether in fact the space in which we are going to find the basis character of a sequence does or does not possess an S.b. Unfortunately, we do not yet have a general solution to this problem. To formulate comparatively more concrete problems in this direction and seek their solutions, let us begin with

Proposition 4.1.1: Every TVS (X,T) possessing a t.b. $\{x_n;f_n\}$ is separable.

Proof: The proof is a direct consequence of the observation that the countable set $\{\sum_{i=1}^{n} r_i x_i : r_i$'s are rationals and $n \in \mathbb{N}\}$ is dense in (X,T). □

Corollary 4.1.2: Every σ-barrelled space with a t.b. is barrelled.

We may now raise the following two problems.

Problem 4.1.3: Does every separable TVS possess a t.b.?

Problem 4.1.4: Does every separable TVS possess an S.b.?

In order to deal with these problems effectively, we would, however, classify the family of all separable TVS as shown in Diagram 4.1.5.

At the outset, we have (cf. [217], p. 454)

Theorem 4.1.6: If (X,T) is a separable F-space with $X^* = \{0\}$, then (X,T) has no t.b. and hence no S.b.

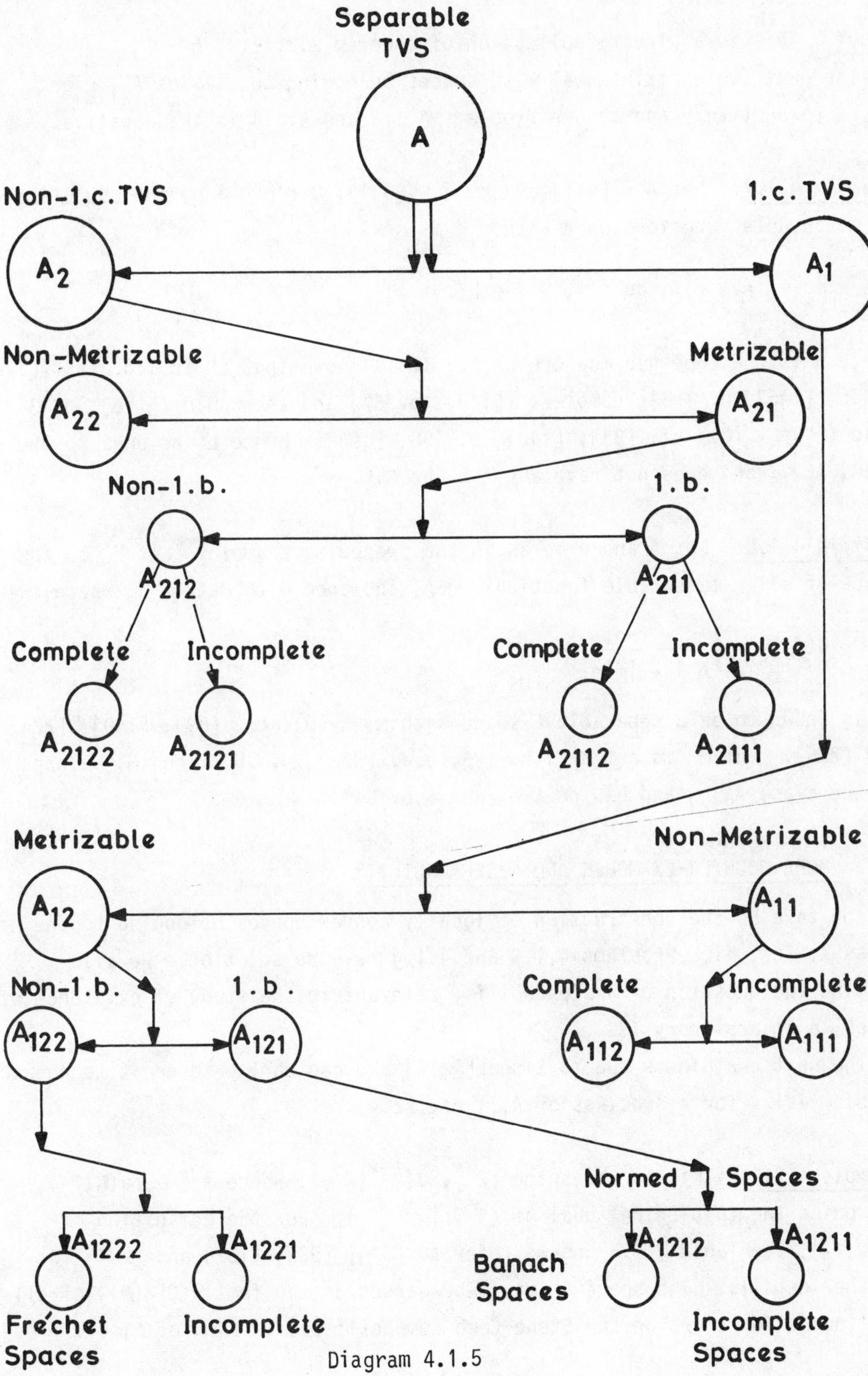
Separable
TVS
A
Non-1.c.TVS
1.c.TVS
A2
A1
Non-Metrizable
Metrizable
A22
A21
Non-1.b.
1.b.
A212
A211
Complete
Incomplete
Complete
Incomplete
A2122
A2121
A2112
A2111
Metrizable
Non-Metrizable
A12
A11
Non-1.b.
1.b.
Complete
Incomplete
A122
A121
A112
A111
Normed Spaces
A1222
A1221
A1212
A1211
Banach
Spaces
Fre'chet
Spaces
Incomplete
Incomplete
Spaces

Diagram 4.1.5

Proof: This is a direct application of Theorem 2.2.12. □

The next two examples deal with spaces belonging to classes A_{2112} and A_{2122} respectively and answer Problems 4.1.3 and 4.1.4 in the negative.

Example 4.1.7: Let $A = [0,1]$ and $L^p \equiv L^p[0,1]$, $0 < p < 1$ be the usual space of measurable functions on A with

$$\|f\|_p = \int_A |f|^p d\mu < \infty, \ \forall \ f \in L^p.$$

where μ is the Lebesgue measure on A. It is known that L^p is l.b. TVS ([140], p. 55; [146], p. 156), complete ([140], p. 67) and separable ([156], p. 34). Also $(L^p)^* = \{0\}$; cf. [33]; [146], p. 158; [193]. Hence L^p belongs to the class A_{2112} and does not have any t.b. or S.b.

Example 4.1.8: Let A and μ be as in the preceding example. Let M_d be the space of all μ-measurable functions on A, equipped with metric d, where

$$d(f,g) = \int_A \frac{|f-g|}{1 + |f-g|} d\mu.$$

M_d is known to be a separable F-space with $M_d^* = \{0\}$; cf. [69], [140], [224] and [225]. But M_d is not an l.b. TVS, e.g. [235], p. 55. Thus M_d belongs to the class A_{2122} and has no t.b. and S.b.

4.2. MORE COUNTER-EXAMPLES AND ω-SEPARABILITY

We now take up the construction of locally convex spaces belonging to the class A_1 for which Problems 4.1.3 and 4.1.4 have no solution. We also examine the question of ω-separability relevant to the study of existence of bases in an arbitrary TVS.

The next example is due to Singer [217] who had earlier used it to negate Problem 4.1.4 for a subclass of A_{11}; cf. [214].

Example 4.2.1: This is the space $(\ell^{\infty*}, \sigma(\ell^{\infty*}, \ell^{\infty}))$, where $\ell^{\infty*} \approx ba(\mathbb{N}, \Phi_\infty)$, $\ell^{\infty*}$ being the topological dual of $(\ell^\infty, \|\cdot\|_\infty)$. For the definition of $ba(\mathbb{N}, \Phi_\infty)$, see Table 1.3.5 and we refer to [51], [90], [132] and [224] for further details. The space ℓ^∞ may be regarded as the family $C[\beta(\mathbb{N})]$ of all continuous functions on the Stone-Čech compactification $\beta(\mathbb{N})$ of $\mathbb{N}$; cf. [140],

p. 209. Let us recall the relevant features of the space in question.

(i) $(\ell^{\infty *}, \sigma(\ell^{\infty *}, \ell^{\infty}))$ is separable. In fact, if f_n in $\ell^{\infty *}$ is defined by $f_n(x) = x_n$, $x \in \ell^{\infty}$, then, using the Hahn-Banach theorem, we find that the set $A = \{\sum_{i=1}^{n} r_i f_i : r_i$'s are rationals and $n \in \mathbb{N}\}$ is dense in the above space.

(ii) $\ell^{\infty *}$ is not $\sigma(\ell^{\infty *}, \ell^{\infty})$-complete, otherwise $\ell^{\infty *} = (\ell^{\infty})'$ (cf. [93], p. 189), and so every $\sigma(\ell^{\infty}, \ell^{\infty *})$-bounded set is finite-dimensional by Theorem 1.2.10. However, this is not true.

(iii) The space in question is nuclear (cf. [226]; (3), p. 62 and (7), p. 75).

(iv) We also recall the following property of $\ell^{\infty *}$ due to Grothendieck [72], p. 168; cf. also [74], p. 229. Consider a sequence $\{g_n\}$ in $\ell^{\infty *}$ with $g_n \to 0$ in $\sigma(\ell^{\infty *}, \ell^{\infty})$; then $g_n \to 0$ in $\sigma(\ell^{\infty *}, \ell^{\infty **})$. The proof of the last statement is a variation of Radon-Nikodym theorem due to Fefferman [56]; see also [35], p. 140, 4(a).

Returning to the basic purpose of Example 4.2.1, suppose on the contrary that $\{f_n\}$ is a t.b. for $(\ell^{\infty *}, \sigma(\ell^{\infty *}, \ell^{\infty}))$. Using (iv), we easily conclude that $\{f_n\}$ is a $\sigma(\ell^{\infty *}, \ell^{\infty **})$-S.b. for $\ell^{\infty *}$ and so it is an S.b. for the Banach space $\ell^{\infty *}$; cf. [162], p. 32 and [218], p. 146, also Theorem 6.1.4. Consequently, ℓ^{∞} is norm separable, which is not true. Thus the space $(\ell^{\infty *}, \sigma(\ell^{\infty *}, \ell^{\infty}))$ which belongs to a subclass of A_{111} does not have a t.b.

Let us now pass on to the concept of ω-separability (cf. [108]).

Definition 4.2.2: A TVS (X,T) is said to be ω-*separable* if there exists a subspace Y of X with dim $Y \leq \aleph_0$ such that each x in X is the T-limit of a sequence in Y.

Proposition 4.2.3: Every TVS(X,T) with a t.b. is ω-separable and each ω-separable TVS(X,T) is separable with $\#(X) = \aleph$.

Proof: It is enough to prove the last part only. Let Y be as in Definition 4.2.2 and $S_Y = Y^{\mathbb{N}}$. Then $\#(S_Y) = \aleph^{\aleph_0} = \aleph$; cf. [119], pp. 38, 47 and 49. Now $\#(X) \leq \#(S_Y)$; also $\#(X) \geq \#(Y)$ and so $\#(X) = \aleph$. □

The fact that the notion of ω-separability is stronger than the separability is exhibited in the following two examples.

Example 4.2.4: Consider $\mathbb{K}^{\aleph}$ equipped with its product topology. It is known (cf. [50], p. 175) that $\mathbb{K}^{\aleph}$ is separable; cf. also [143], p. 37 where a more simplified argument is given for the separability of $\mathbb{K}^{\aleph}$. If $\mathbb{K}^{\aleph}$ were ω-separable, then $2^{\aleph} = \#(\mathbb{K}) = \aleph$, which is not true.

Example 4.2.5: The space $(\ell^{\infty *}, \sigma(\ell^{\infty *}, \ell^{\infty}))$ is not ω-separable; indeed $\ell^{\infty *}$ contains a copy of $\beta(\mathbb{N})$ and so $\#(\ell^{\infty *}) \geq 2^{\aleph}$; cf. [50], p. 244.

Finally we come to the example of Kalton [108] where a space is constructed not to have a t.b.

Example 4.2.6: This is the space $X = \mathbb{K}^{\aleph}$ of Example 4.2.4 where the product topology T on $\mathbb{K}$ is easily seen to be equivalent to $\sigma(X,X^*)$ with $X^* = \bigoplus_{\aleph} \mathbb{K}$; cf. [192], p. 93. By Proposition 1.2.18, (X,T) is nuclear; further, this space is bornological ([146], p. 390), complete ([140], p. 57), barrelled and Pták ([205], p. 162). By Proposition 4.2.3 and Example 4.2.4, this space does not have a t.b.

The foregoing discussion leads to the following

Problem 4.2.7: Do all ω-separable non-metrizable spaces have t.b. or S.b.?

4.3. THE AMEMIYA-KŌMURA EXAMPLE

The purpose of this section is two-fold: to construct another rich space having no S.b. and secondly to answer in the affirmative a question raised in [66], related to Example 3.2.5 in the form of

Problem 4.3.1: Does there exist a separable barrelled ω-complete but not complete space?

The next example serves both these purposes. Originally this example was given by Amemiya and Kōmura [1] and subsequently a major correction to it was provided by Knowles and Cook [145].

Example 4.3.2: Let A be an indexing set with $\#(A) = \aleph$ and such that

$$A = \bigcup_{1 \leq \alpha < Q} A_\alpha; \quad A_\alpha \cap A_\beta = \emptyset, \quad \alpha \neq \beta; \quad \#(A_\alpha) = \aleph, \quad \alpha \in Q,$$

where Q denotes the smallest transfinite number having the smallest uncountable cardinality $\aleph_1$. Recall the subspaces $\omega_0(A)$ and $\phi(A)$ as introduced in

Example 3.2.5; for simplicity, we write X for $\omega_o(A)$ and equip X with the topology $\sigma(X,\phi(A))$. The system $\{e^\alpha:\alpha \in A\}$ discussed in Example 3.2.5 will henceforth be referred to as the *normal base* (n.b.) for $\omega(A) = \mathbb{R}^A$. Before we actually construct a suitable space for the purpose outlined earlier, let us recall the following general definition: let two vector spaces X and Y form a dual system $\langle X,Y\rangle$; X is called *finite relative to* Y if each $\sigma(X,Y)$-bounded subset of X is finite-dimensional.

Let Y and Z be subspaces of $X' \equiv (\omega(A))'$ such that (i) X is finite relative to Y, (ii) Y is finite relative to X, (iii) for each y in Y, the set $\{a \in A:y_a = \langle y,e^a\rangle \neq 0\}$ is countable, (iv) $\dim Z = \aleph_o$, (v) Z is dense in $(X',\sigma(X',X))$ and (vi) for $A_1 \subset A$ with $A \setminus A_1$ being countable, Z is finite relative to $\phi(A_1)$, where $\phi(A_1) = \{x \in \phi(A):x(a) = 0, a \in A \setminus A_1\}$. With all this background, we formulate the space $(Y + Z, \sigma(Y + Z,X))$ and claim that it is an incomplete separable Montel space having no S.b.

We will return to the precise construction of Y and Z at a later stage; in the meantime, let us temporarily take as granted the existence of Y and Z satisfying conditions (i) through (vi) and prove the desired conclusion about Y + Z in the form of

<u>Proposition 4.3.3</u>: The space $(Y + Z, \sigma(Y + Z,X))$ is an incomplete separable Montel space having no S.b.

<u>Proof</u>: By virtue of Theorem 3.2.2, it suffices to establish that the space in question is (a) incomplete separable and (b) Montel. To prove (b), we need to show that each of the spaces is finite relative to the other.

(a) For proving incompleteness, assume the contrary. Then $\dim (Y + Z) = \dim (X')$; cf. [93], p. 188. However, $\dim X < \dim X'$ and so, using properties (iii) and (iv), we arrive at a contradiction.

The separability of $(Y + Z, \sigma(Y + Z,X))$ is a consequence of the properties (iv) and (v) of Z.

(b) In view of (i), we need show that Y + Z is finite relative to X. If this were not true, there would exist a $\sigma(Y + Z,X)$-bounded subset B which is not finite-dimensional. B therefore contains a linearly independent sequence $\{x^n\}$, where $x^n = y^n + z^n$ with $y^n \in Y$ and $z^n \in Z$, $n \geqslant 1$. Put $A_1 = \{a \in A: \langle e^a,y^n\rangle = 0, n \geqslant 1\}$. By (iii), $A \setminus A_1$ is countable. The sequence $\{z^n\}$ is easily seen to be $\sigma(Z,\phi(A_1))$-bounded and so, using (vi), we find that $\{z^n\}$

is $\sigma(Z,X)$-bounded and consequently, by (ii), $\{y^n\}$ is finite-dimensional. Therefore $\{x^n\}$ is finite-dimensional. □

We now take up the desired construction of Y and Z.

Construction of Y

In the course of discussion, we require the concept of an $\aleph_0$-*dual* N of X relative to a Hamel base $\mathcal{H}$ of X containing the normal base $\{e^a : a \in A\}$; that is, $N = \{f \in X' : f_a \equiv \langle e^a, f\rangle = 0$, except for countable a's in A$\}$. Further, we write $A(< \alpha) = \cup\{A_\beta : \beta < \alpha\}$, $A(> \alpha) = \cup\{A_\beta : \beta > \alpha\}$ and $X(< \alpha)$, $X(> \alpha)$, $X(\alpha)$ to mean respectively the spaces $\omega_0(A(< \alpha))$, $\omega_0(A(> \alpha))$, $\omega_0(A_\alpha)$ for $1 \leqslant \alpha < Q$. We use the notation $N(< \alpha)$ to mean the $\aleph_0$-dual of $X(< \alpha)$ relative to a Hamel base $\mathcal{H}_\alpha$ of $X(< \alpha)$ containing the normal base $\{e^a : a \in A(< \alpha)\}$.

For α with $1 \leqslant \alpha < Q$, let Y_α denote a subspace of X' satisfying (I), (II) and (III) below.

(I) For x in Y_α, if $y = x|X(< \alpha)$, then $y_a \equiv \langle e^a, y\rangle = 0$ except for countable values of a in $A(< \alpha)$ and the natural mapping $x \to y$ from Y_α to $N(< \alpha)$ is 1-1.

(II) For x in Y_α and $y = x|X(\alpha)$, $y_a = 0$ except for a finite subset of a's in A_α and the mapping $x \to y$ from Y_α to $\phi(A_\alpha)$ is 1-1.

(III) For x in Y_α, $x|X(> \alpha) = 0$.

The choice of Y_α satisfying (I), (II) and (III) above is possible since $\dim N(< \alpha) = \aleph = \dim \phi(A_\alpha)$ and $X = X(< \alpha) \oplus X(\alpha) \oplus X(> \alpha)$. Put $Y = \mathrm{sp}\,\{\cup\, Y_\alpha : 1 \leqslant \alpha < Q\}$. We have

Proposition 4.3.4: The space Y constructed above satisfies (i), (ii) and (iii).

Proof: (i) Let $\{x^n\}$ be a linearly independent $\sigma(X,Y)$-bounded sequence in X. Then each $B_n = \{a \in A : x^n(a) \neq 0\}$ and so $B = \cup\{B_n : n \geqslant 1\}$ is countable. Hence we may write $B = \{a_n\}$ and consequently, for some $\alpha < Q$,

$$B \subset \cup\{A_\beta : \beta < \alpha\}; \quad B \cap A_\alpha = \emptyset.$$

If $S = \mathrm{sp}\,\{x^n; e^{a_n} : n \geqslant 1\}$, then, defining f by $f(x^n) = n$ with $f(e^{a_n})$ being chosen appropriately and extending f to S linearly, we get g in Y_α so that $g(x^n) = f(x^n)$. Hence $\{x^n\}$ is not $\sigma(X,Y)$-bounded.

(ii) To prove this assertion, consider a linearly independent and $\sigma(Y,X)$-

bounded sequence $\{y^n\}$ in Y. Put $D = \{a \in A: y^n(e^a) \neq 0, n \geq 1\}$. Then D is countable. Consequently, there exists an index $\alpha < Q$ such that $D \subset \bigcup_{\beta<\alpha} A_\beta$ and

$$y^n|X(<\alpha) = z^n \in N(<\alpha);$$

$$y^n|X(\alpha) = u^n \in \phi(A_\alpha);$$

$$y^n|X(>\alpha) = 0.$$

Clearly, $\{u^n\}$ is linearly independent and $\sigma(\phi(A_\alpha), \omega_o(A_\alpha))$-bounded. Following Example 3.2.5 (ii), we find that $\{u^n\}$ is finite dimensional and therefore so is $\{y^n\}$.

(iii) This is a trivial consequence of the construction of Y. □

Construction of Z

The following lemma due to Amemiya and Kōmura [1], subsequently corrected in [145], plays a key role in the construction of Z.

Lemma 4.3.5: There exists a metrizable barrelled subspace ω_o of $\omega \equiv (\omega, \sigma(\omega,\phi))$ with dimension $\aleph$ such that (i) codimension of ω_o is $\aleph$ and (ii) ω_o has a Hamel base $\{e_a^*\}$ so that each subspace λ of ω_o with codimension $\aleph_o$ and spanned by a subset of the Hamel base is total and barrelled.

Proof: Let $\omega_o = \{\alpha x: \alpha \in \mathbb{O}, x \in \omega\}$, where $\mathbb{O} = \{\alpha \in \omega: \alpha_n = 0 \text{ or } 1 \text{ and } (\sum_{i=1}^{n} \alpha_i)/n \to 0\}$. Then $(\omega_o, \sigma(\omega_o, \phi))$ is metrizable and barrelled (cf. [132] p. 200). If we define $g: \omega \to \omega_o$ by

$$g(a) = \sum_{n\geq 1} a_n e^{2^n+n} = a\alpha$$

where $a \in \omega$, $\alpha \in \mathbb{O}$ with $\alpha_n = 1$, $n = 2^k+k$, $k \geq 1$ and zero elsewhere, then g is 1-1 and linear. Hence $\dim \omega_o = \aleph$.

(i) If c_k in ω is defined by

$$c^k = \sum_{n\geq 1} e^{(2n-1)2^{k-1}}, \quad k \geq 1,$$

then $\{c^k\}$ is linearly independent. Hence the map $h: \omega \to \omega$, $h(\alpha) = \sum_{n\geq 1} \alpha_n c^n$

is a 1-1 linear map. Put $\lambda = h[\omega]$. Then $\dim \lambda = \aleph$ and $\lambda \cap \omega_0 = \{0\}$. But the map $F:\lambda \to \omega/\omega_0$, $F(a) = a + \omega_0$, is 1-1 linear map and so $\dim (\omega/\omega_0) = \aleph$.

(ii) Consider an arbitrary set A^* with $\#(A^*) = \aleph$. For a in A^*, define f_a^1 in ω_0 so that the first two coordinates of f_a^1 are zeros and $\{f_a^1; e^1, e^2, \ldots\}$ is linearly independent. Next, choose f_a^2 in ω_0 such that the first three coordinates of it are zeros and $\{f_a^2, f_a^1; e^1, e^2, \ldots\}$ is linearly independent. We thus choose $\{f_a^k\}$ inductively and conclude that $\{f_a^k + e^k\}$ is linearly independent and total in $\omega \equiv \omega(\omega, \sigma(\omega, \phi))$.

Put $B = \{e^*_{a,k} = f_a^k + e^k : a \in A^*, k \geq 1\}$. Then B is linearly independent and total in ω. If possible, extend B to a Hamel base $\mathcal{B}^* = \{e^*_a : a \in A^*\}$ for ω_0. Further, if $\lambda = \text{sp}\,\{\mathcal{B}^*_1\}$ with $\omega_0 = \lambda \oplus \mu$, $\mathcal{B}^*_1 \subset \mathcal{B}^*$ and $\dim \mu = \aleph_0$, then, for some a in A^*, $\{e^*_{a,k} : k \in \mathbb{N}\} \subset \mathcal{B}^*_1$. Hence λ is total. It is barrelled by Proposition 1.2.11(ii). □

Returning to the construction of Z, consider A^* appearing in the construction of the Hamel base $\{e^*_a : a \in A^*\}$ of ω_0 of Lemma 4.3.5 to be the set A of the current example. Define a one-to-one and onto map $T:\phi(A) \to \omega_0$ by

$$T\Big(\sum_{i\in\sigma} \alpha_i e^{a_i}\Big) = \sum_{i\in\sigma} \alpha_i e^*_{a_i}, \quad \sigma \in \Phi .$$

By Lemma 4.3.5, there exists a subspace λ of ω with $\dim \lambda = \dim \omega/\omega_0 = \aleph$. Also, let G be a subspace of $\omega_0(A)$ such that $\dim G = \dim \omega_0(A)/\phi(A)$. If S is the natural isomorphism from G onto λ, then the map $\tilde{T}:\omega_0(A) = \phi(A) \oplus G \to \omega_0 \oplus \lambda = \omega$, $\tilde{T}(x+y) = T(x) + S(y)$, is a 1-1 linear extension of T. If $\tilde{T}'$ is the algebraic adjoint of $\tilde{T}$, let us write $Z = \tilde{T}'[\phi]$ and there remains to prove finally the following

<u>Proposition 4.3.6</u>: The space Z satisfies (iv), (v) and (vi).

<u>Proof</u>: (iv) This follows from the construction.

(v) Suppose $x' \in X'$ and $x' \notin \bar{Z}$, the $\sigma(X',X)$-closure of Z. By the Hahn-Banach theorem, $\langle \tilde{T}(x), \alpha \rangle = 0$ for all α in ϕ and some nonzero x in X. This leads to a contradiction.

(vi) Consider $A_1 \subset A$ such that $A \setminus A_1$ is countable. By Lemma 4.3.5, the space $\omega_0 = \tilde{T}[\phi(A)]$ has a Hamel base $\{e^*_a : a \in A\}$. Put $\omega_1 = \text{sp}\,\{e^*_a : a \in A_1\}$. Then by Lemma 4.3.5 and Theorem 1.2.10, ϕ is finite relative to ω_1. If B is any $\sigma(Z, \phi(A_1))$-bounded subset of Z, it follows that $B^* = \{\alpha \in \phi : \tilde{T}'(\alpha) \in B\}$

is finite-dimensional and so is B. □

4.4 ENFLŐ'S EXAMPLE

Earlier we have seen different types of spaces having no t.b. or S.b. and we discovered that the rich structure of the space is no guarantee of the existence of a t.b. or an S.b. in that space. Since all separable Hilbert spaces have S.b., the only extreme case which we have yet to consider is that of Banach spaces. However, in this case, Problems 4.1.3 and 4.1.4 are the same and consequently we seek the solution of

Problem 4.4.1: Does every separable Banach space possess a t.b. or, equivalently, an S.b.?

The above problem was first raised by Banach [5] in 1932. It remained unsolved for forty years and was finally negated by Enflő [55] in 1972. In fact, Enflő's example of a separable Banach space is devoid of the so-called approximation property and hence it does not have an S.b.; cf. Proposition 4.4.4 below.

In this last section, we first discuss briefly the approximation property and then turn our attention to the precise construction of Enflő's example.

Approximation property

The reason for introducing this notion basically comes from the well known theorem on the characterization of compact operators from Hilbert spaces to themselves in terms of finite dimensional operators, due to Grothendieck [73].

Definition 4.4.2: A TVS (X,T) is said to have the *approximation property* (a,p.) if, for every u in $\mathcal{B}_T$ and every T-precompact subset P in X, there exists a linear operator $A:X \to X$ with $\dim A[X] < \infty$ such that $A(x) - x \in u$ for all x in P; or equivalently, the identity operator on X can be approximated uniformly on precompact sets by finite dimensional operators A.

Note: For classification of a.p. in Banach (resp. Fréchet) spaces, one may see [154] (resp. [46]) and for results on a.p. in locally convex spaces, one may consult [147]; cf. also [205].

Before we present a result on the relationship of a.p. of an l.c. TVS with its S.b., let us introduce

Definition 4.4.3: An S.b. for a TVS (X,T) is called e-*Schauder* if $\{S_n\}$ is T-T equicontinuous on X.

Note: Every S.b. in a barrelled space is e-Schauder, while the S.b. $\{e^n;e^n\}$ for $(\phi,\sigma(\phi,\omega))$ is not e-Schauder; cf. [132], p. 65.

Proposition 4.4.4: Every TVS(X,T) having e-Schauder base $\{x_n;f_n\}$ possesses the a.p.

Proof: Let u and P be as in Definition 4.4.2. There exist v and w_1 in $\mathcal{B}_T$ such that $v + v + v \subset u$ and

$$w_1 \subset \bigcap_{n \geq 1} S_n^{-1}[v].$$

Set $w = w_1 \cap v$, then there exist $\{y_1,\ldots,y_k\}$ in P so that

$$P \subset \bigcap_{i=1}^{k} (y_i + w).$$

Also, there exists N in $\mathbb{N}$ such that $y_i - S_n(y_i) \in v$ for all $n \geq N$ and $1 \leq i \leq k$. Thus $x - S_n(x) \in u$ for all $n \geq N$ and all x in P. □

The converse of the above result is not true, for we have

Example 4.4.5: Here we consider the space $\mathbb{K}^{\aleph}$ of Example 4.2.6. This space does have the a.p. as follows from the more general

Proposition 4.4.6: Every nuclear l.c. TVS (X,T) has the a.p.

Proof: Let u and P be as in Definition 4.4.2 and $\varepsilon > 0$. Therefore there exist α in ℓ^1 with $\alpha_n > 0$, an equicontinuous sequence $\{f_n\}$ in X* and a sequence $\{y_n\}$ in X with $\{p_u(y_n)\}$ being bounded such that

$$p_u(x - \sum_{i=1}^{n} \alpha_i f_i(x) y_i) \to 0, \quad \forall\, x \in X. \qquad (*)$$

Choose w in $\mathcal{B}_T$ and positive constants M,R with $MR > 1$ such that $|f_n(x)| \leq p_w(x)$ for all $n \geq 1$ and x in X, $\sum_{n \geq 1} \alpha_n = R$ and $p_u(y_n) \leq M$ for $n \geq 1$. Put $v = u \cap w$. There exists a finite subset $\{z_1,\ldots,z_m\}$ such that $p_v(x-z_j) < \varepsilon/3MR$ for each x in P and some z_j. Determine an N with

$$p_u(z_j - \sum_{i=1}^{n} \alpha_i f_i(z_j)y_i) < \frac{\varepsilon}{3MR}; \quad \forall\, n \geqslant N,\ 1 \leqslant j \leqslant m.$$

With all these considerations, it is easy to show that the left-hand side in (*) converges to zero uniformly in $x \in P$. □

Proposition 4.4.7: For every dual pair $\langle X,Y\rangle$, the space $(X,\sigma(X,Y))$ or $(Y,\sigma(Y,X))$ has the a.p.

Proof: From [226], p. 62 and 75, $(X,\sigma(X,Y))$ is nuclear; now apply Proposition 4.4.6. Similarly the other part follows. □

Alternative proof: This is independent of Proposition 4.4.6. Consider the space $(X,\sigma(X,Y))$ and choose u in $\mathcal{B}_\sigma$. We may then find $y_1,\ldots,y_n$ in Y and $k > 0$ so that $\{x \in X: |\langle x,y_i\rangle|<k,\ 1 \leqslant i \leqslant n\} \subset u$. Put $Z = sp\,\{y_1,\ldots,y_n\}$. Using the idea of topological supplements (cf. [9], p. 137; [192], p. 97), we can find a $\sigma(Y,X)$-$\sigma(Y,X)$ continuous projection A on Y with $A[Y] = Z$. Hence the adjoint $A^*;X \to X$ is continuous. By the Hahn-Banach theorem, we can find $x_1,\ldots,x_n$ in X with $\langle x_i,y_j\rangle = \delta_{ij}$ and so

$$A^*x = \sum_{i=1}^{n} \langle x,y_i\rangle\, A^*x_i,\ x \in X.$$

Therefore $x-A^*x \in u$ for all x in X and, in particular, in any precompact subset P of X. □

Exercise 4.4.8: Show that $(\ell^\infty{}^*, \sigma(\ell^\infty{}^*,\ell^\infty))$ has no t.b. but has the a.p. [Hint: cf. Example 4.2.1 for first part].

Note: The nuclearity of an l.c. TVS (resp. the e-Schauder character of the S.b. of the space) is a sufficient condition to guarantee the a.p. in that space. None of these is a necessary condition; for instance, consider $(\ell^\infty, \|\cdot\|_\infty)$; cf. [147], p. 258. A still better example is

Example 4.4.9: Consider the uncountable product of $(\ell^\infty, \|\cdot\|_\infty)$, which is clearly non-normable and has no t.b. Also, it has a.p. ([147], p. 245) and is non-nuclear, otherwise a subspace isomorphic to ℓ^∞ would be nuclear.

The main example:

Let us now return to the main theme of this section: to construct a separable Banach space having no t.b. As pointed out earlier, we present an example of a separable Banach space having no a.p. and hence, by Proposition 4.4.4, this space has no t.b. or S.b. Enflö's original construction of this example is quite complicated. There is a substantial simplification of it in [57]; however, a still more simplified version of the example is due to Davie [29], whom we follow for the rest of this section. In essence, the desired space is the sup norm closure of a linear span of a certain sequence of linearly independent complex valued bounded functions defined on a set which is itself a union of mutually disjoint sequence of abelian groups of finite order.

Intermediary lemmas:

In order to construct the precise space, we need a number of lemmas from probability theory and refer to [85] or [208] for several unexplained terms which follow hereafter.

Lemma 4.4.10: Let (Q,S,P) be a probability space and $f_1,\ldots,f_N$ any N random variables on Q with values in $\mathbb{R}$ such that each f_i is either -1 or 2. Assume that $P\{x \in Q: f_i(x) = -1\} = 2/3$ and $P\{x \in Q: f_i(x) = 2\} = 1/3$ for each i. Then, for some $A > 0$,

$$P\{x \in Q: |\sum_{j=1}^{N} \alpha_j f_j(x)| > A(\sum_{j=1}^{N} |\alpha_j|^2 \log N)^{1/2}\} \leq \frac{A}{N^3}, \tag{4.4.11}$$

for all $\alpha_1,\ldots,\alpha_N$ in $\mathbb{C}$.

Proof: It is enough to prove (4.4.11) when all α_i's are real and $\sum_{j=1}^{N} \alpha_j^2 = 1$. Now for any real λ, we have (cf. [85], p. 193)

$$I \equiv \int_Q \exp\{\lambda \sum_{i=1}^{N} \alpha_i f_i(x)\}dP = \prod_{i=1}^{N} \{\tfrac{1}{3}\exp(2\lambda\alpha_i) + \tfrac{2}{3}\exp(-\lambda\alpha_i)\}.$$

But, for any real θ, $\exp(2\theta) + 2\exp(-\theta) \leq 3\exp(2\theta^2)$. Hence $I \leq \exp(2\lambda^2)$. Therefore

$$\int_Q \exp\{\lambda|\sum_{i=1}^{N} \alpha_i f_i(x)|\}dP \leq 2e^{2\lambda^2}.$$

Define for x in Q,

$$F(x) = \exp \{\lambda|\sum_{i=1}^{N} \alpha_i f_i(x)|-2\lambda^2 - 3 \log N\}$$

for a suitable real $\lambda > 0$ to be chosen later and consider a random variable g on Q defined by $g(x) = |\sum_{i=1}^{N} \alpha_i f_i(x)|-4\lambda$. Then

$$P\{x \in Q:g(x) > 0\} \leq \int_Q \exp(\lambda g(x))dP.$$

Note that

$$\int_Q F(x)dP \leq \frac{2}{N^3},$$

and consequently for $\lambda = \sqrt{(3 \log N)/2}$, $F(x) = \exp\ \{\lambda g(x)\}$ and this gives $P\{x \in Q:g(x) > 0\} \leq 2/N^3$. Therefore

$$P\{x \in Q:|\sum_{i=1}^{N} \alpha_i f_i(x)| > 2\sqrt{6 \log N}\} \leq \frac{2}{N^3} < \frac{2\sqrt{6}}{N^3},$$

and the required proof is complete. □

Lemma 4.4.12: Let (Q,S,P) be a probability space and $f_1,\ldots,f_N$ any N random variables on Q, taking values either -1 or 1. Let $P\{x \in Q:f_i(x) = -1\} = P\{x \in Q:f_i(x) = 1\} = 1/2$, for $1 \leq i \leq N$. Then there exists $A > 0$ independent of N such that

$$P\{x \in Q:|\sum_{i=1}^{N} \alpha_i f_i(x)| > A(\sum_{i=1}^{N} |\alpha_i|^2)^{1/2}\ \sqrt{\log N}\} < \frac{A}{N^3},$$

for all scalars $\alpha_1,\ldots,\alpha_N$.

Proof: As in the preceding lemma and use the fact that $\exp(\theta) + \exp(-\theta) \leq 2 \exp(\theta^2)$ for all real θ. □

Lemma 4.4.13: Let G_k be an Abelian group of order 3.2^k, k being a non-negative integer. Suppose $\phi_1^k,\ldots,\phi_{3.2^k}^k$ are the elements of the character group $\hat{G}_k$ of G_k. Then there exist reals $r_1^k,\ldots,r_{3.2^k}^k$, r_i^k being equal to 2 or -1 with

$$\sum_{i=1}^{3.2^k} r_i^k = 0, \tag{4.4.14}$$

and a constant A independent of k such that

$$|\sum_{i=1}^{3.2^k} r_i^k \phi_i^k(g)| \leq A2^{k/2}\sqrt{k+1}, \quad \forall\, g \in G_k. \tag{4.4.15}$$

<u>Proof</u>: For simplicity, put $K = 3.2^k$ and recall $Q, f_1,\dots,f_K$ from Lemma 4.4.10. For each g in G_k, let

$$A_g = \{x \in Q: |\sum_{i=1}^{K} f_i(x)\phi_i^k(g)| > 2\sqrt{6}\,\overline{K\log K}\}.$$

Then $P(A_g) < 4/K^3$ and, as

$$P(\cap\{\tilde{A}_g : g \in G_k\}) = 1-P(\cup A_g : g \in G_k) > 0, \quad (\tilde{A}_g = Q \setminus A_g)$$

there exists $x \notin A_g$ for each g in G_k such that

$$|\sum_{i=1}^{K} f_i(x)\phi_i^k(g)| \leq (2^k(1+k))^{1/2}A;\ \forall\, g \in G_k, \tag{$*$}$$

where $A = 6\sqrt{2\log 3}$. Let $I = \{i \in \mathbb{N}: f_i(x) = 2\}$, $J = \{i \in \mathbb{N}: f_i(x) = -1\}$, and $m = \#(I)$, $n = \#(J)$. Then $m+n = K$. The required result is proved by using (*) and considering cases (a), (b) and (c).

(a) Suppose $m = 2^k$ and $n = 2^{k+1}$. Assume $r_i^k = f_i(x)$, $1 \leq i \leq K$. Note that (4.4.14) is clearly satisfied and (*) results in (4.4.15).

(b) Suppose $m > 2^k$. Let $M \subset I$ be such that it contains exactly 2^k elements. If $r_i^k = 2$ for $i \in M$ and $= -1$ if $i \in I \cup J \setminus M$, then $\{r_i^k : 1 \leq i \leq K\}$ satisfies (4.4.14). Now, for every g in G_k,

$$|\sum_{i=1}^{K} (r_i^k - f_i(x))\phi_i^k(g)| \leq 3\ \#(I\setminus M)$$

$$= 2m-n \leq |\sum_{i=1}^{K} f_i(x)| \leq A(2^k(1+k))^{1/2},$$

using (*) and putting g = e, the identity element of G_k. This last inequality, when coupled with (*) again, results in (4.4.15).

(c) Finally, let $m < 2^k$. Then $n > 2^{k+1}$. Choose a subset N of J having exactly 2^{k+1} elements. By introducing r_i^k as equal to -1 when $i \in N$ and equal to 2 when $i \in I \cup J \setminus N$ and proceeding as in (b), we can easily establish the truth of (4.4.14) and (4.4.15). □

Construction of the example:

Invoking the contents, notation and proof of Lemma 4.4.13, let $\Sigma_1 = \{i \in \mathbb{N}: r_i^k = 2\}$ and $\Sigma_2 = \{i \in \mathbb{N}: r_i^k = -1\}$; then $\#(\Sigma_1) = 2^k$ and $\#(\Sigma_2) = 2^{k+1}$. So we can rearrange Σ_1 and Σ_2 as $\{1,\dots,2^k\}$ and $\{1,\dots,2^{k+1}\}$ respectively. Breaking ϕ_i^k as $\eta_i^k (i \in \Sigma_1)$ and $\mu_i^k (i \in \Sigma_2)$, we find from (4.4.15) the following inequality

$$|2 \sum_{i=1}^{2^k} \sigma_i^k(g) - \sum_{i=1}^{2^{k+1}} \tau_i^k(g)| \leq A2^{k/2}\sqrt{1+k}; \ \forall g \in G_k, \tag{4.4.16}$$

where we have rewritten η_i^k as σ_i^k and μ_i^k as τ_i^k according to the new arrangement of Σ_1 and Σ_2.

Consider the sequence $\{G_k : k \geq 0\}$ of Abelian groups mentioned in Lemma 4.4.13 with $G_i \cap G_j = \emptyset$ for $i \neq j$, and let $G = \cup\{G_k : k \geq 0\}$. For each $k \geq 0$ and i with $1 \leq i \leq 2^k$, define $e_i^k : G \to \{z \in \mathbb{C}: |z| = 1\}$ by

$$e_i^k(g) = \begin{cases} \tau_i^{k-1}(g), & g \in G_{k-1} (k \geq 1); \\ \varepsilon_i^k \sigma_i^k(g), & g \in G_k (k \geq 0) \\ 0, & \text{otherwise,} \end{cases}$$

where $\varepsilon_i^k (1 \leq i \leq 2^k, k \geq 0)$ are to be specified later. Put

$$E = \overline{\text{sp}}\, \{e_i^k : 1 \leq i \leq 2^k, k \geq 0\},$$

where the closure is taken relative to the sup norm topology with which the space of all $\mathbb{C}$-valued bounded functions on G is supposed to be equipped. The principal result of this section may now be stated as follows:

Theorem 4.4.17: For a suitable choice of ε_i^k (= ± 1) for $1 \leq i \leq 2^k$, $k \geq 0$, the space E is a separable Banach space having no a.p.

The proof that E has no a.p. is quite involved (the rest of the proof is obvious). At the outset, let us recall the familiar notation for the sake of brevity. Let $\mathcal{L}(E)$ denote the space of all continuous operators on E, equipped with sup norm topology. The basic idea of proof lies in constructing β in $(\mathcal{L}(E))^*$ as the limit of a sequence $\{\beta_k\} \subset (\mathcal{L}(E))^*$ such that

$$\beta \text{ is continuous on } \mathcal{L}(E) \text{ relative to the topology of uniform convergence on compact subsets of } E, \tag{4.4.18}$$

$$\beta(I_E) = 1, \; I_E \text{ being the identity operator of } E \tag{4.4.19}$$

$$\beta(T) = 0 \text{ for any } T \text{ in } \mathcal{L}(E) \text{ with } \dim T[E] < \infty . \tag{4.4.20}$$

These statements, once established, clearly imply that E has no a.p.

Construction of functionals:

We now proceed to find functionals on $\mathcal{L}(E)$ satisfying (4.4.18), (4.4.19) and (4.4.20).

Define α_i^k and β_i^k in E* as follows:

$$\alpha_i^k(f) = \frac{1}{3.2^k} \sum_{g \in G_k} \varepsilon_i^k \sigma_i^k(g^{-1}) f(g); \quad 1 \leqslant i \leqslant 2^k, \; k \geqslant 0 \tag{4.4.21}$$

and

$$\beta_i^k(f) = \frac{1}{3.2^{k-1}} \sum_{g \in G_{k-1}} \tau_i^{k-1}(g^{-1})\, f(g); \quad 1 \leqslant i \leqslant 2^k, \; k \geqslant 1. \tag{4.4.22}$$

Proposition 4.4.23: $\alpha_i^k = \beta_i^k$ for $1 \leqslant i \leqslant 2^k$, $k \geqslant 1$.

Proof: It is sufficient to show that $\alpha_i^k(e_n^m) = \beta_i^k(e_n^m)$ for $m \geqslant 0$ and $n \geqslant 1$. From a known theorem on Haar measure ([89], p. 363) we recall the following simple consequence ([95], p. 18-20): if σ_1 and σ_2 are two continuous characters on a compact Abelian group G equipped with a normalized Haar measure μ, then

$$\int_G \sigma_1(g)\overline{\sigma_2(g)}\,d\mu = \begin{cases} 1, & \text{if } \sigma_1 = \sigma_2; \\ 0, & \text{otherwise.} \end{cases} \tag{4.4.24}$$

Since

$$\alpha_i^k(e_n^m) = \varepsilon_i^k \int_{G_K} \overline{\sigma_i^k(g)}\, e_n^m(g)d\mu,$$

μ being the Haar measure on G_k as specified above, $\alpha_i^k(e_n^m) = \delta_{km}\delta_{in}$. Similarly, $\beta_i^k(e_n^m) = \delta_{km}\delta_{in}$. □

Next, we have

<u>Proposition 4.4.25</u>: If β_k in $(\mathcal{L}(E))^*$ is defined by

$$\beta_k(T) = \frac{1}{2^k} \sum_{i=1}^{2^k} \alpha_i^k(T(e_i^k)); \quad \forall T \in \mathcal{L}(E),\ k \geq 0, \tag{4.4.26}$$

then (i) $\beta_k(I_E) = 1$, $k \geq 0$; and (ii) for $k \geq 0$,

$$|(\beta_{k+1}-\beta_k)(T)| \leq \sup \{\, \|T(\psi_g^k)\| : g \in G_k\},$$

where

$$\psi_g^k = \frac{1}{2^{k+1}} \sum_{i=1}^{2^{k+1}} \tau_i^k(g^{-1})e_i^{k+1} - \frac{1}{2^k} \sum_{i=1}^{2^k} \varepsilon_i^k \sigma_i^k(g^{-1})e_i^k.$$

<u>Proof</u>: (i) This follows from (4.4.24).

(ii) Make use of (4.4.21), (4.4.22) and (4.4.26). □

In order to infer the convergence of $\{\beta_k\}$, we have to estimate $\|\psi_g^k\|$ suitably and this is shown in

<u>Proposition 4.4.27</u>: There exist a large k_o in $\mathbb{N}$ and $B > 0$ such that

$$\|\psi_g^k\| = \sup_{h\in G} |\psi_g^k(h)| \leq B\sqrt{1+k}\ 2^{-k/2}; \quad \forall g \in G_k,\ k \geq k_o.$$

<u>Proof</u>: Let $h \in G$ and $A_1 = 3\sqrt{2 \log 3}$. If $h \notin G_{k-1}, G_k, G_{k+1}$, then $\psi_g^k(h) = 0$ for every g in G_k and $k \geq 1$. For g in G_k and h in G_{k-1} with $k \geq 1$, define

$$A_{g,h} = \{x \in Q: |\sum_{i=1}^{2^k} f_i(x)\sigma_i^k(g^{-1})\tau_i^{k-1}(h)| > A_1 2^{k/2} k^{1/2}\}.$$

By Lemma 4.4.12, there exists $B > 0$ independent of k such that $P(A_{g,h}) \leq B/2^{3k}$ $(k \geq 1)$. Hence, for large $k \geq k_o$ (say),

$$P(\cup \{A_{g,h}: h \in G_{k-1}, g \in G_k\}) < 1, \quad \forall k \geq k_o.$$

Therefore there exists x_o in Q with $x_o \notin A_{g,h}$ for all h in G_{k-1}, g in G_k and $k \geq k_o$. Setting $f_i(x_o) = \varepsilon_i^k = \pm 1$, $1 \leq i \leq 2^k$, $k \geq k_o$, we conclude that

$$|\psi_g^k(h)| \leq A_1 2^{-k/2} k^{1/2}; \quad \forall g \in G_k,\ h \in G_{k-1},\ k \geq k_o.$$

Next, for $g \in G_k$ and $h \in G_{k+1}$, observe that $\psi_g^k(h) = -\psi_{h^{-1}}^{k+1}(g^{-1})$ and therefore

$$|\psi_g^k(h)| \leq A_1 2^{-(k+1)/2} (k+1)^{1/2}, \quad \forall g \in G_k,\ h \in G_{k+1},\ k \geq k_o.$$

Finally, for all g,h in G_k, we have from (4.4.16)

$$|\psi_g^k(h)| \leq A_1\ 2^{-k/2}\ \sqrt{1+k}.$$

The above estimates of $|\psi_g^k(h)|$ collectively yield the desired inequality. □

<u>Proposition 4.4.28</u>: There exists a relatively compact subset A of E such that for T in $\mathcal{L}(E)$

$$|(\beta_{k+1} - \beta_k)(T)| \leq (1+k)^{-2} \sup_{x \in A} \|Tx\|, \quad \forall k \geq 0. \tag{4.4.29}$$

Consequently $\beta_k \to \beta$ in $\sigma((\mathcal{L}(E))^*, \mathcal{L}(E))$ with $|\beta(T)| \leq \lambda \sup\{\|Tx\| : x \in A\}$, where $\lambda = 2 + \sum_{n \geq 1} 1/n^2$.

<u>Proof</u>: Let $A = \{e_i^o\} \cup \{(1+k)^2\ \psi_g^k : g \in G_k,\ k \geq 0\}$, then A is relatively compact by Proposition 4.4.27. The inequality (4.4.29) is a consequence of Proposition 4.4.25 and so the existence of β is guaranteed by the Banach-Steinhaus theorem. For the last inequality, note that

$$|\beta(T)| \leq \left(\sum_{i=1}^{k_1} i^{-2}\right) \sup_{x \in A} \|Tx\| + |\beta_o(T)| + \sup_{x \in A} \|Tx\|$$

for some $k_1 \geq 1$. This, on making use of (4.4.21), (4.4.26) and the fact that $e_1^o \in A$, yields the result. □

Finally, we return to

Proof of Theorem 4.4.17: In view of Propositions 4.4.25 and 4.4.28, we have to establish only (4.4.20). Choose, therefore, a finite-dimensional operator T in $\mathcal{L}(E)$. Then, for each x in E, we may express Tx as a finite linear combination of $\langle x,\lambda_i^k\rangle e_i^k$, where $\lambda_i^k \in E^*$; $1 \leqslant i \leqslant 2^k$, $k \geqslant 0$. If $S_i^k: E \to E$ with $S_i^k(x) = \langle x,\lambda_i^k\rangle e_i^k$, then from (4.4.26), we deduce

$$\beta_m(S_i^k) = \frac{1}{2^m} \sum_{j=1}^{2^m} \langle e_j^m,\lambda_i^k\rangle \, \delta_{km}\delta_{ij};$$

cf. the proof of Proposition 4.4.23. This shows that $\beta_m(T) \to 0$ and therefore $\beta(T) = 0$. □

5 Characterization of Schauder bases

5.1 THE MAIN RESULT

In this chapter we deal with the problem of characterizing a Schauder base in an arbitrary TVS. Needless to say, we do not restrict our discussion to those spaces which are devoid of S.b. In the process of investigation, several sufficient conditions are discovered that ensure the Schauder character of a sequence and the importance of these conditions is exhibited by a set of examples. As one aspect of application of the main result, we also deal with Haar and Schauder systems and prove their S.b. character.

At this stage, it will be convenient to introduce the following

Definition 5.1.1: A sequence $\{x_n\}$ in a TVS (X,T) is called (ii) *complete* if $[x_n] \equiv [x_n]^T = X$; and (ii) *Schauder basic* (S. *basic*) [resp. *basic*] if $\{x_n\}$ is an S.b. [resp. t.b.] for $[x_n]$.

Theorem 5.1.2: Suppose that there is a sequence $\{x_n\}$ in a TVS (X,T) with $x_n \neq 0$ $(n \geqslant 1)$. If for each p in $D \equiv D_T$, there exist q in D and $M > 0$ such that

$$p\left(\sum_{i=1}^{m} \alpha_i x_i\right) \leqslant Mq\left(\sum_{i=1}^{n} \alpha_i x_i\right), \tag{5.1.3}$$

for all m,n in $\mathbb{N}$ with $m \leqslant n$ and arbitrary $\alpha_1,\ldots,\alpha_n$ in $\mathbb{K}$, then $\{x_n\}$ is S. basic in (X,T). In particular, if $\{x_n\}$ is also complete, then $\{x_n\}$ is an S.b. for (X,T).

Proof: For each $n \geqslant 1$, let $X_n = \text{sp}\,\{x_i : 1 \leqslant i \leqslant n\}$, and introduce the maps $P_{nm} : X_m \to X_n$, where P_{nm} is the identity operator of X_m if $m \leqslant n$, otherwise P_{nm} is the projection of X_m onto X_n defined by $P_{nm}(x+y) = x$; $x \in X_n$ and $y \in \text{sp}\,\{x_i : n < i \leqslant m\}$. By (5.1.3), P_{nm} is continuous for all $m,n \geqslant 1$. Let $Y = \{x \in X_m : 1 \leqslant m < \infty\}$ and $P_n : Y \to X_n$ $(n \geqslant 1)$ with $P_n(x) = P_{nm}(x)$, $x \in X_m$ $(1 \leqslant m < \infty)$. Let us denote by P_n itself the unique continuous extension to $\bar{Y} = [x_n]$. Using (5.1.3), it is readily verified that $\{P_n\}$ is equicontinuous on $\bar{Y}$ and hence, for each σ in D, there exist σ_1 in D and $M > 0$ such that

$$\sigma(P_n(x)) \leq M\sigma_1(x); \ \forall n \geq 1, x \in \bar{Y}. \tag{5.1.4}$$

Let p in D, x in $\bar{Y}$ and $\varepsilon > 0$ be chosen arbitrarily. There exists σ in D so that $p(P_n(x)-x) \leq \sigma(P_n(x)-y) + \sigma(x-y)$ for all y in Y and $n \geq 1$. Put $r = \max(\sigma,\sigma_1)$, then $r(x-y) \leq \varepsilon$ for some y in $Y_m \subset Y$. Hence using (5.1.4),

$$p(P_n(x)-x) \leq (1+M)\varepsilon, \ \forall n \geq m. \tag{5.1.5}$$

Finally, let us define f_i in $(\bar{Y})^*$ by $f_i(x)x_i = (P_i-P_{i-1})(x)$, where $P_o = 0$. Then $f_i(x_j) = \delta_{ij}$. Since

$$P_n(x) = \sum_{i=1}^{n} f_i(x)x_i, \ x \in \bar{Y}$$

and $\{x_n\}$ is ω-linearly independent, the required result follows from (5.1.5).□

Exercise 5.1.6: Prove that $\{e^n\}$ is an S.b. for $(\lambda,\eta(\lambda,\lambda^\times))$ by using Theorem 5.1.2.

A partial converse of Theorem 5.1.2 is contained in

Theorem 5.1.7: Let (X,T) be either a TVS containing a set of second category or a barrelled space with an S.b. $\{x_n;f_n\}$. Then $\{x_n\}$ is complete in X and (5.1.3) holds good.

Proof: By the condition on (X,T), the sequence $\{S_n\}$ is equicontinuous and this results in (5.1.3). The other part is trivially true. □

As a special case of Theorems 5.1.2 and 5.1.7, we have

Theorem 5.1.8: Let (X,ρ) be an F-space containing a nonzero sequence $\{x_n\}$ with $[x_n] = X$, where ρ is the F-norm on X. Then $\{x_n\}$ is an S.b. for (X,ρ) if and only if there exists $M \geq 1$ such that

$$\rho\Big(\sum_{i=1}^{m} \alpha_i x_i\Big) \leq M\rho\Big(\sum_{i=1}^{n} \alpha_i x_i\Big) \tag{5.1.9}$$

for all m,n in $\mathbb{N}$ with $m \leq n$ and all scalars $\alpha_1,\ldots,\alpha_n$.

Note: If ρ is a norm in Theorem 5.1.8, this result is usually known as the *Nikolśkii theorem* and the inequality (5.1.9) is termed the Nikolškii inequality; cf. [178] and also [162], p. 57. The locally convex analogue of Theorem 5.1.2 is given in [191] where one finds a different proof of this result depending upon the technique of constructing new spaces from the old ones; cf. also [26]. Theorems 5.1.2 and 5.1.7 in their present form are given in [127] and their extensions to Schauder decompositions are to be found in [134] and [197].

5.2 EXAMPLES AND COUNTER-EXAMPLES

In the next few pages, we examine how far the conditions on $\{x_n\}$ in Theorem 5.1.2 are indispensable if its S.b. character for (X,T) is to be inferred. Also, we examine the usefulness of the barrelled property of (X,T) in Theorem 5.1.7. The examples constructed for this purpose, will also justify the extension of Theorem 5.1.8 to Theorems 5.1.2 and 5.1.7.

To begin with, let us mention that completeness in Theorem 5.1.2 cannot be dropped.

Example 5.2.1: Consider the sequence $\{e^n\}$ in $(\ell^\infty, \|\cdot\|_\infty)$. Then $[e^n] = c_o \subsetneqq \ell^\infty$ and $\{e^n\}$ is not an S.b. for ℓ^∞.

At the same time, completeness of $\{x_n\}$ alone is not sufficient to ensure its S.b. character for (X,T), for we have

Example 5.2.2: Let $C_{2\pi}$ denote the Banach space of all continuous periodic functions $f:\mathbb{R} \to \mathbb{R}$ with period 2π and $\|f\| = \sup\{|f(x)| : x \in \mathbb{R}\}$. Define f_n in $C_{2\pi}$ for n = 0,1,... by $f_o(x) = 1/2$, $f_{2n}(x) = \cos nx$ and $f_{2n-1}(x) = \sin nx$ for $n \geqslant 1$. From [239], p. 90 or [227], §13.33, $[f_n] = C_{2\pi}$. However, by the well known Fejér's Theorem (cf. [227], p. 416), $\{f_n\}$ cannot be an S.b. for $C_{2\pi}$.

Turning to the other condition, we now claim that (5.1.3) is indispensable for the S.b. character of $\{x_n\}$; indeed, consider (cf. [82])

Example 5.2.3: Recall the sequence $\{x^n\}$ in the space $(\phi, \|\cdot\|)$ with $\|\cdot\| \equiv \|\cdot\|_\infty$ of Example 2.2.3. Then $[x^n] = \phi$ and $\{x^n\}$ is not an S.b. for ϕ. In this case, we need to consider (5.1.9) in place of (5.1.3) and if this were true in the present case, then (5.1.9) would hold with m = N, n = 2N,

$N > \max(3, 1 + M/3)$ and $\alpha_1 = 1$, $\alpha_i = N (2 \leq i \leq N)$, $\alpha_i = -(N-1)$ for $N + 1 \leq i \leq 2N$. But

$$\left\| \sum_{i=1}^{N} \alpha_i x^i \right\| = (N-1)N; \quad \left\| \sum_{i=1}^{2N} \alpha_i x^i \right\| = \frac{N}{3}$$

and consequently $M \geq 3(N-1)$, a contradiction.

Next, we show the importance of barrelledness in Theorem 5.1.7 in the form of (cf. [82])

Example 5.2.4: Here we recall the space $(c_o, \sigma(c_o, \ell^1))$ which is not even infrabarrelled ([132], p. 120) and has $\{e^n; e^n\}$ as its S.b. If (5.1.3) were true, then, for every y in ℓ^1, there would exist z in ℓ^1 and $M > 0$ such that for all m,n in $\mathbb{N}$ with $m \leq n$ and any choice of scalars $\alpha_1, \ldots, \alpha_n$, we would have

$$\left| \sum_{i=1}^{m} \alpha_i y_i \right| \leq M \left| \sum_{i=1}^{n} \alpha_i z_i \right|. \tag{5.2.5}$$

We proceed to show that (5.2.5) is not satisfied for $y = \{n^{-2}\}$. We consider two cases:

Case 1: Consider any z in ϕ and N in $\mathbb{N}$ and let ℓ be the length of z. Put

$$R = \sum_{i=1}^{\ell} i|z_i|,$$

and choose m_o in $\mathbb{N}$ with $m_o > \ell$ such that

$$\sum_{i=1}^{m_o} \frac{1}{i} > NR.$$

Choose α in ω with $\alpha_i = i (1 \leq i \leq 2m_o)$ and arbitrary elsewhere. We then obtain

$$\left| \sum_{i=1}^{m_o} \alpha_i y_i \right| > N \left| \sum_{i=1}^{2m_o} \alpha_i z_i \right|,$$

and this contradicts (5.2.5).

Case 2: Fix N in $\mathbb{N}$ and z in $\ell^1 \setminus \phi$. Put

$$R = \sum_{n \geq 1} |z_n| < \infty ,$$

and choose n_o so that $\sum_{i=1}^{n_o} i^{-1} > NR$. We can find $n_1, \ldots, n_{n_o}$ such that (i) $n_o < n_1 < \ldots < n_{n_o}$, (ii) $z_{n_j} \neq 0$ for $j = 1, \ldots, n_o$ and (iii) $z_i = 0$ for $n_j < i < n_{j+1}$ $(j = 0, \ldots, n_o - 1)$. Define α in ω by

$$\alpha_i = \begin{cases} i, \ 1 \leq i \leq n_o; \\ (1-j)\, z_j\, z_i^{-1}, \ i = n_j (1 \leq j \leq n_o) \\ \text{arbitrary}, \ n_j < i < n_{j+1} \ (0 \leq j \leq n_o - 1), \ i > n_o. \end{cases}$$

Write $\ell = n_{n_o}$, then

$$N \left| \sum_{i=1}^{\ell} \alpha_i z_i \right| < NR < \left| \sum_{i=1}^{n_o} \alpha_i y_i \right| ,$$

and this again contradicts (5.2.5).

Remark: The invalidity of (5.1.3) in Example 5.2.4 is also a simple consequence of a known result in the sequence space theory. In fact, the truth of (5.1.3) for an l.c.s.s. $(\lambda, \sigma(\lambda, \lambda^\times))$ is equivalent to the fact that $\lambda^\times = \phi$; cf. [132], p. 65. However, the construction of Example 5.2.4 is of independent interest and of constructive nature.

Applications and examples

We apply here the previous results to ascertain the S.b. nature of certain concrete sequences in some more familiar spaces.

Example 5.2.6: Let C[0,1] denote the real Banach space of all continuous functions $f : [0,1] \to \mathbb{R}$ equipped with the usual sup norm $\|f\| = \sup \{|f(x)| : 0 \leq x \leq 1\}$. It is natural to enquire the S.b. nature of the sequence $\{f_n\}$ in C[0,1] given by $f_n(x) = x^n$, $n = 0,1,\ldots$. By an application of Theorem 5.1.8, it will turn out that $\{f_n\}$ is not an S.b. for C[0,1]. In this direction, let us observe first of all that $\{f_n\}$ is complete in C[0,1]; cf. [69], p. 32 or [70], p. 261. On the other hand, if (5.1.9) were true for some $M \geq 1$, then choosing $m = N$, $n = N+1$ with $M < (1+N)(1+N^{-1})^N$ for some N in $\mathbb{N}$ and $\alpha_o = \alpha_1 = \ldots = \alpha_{N-1} = 0$, $\alpha_N = N$, $\alpha_{N+1} = -N$, we find that

$$N \sup_{0 \leq x \leq 1} |x^N| \leq MN \sup_{0 \leq x \leq 1} |x^N - x^{N+1}| = M\{\frac{N}{N+1}\}^{N+1}.$$

Hence $(1+N)(1+N^{-1})^N \leq M$ and this contradicts the choice of N.

Example 5.2.7: Let $\mathfrak{P} \equiv \mathfrak{P}_R$ denote the subspace of C[0,1] consisting of all polynomials. It follows from the preceding example that $\{f_n\}$ is not an S.b. for $\mathfrak{P}$. However, this sequence is a t.b. for $\mathfrak{P}$. We may also demonstrate the non-S.b. character of $\{f_n\}$ for $\mathfrak{P}$ as follows. At the outset, let $\{F_i : i \geq 0\}$ denote the s.a.c.f. corresponding to $\{f_i : i \geq 0\}$.

Define $\Psi_n (n \geq 1)$ in C[0,1] by

$$\psi_n(x) = \begin{cases} 0, & 0 \leq x \leq 1/n; \\ (1-nx)/nx^2, & 1/n \leq x \leq 1. \end{cases}$$

There exist $g_n (n \geq 1)$ in $\mathfrak{P}$ such that $\|\psi_n - g_n\| < 1/n$. Put $P_n(x) = x + x^2 g_n(x)$; $0 \leq x \leq 1$, $n \geq 1$. Then each $P_n \in \mathfrak{P}$ and has degree at least 2. Also $\|P_n\| \leq 2/n$ and so $P_n \to 0$. But $F_1(P_n) = 1$, $n \geq 1$ and this shows the discontinuity of F_1.

Note: The preceding three examples are given in [82].

The following notation will be helpful in considering further examples. For $r > 0$, let

$$D_r = \{z \in \mathbb{C} : |z| < r\},$$

$$\Gamma_r = \{z \in \mathbb{C} : |z| = r\}$$

$$\bar{D}_r = D_r \cup \Gamma_r.$$

By a trignometric polynomial P_n we mean the mapping $P_n : \Gamma_1 \to \mathbb{C}$ with

$$P_n(e^{it}) = \sum_{p=-n}^{n} \alpha_p \varepsilon_p(e^{it}),$$

where $\alpha_p \in \mathbb{C}$, $\varepsilon_p(e^{it}) = \exp(ipt)$ with $p = 0, \pm 1, \ldots$ and $0 \leq t \leq 2\pi$. Also, we write Γ for Γ_1.

Let $L^p(\Gamma)$, $0 < p < \infty$, denote the space of all $\mathbb{C}$-valued Lebesgue measurable ($= \mu$-measurable) functions f on Γ such that

$$\|f\|_p = [\frac{1}{2\pi}\int_\Gamma |f|^p\, d\mu]^{1/p} \equiv [\frac{1}{2\pi}\int_0^{2\pi} |f(e^{it})|^p\, dt]^{1/p} < \infty,$$

and let $C(\Gamma)$ be the subspace of $L^p(\Gamma)$, consisting of all continuous functions on Γ. $C(\Gamma)$ is also equipped with its own norm $\|f\| = \sup\{|f(e^{it})| : 0 \leq t \leq 2\pi\}$. If $1 \leq p < \infty$, then $C(\Gamma)$ is dense in $L^p(\Gamma)$; cf. [199], p. 68. Hence, using a well known result of Fejér on (C,1)-summability (cf. [227], p. 414), the family of all trigonometric polynomials is dense in $L^p(\Gamma)$ for $1 \leq p < \infty$. In particular, $[\varepsilon_n : n = 0, \pm 1, \ldots] = L^p(\Gamma)$, for $1 \leq p < \infty$.

Example 5.2.8: Consider $L^2(\Gamma)$ and define $\{\eta_n : n \geq 0\}$ by $\eta_{2n}(t) = \exp(int)$, $n \geq 0$ and $\eta_{2n-1}(t) = \exp(-int)$ for $n \geq 1$. From the preceding paragraph and the fact that

$$\|\sum_{m=0}^{N} \alpha_m\eta_m\|_2^2 = \sum_{m=0}^{N} |\alpha_m|^2 \leq \|\sum_{m=0}^{N+1} \alpha_m\eta_m\|_2^2,$$

we easily conclude that $\{\eta_m\}$ is an S.b. for $L^2(\Gamma)$.

Example 5.2.9: The sequence $\{\eta_m : m \geq 0\}$ is complete but not an S.b. for $(C(\Gamma), \|\cdot\|)$. On the other hand, if it were an S.b., then for each f in $C(\Gamma)$,

$$f(1) = \lim_{n\to\infty} \frac{1}{2\pi}\int_0^{2\pi} [\sum_{k=-n}^{n} e^{-ik\theta}]\, f(e^{i\theta})d\theta. \qquad (*)$$

Put

$$F_n(f) = \frac{1}{2\pi}\int_0^{2\pi} D_n(\theta)\, f(e^{i\theta})d\theta,\ f \in C(\Gamma)$$

where

$$D_n(\theta) = \sum_{k=-n}^{n} e^{-ik\theta} = \frac{\sin(n+\frac{1}{2})\theta}{\sin\frac{\theta}{2}}.$$

Then $F_n \in (C(\Gamma))^*$, $n \geq 0$. Let $y_n(\theta) = 1$ if $D_n(\theta) \geq 0$, and $= -1$ if $D_n(\theta) < 0$. Then $y_n \in L^1(\Gamma)$ for $n \geq 0$ and so for $\varepsilon > 0$ (given), we find g_n in $C(\Gamma)$ with $\|g_n\| = 1$ and

$$\frac{1}{2\pi}\int_0^{2\pi} |y_n(\theta) - g_n(\theta)|d\theta < \frac{\varepsilon}{\sup\{|D_n(\theta)| : 0 \leq \theta \leq 2\pi\}}.$$

Therefore

$$\|F_n\| \geq |F_n(g_n)| \geq \frac{1}{\pi^2} \sum_{k=0}^{2n} \frac{1}{1+k} \int_{k\pi}^{(1+k)\pi} |\sin x|dx - \varepsilon.$$

Hence, by the Banach-Steinhaus theorem, this yields an f in $C(\Gamma)$ such that $\{F_n(f)\}$ is unbounded, thereby contradicting (*).

Exercise 5.2.10: Prove that $\{\eta_n : n \geq 0\}$ is complete but not an S.b. for $(L^1(\Gamma), \|\cdot\|_1)$. [Hint: Assume the contrary and for some f in $L^1(\Gamma)$ with $\|f\|_1 = 1$, show that $\{S_{2n}(f)\}$ is unbounded, where

$$S_{2n}(f)(t) = \frac{1}{2\pi} \int_0^{2\pi} \frac{\sin (n + \frac{1}{2})\theta}{\sin \frac{\theta}{2}} f(e^{i(t-\theta)})d\theta.]$$

Example 5.2.11: This is the classical example of Schauder [205], consisting of a sequence of functions from C[0,1], usually referred to as the *Schauder system*, which forms an S.b. for C[0,1]. Accordingly, let us introduce a linearly independent sequence $\{f_n : n \geq 0\}$ in C[0,1] by $f_o = 1$, $f_1(x) = x$ and

$$f_{2^n+i}(x) = \begin{cases} 2(2^n x - i + 1), & (2i-2)/2^{n+1} \leq x \leq (2i-1)/2^{n+1}, \\ 1, & x = (2i-1)/2^{n+1}; \\ 2(i - 2^n x), & (2i-1)/2^{n+1} \leq x \leq 2i/2^{n+1}, \\ 0, & \text{elsewhere}, \end{cases}$$

where $i = 1,\ldots,2^n$; $n = 0,1,\ldots$.

To appreciate the geometry of the functions $f_2, f_3, \ldots$, divide [0,1] successively into even number of subintervals by the points $1/2$, $1/2^2$, $3/2^2, \ldots; 1/2^n, \ldots, (2^n-1)/2^n; \ldots$ (at the n-th stage there are 2^{n-1} division points) and draw isosceles triangles with bases as the preceding undivided subintervals, each triangle having its vertex height equal to unity. The successive triangles obtained in this way represent $f_2, f_3, f_4, \ldots, f_{2^{n-1}+1}, \ldots, f_{2^n}, \ldots$ respectively. The situation corresponding to divisions n = 1,2 and 3 is shown in Figures 5.2.12, 5.2.13 and 5.2.14.

Let us first show that $\{f_n : n \geq 0\}$ is complete in C[0,1]. Consider there-

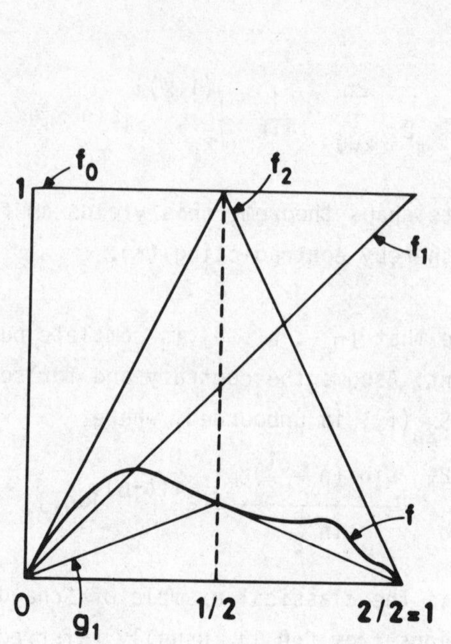

Fig. 5.2.12

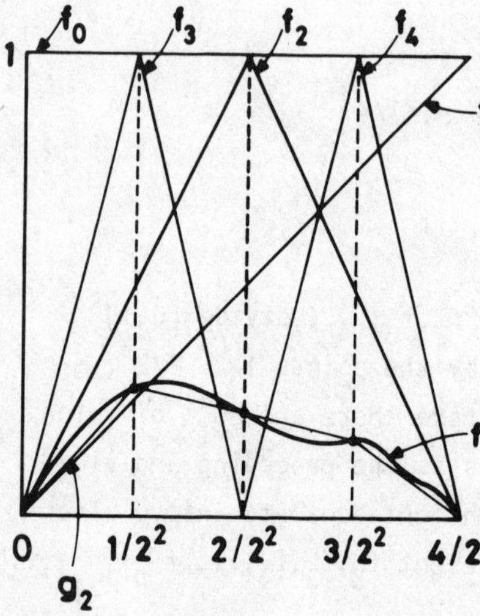

Fig. 5.2.13

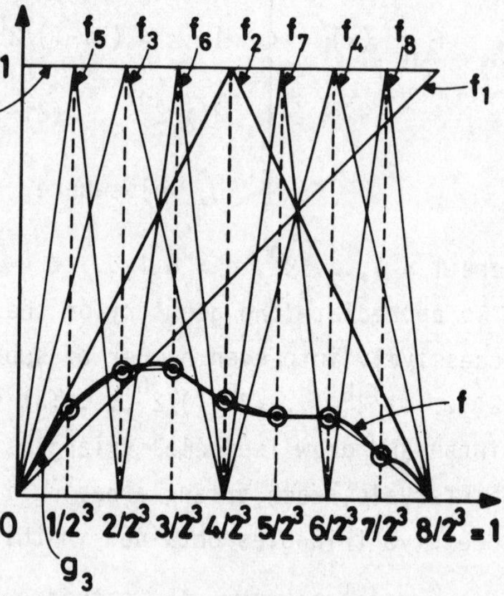

Fig. 5.2.14

Schauder System $\{f_o, f_1, \ldots\}$

fore an f in C[0,1] with $f \neq 0$, $f(0) = f(1) = 0$. Consider the polygonal arcs $g_1, \ldots, g_n$ obtained by joining linearly the consecutive points where f coincides with the vertex heights of the successive isosceles triangles representing $f_2, f_3, f_4, \ldots, f_{2^{n-1}+1}, \ldots, f_{2^n}$. For the case when n = 1,2,3, see the figures referred to earlier. By choosing n large enough, g_n approximates f as closely as we wish. Further, it is easily seen that $\{g_n, f_2, \ldots, f_{2^n}\}$ is linearly dependent and it follows that $[f_n : n \geqslant 0] = C[0,1]$.

Finally, choose an arbitrary set $\alpha_0, \ldots, \alpha_m, \ldots, \alpha_n$ of scalars and let

$$\phi_k = \sum_{i=0}^{k} \alpha_i f_i \quad (0 \leqslant k \leqslant n).$$

The maximum of $|\phi_m(x)|$ lies at $x = x_0$, where $x_0 = 0$ or 1 or one of the points in (0,1) intersected by the vertex heights of triangles representing $f_2, \ldots, f_m$. If $n > m$, then

$$\|\phi_n\| \geqslant |\phi_n(x_0)| = |\phi_m(x_0) + \sum_{i=m+1}^{n} \alpha_i f_i(x_0)|$$

$$= \|\phi_m\| .$$

Hence (5.1.9) holds for the sequence $\{f_n : n \geqslant 0\}$ and therefore it is an S.b. for C[0,1].

Let us now consider an S.b. usually referred to as the *Haar system* for $L^p[0,1]$ where $1 \leqslant p < \infty$. Accordingly, we have [207].

Example 5.2.15: Introduce the linearly independent sequence $\{y_1, y_n, \ldots\}$ in $L^p[0,1]$, where $y_1 = 1$ and

$$y_{2^n+m}(t) = \begin{cases} 2^{n/2}, & (2m-2)/2^{n+1} \leqslant t < (2m-1)/2^{n+1}, \\ -2^{n/2}, & (2m-1)/2^{n+1} \leqslant t < 2m/2^{n+1}; \\ 0 & , \text{elsewhere}, \end{cases}$$

where $m = 1, \ldots, 2^n$; $n = 0, 1, \ldots$. The computation of $y_1, \ldots, y_8$ is shown in Figure 5.2.16.

To show that $\{y_1, \ldots, y_n, \ldots\}$ is complete in $L^p[0,1]$, let us choose g in

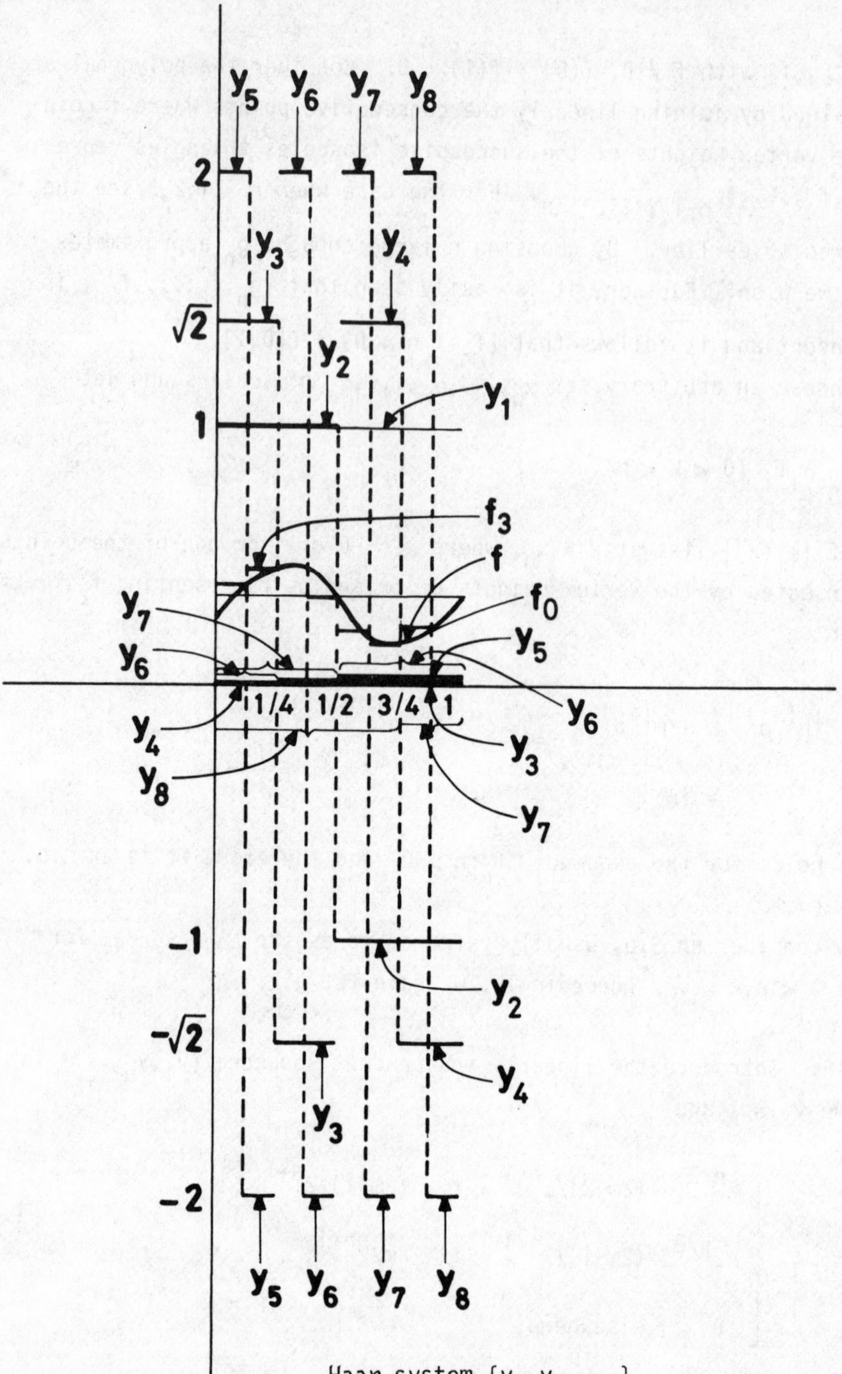

Haar system $\{y_1, y_2, \ldots\}$

Fig. 5.2.16

$L^p[0,1]$ and $\varepsilon > 0$ arbitrarily. There exists f in C[0,1] such that $\|f-g\|_p < \varepsilon/2$. Also, there exists $\delta > 0$ such that, for $|t_1-t_2| < \delta$, $|f(t_1)-f(t_2)| < \varepsilon/2$. Divide [0,1] into 2^n subintervals each of length $1/2^n < \delta$ (this is possible for large n). For this choice of n, let us define the step function f_n on [0,1] by

$$f_n(t) = \sum_{i=1}^{2^n} \alpha_i y_i(t),$$

where $\alpha_1,\ldots,\alpha_{2^n}$ are chosen in such a way that $f_n(t) = f(t)$ for $t = 1/2^n$, $2/2^n,\ldots,1$. Then $|f_n(t)-f(t)| < \varepsilon/2$ for all t in [0,1]. Thus $\|f_n - g\|_p < \varepsilon$, giving $L^p[0,1] = [y_n]$.

Next, choose arbitrary (m+1) real scalars $\alpha_1,\ldots,\alpha_{m+1}$ and let

$$f(t) = \sum_{i=1}^{m} \alpha_i y_i(t),\ g(t) = f(t) + \alpha_{m+1}\, y_{m+1}(t).$$

Let α (resp. β) be the inf (resp. sup) of t in [0,1] for which $y_{m+1}(t) \neq 0$. Then

$$y_{m+1}(t) = \begin{cases} 2^{n/2}, & \alpha \leq t < (\alpha+\beta)/2, \\ -2^{n/2}, & (\alpha+\beta)/2 \leq t < \beta, \end{cases}$$

for some n depending upon m and f(t) = k (= a constant) for t in $[\alpha,\beta)$. Thus

$$\int_\alpha^\beta |g(t)|^p\, dt = \int_\alpha^{(\alpha+\beta)/2} |k+2^{n/2}\alpha_{m+1}|^p dt + \int_{(\alpha+\beta)/2}^{\beta} |k-2^{n/2}\alpha_{m+1}|^p\, dt$$

$$\geq (\beta-\alpha)\ |k|^p = \int_\alpha^\beta |f(t)|^p\, dt.$$

Consequently $\|f\|_p \leq \|g\|_p$ and hence, by Theorem 5.1.8 again, $\{y_n\}$ is an S.b. for $L^p[0,1]$, $1 \leq p < \infty$.

Example 5.2.17: Let E_R denote the class of all functions analytic in D_R, equipped with the compact open topology T_R generated by $\{p_r : 0 < r < R\}$, where $p_r(f) = \sup\{|f(z)| : z \in \bar{D}_r\}$. As before, let $\delta_n(z) = z^n$ for $n = 0,1,\ldots$ and $z \in \mathbb{C}$. It is well known (cf. [96], [97], [120], [121]) that (E_R,T_R) is a non-normable Fréchet space (for similar results in several complex variables and infinite-dimensional holomorphy, one may refer respectively to [122]; [123] and [15]; [16]). For several reasons of analysis, it is convenient to introduce another equivalent topology S_R on E_R generated by $\{q_r : 0 < r < R\}$, where, for f in E_R,

$$q_r(f) = \sum_{n \geq 0} |a_n| r^n, \quad f(z) = \sum_{n \geq 0} a_n z^n$$

with

$$\limsup_{n\to\infty} |a_n|^{1/n} = \frac{1}{R}.$$

From the classical theory of analytic functions, it is easily seen that $E_R = [\delta_n]^{S_R}$. Also, for any $r(0 < r < R)$ and each set $\alpha_1, \ldots, \alpha_{n+1}$ of complex scalars,

$$q_r\left(\sum_{i=0}^{n} \alpha_i \delta_i\right) \leq \sum_{i=0}^{n+1} |\alpha_i| \, r^i = q_r\left(\sum_{i=0}^{n+1} \alpha_i \delta_i\right),$$

and so, by Theorem 5.1.2, $\{\delta_n : n \geq 0\}$ is an S.b. for (E_R, S_R) and hence for (E_R, T_R). [By using Taylor series expansion for each f in E_R and the uniqueness theorem for power series, it is also easily verified that $\{\delta_n\}$ is an S.b. for (E_R, T_R) and hence for (E_R, S_R).]

Finally, let us consider the example of *disc algebra* $A_c = \{f : f$ is analytic in D_1 and continuous on $\bar{D}_1\}$. A_c is a Banach space under the norm $\|f\| = \sup\{|f(z)| : |z| \leq 1\} = \sup\{|f(z)| : |z| = 1\}$. This is again a problem of Banach [5], who asked whether or not A contains an S.b. The question was answered in the affirmative by Bočkarev [14], who constructed an S.b. for A out of the so-called Franklin system to which we will return at the end of this section. In the meantime, let us discuss the behaviour of $\{\delta_n : n \geq 0\}$ with regard to its S.b. character of A_c. At the outset, let us observe that by using the notion of (C,1) summability, the complex form of the Fourier series for f in A_c (cf. [150], p. 450; [239], p. 7), the maximum modulus principle and Fejér's theorem, we have (see, for instance, [159], p. 329 or [224], p. 94)

$$f(z) = \lim_{n\to\infty} \sum_{k=0}^{n-1} \left(1 - \frac{k}{n}\right) a_k(f) z^k; \quad \liminf_{n\to\infty} |a_n(f)|^{1/n} \leq 1,$$

where the a_n's are Taylor's coefficients corresponding to f and the series on the right converges uniformly in $\bar{D}_1$. Hence $[\delta_n] = A_c$; this fact may also be proved by using Walsh's theorem on approximation by polynomials (cf. [161], p. 98). Whether or not $\{\delta_n\}$ is an S.b. for A_c can be established by proving or disproving the following statement:

> There exists an $M \geq 1$ such that, for all integers m and n with $0 \leq m \leq n$ and arbitrary scalars $\alpha_0,\ldots,\alpha_m,\ldots,\alpha_n$, the following inequality is satisfied:
>
> $$\sup_{0\leq\theta\leq 2\pi} \left|\sum_{i=0}^{m} \alpha_i e^{i\theta}\right| \leq M \sup_{0\leq\theta\leq 2\pi} \left|\sum_{i=0}^{n} \alpha_i e^{i\theta}\right|.$$

This statement is equivalent to the question: Does the Fourier series of each f in A_c converge to f itself? However, it is not true. For, consider an f in $C[0,2\pi]$ such that its Fourier series diverges in $[0,2\pi]$, cf. [239], p. 298. Then the function g(z) defined by

$$g(z) = \frac{1}{2\pi}\int_0^{2\pi} f(t)\,\frac{e^{it}+z}{e^{it}-z}\,dt, \quad |z| < 1$$

and $g(e^{it}) = f(t)$, $0 \leq t \leq 2\pi$, is the required function; cf. [88], p. 295.

Bočkarev's example

Toward the end of this chapter we use the example of Bočkarev [14] to show the existence of an S.b. for the disc algebra A_c. Let us first provide the necessary background before we consider the desired result in the form of a theorem.

If $f \in A_c$, then (cf. [88], p. 77)

$$f(z) = \frac{1}{2\pi}\int_{-\pi}^{\pi} u(e^{it})\,\frac{e^{it}+z}{e^{it}-z}\,dt + i\,\mathrm{Im}\, f(0), \tag{5.2.18}$$

where $|z| < 1$ and u is the real part of f(z). Whenever an arbitrary g is defined on Γ_1, we will write interchangeably g(t) for $g(e^{it})$, $-\pi \leq t \leq \pi$; similarly, if g is defined on $[-\pi,\pi]$ and has period 2π, g(t) will mean $g(e^{it})$.

Let g be any continuous function on Γ_1, then it is known (cf. [239], p.103 and 131) that

$$\tilde{g}(x) \equiv -\frac{1}{\pi}\int_0^{\pi} \frac{g(x+t) - g(x-t)}{2\tan(t/2)}\,dt \tag{5.2.19}$$

exists for $-\pi < x < \pi$. Thus, for f in A_c, we have from (5.2.18) and (5.2.19)

$$f(\theta) = u(\theta) + iv(\theta) + i \operatorname{Im} f(0); \quad -\pi < \theta < \pi, \tag{5.2.20}$$

and

$$u(\theta) = -\tilde{v}(\theta) + u(0); \quad -\pi < \theta < \pi, \tag{5.2.21}$$

where $v(\theta) = \tilde{u}(\theta)$. Observe that

$$\frac{1}{2\pi} \int_{-\pi}^{\pi} v(\theta)\, d\theta = 0. \tag{5.2.22}$$

Suppose G is a $\mathbb{C}$-valued function continuous on Γ_1, the function

$$G(z) = \frac{1}{2\pi} \int_{-\pi}^{\pi} \mathrm{Rl}\Big[\frac{e^{it} + z}{e^{it} - z}\Big]\, G(t)\, dt \tag{5.2.23}$$

is analytic in D_1 and so $G \in A_c$; cf. [88], p. 295, [92], p. 32, [160], p. 155.

Let us now turn to the Franklin system $\{f_n\} \equiv \{f_n : n = 0,1,\ldots\}$ defined over $[0,2\pi]$; cf. [20]; [60]. This is obtained by the Schmidt orthonormalizatio process applied to the Schauder system $\{g_n\}$ over $[0,2\pi]$:

$$f_n(t) = \sum_{i=0}^{n} \lambda_{in}\, g_i(t); \quad 0 < t < 2\pi,$$

where (λ_{in}) is the uniquely determined infinite matrix with $\lambda_{nn} > 0$, $n \geq 0$. Here $f_0(t) = 1/\sqrt{2\pi}$ and $f_1(t) = \sqrt{3}\,(t/\pi - 1)/\sqrt{2\pi}$, $0 < t < 2\pi$. In general, when $n \geq 2$, there exists an unique integer $m \geq 0$ with $2^m < n \leq 2^{m+1}$ such that $n = 2^m + p$ for some p in $\mathbb{N}$ with $1 \leq p \leq 2^m$. The function f_n is chosen suitably so as to be continuous on $[0,2\pi]$ and linear piecewise in between the nodal points $\pi\ell/2^{m+1}$, where $\ell = 0,1,\ldots,2p;\ 2p+2,\ 2p+4,\ldots,2^{m+1}$, e.g.,

$$f_2(t) = \begin{cases} \sqrt{\dfrac{3}{2\pi}}\Big(\dfrac{2t}{\pi} - 1\Big), & 0 < t < \pi, \\[2ex] \sqrt{\dfrac{3}{2\pi}}\Big(3 - \dfrac{2t}{\pi}\Big), & \pi < t < 2\pi. \end{cases}$$

Define

$$F_n(t) = \begin{cases} f_n(2t), & 0 \leqslant t \leqslant \pi ; \\ f_n(-2t), & -\pi \leqslant t \leqslant 0. \end{cases}$$

Then $F_n \in C[-\pi,\pi]$ and is even, indeed, $F_n \in \text{Lip } 1$, $n \geqslant 0$. Hence $\tilde{F}_n$ exists and is an odd continuous function over $[-\pi,\pi]$. In particular,

$$\int_{-\pi}^{\pi} F_n(x)dx = 0, \quad n \geqslant 1; \qquad \int_{-\pi}^{\pi} \tilde{F}_n(x)dx = 0, \quad n \geqslant 0, \tag{5.2.24}$$

and hence, using a result of [239], p. 128, we conclude the orthonormality of $\{F_n\}$ and $\{\tilde{F}_n\}$, also $\{F_n\}$ is orthogonal to $\{\tilde{F}_n\}$. Let

$$G_o(t) = \frac{1+i}{2\sqrt{\pi}}, \quad G_n(t) = \frac{1}{\sqrt{2}} (F_n(t) + i\tilde{F}_n(t)), \tag{5.2.25}$$

where $n \geqslant 1$ and $-\pi \leqslant t \leqslant \pi$. Then, from what precedes, $\{G_n\}$ is an orthonormal family on $[-\pi,\pi]$ of which the extensions $G_n(z)$ are members of A_c, cf. (5.2.23) Further for f in A_c with u and v as before

$$(u,F_n) = (v,\tilde{F}_n); \quad (v,F_n) = -(u,\tilde{F}_n), \quad n \geqslant 1, \tag{5.2.26}$$

where $(u,F_n) = \int_{-\pi}^{\pi} u(t)F_n(t)dt$, etc., cf. (5.2.21) and (5.2.24).

Observe that $\{\sqrt{2}\, F_n\}$ is the Franklin system over $[0,\pi]$ and so, if $g \in C[0,\pi]$, then

$$g(x) = \sum_{n \geqslant 0} \left(2 \int_o^{\pi} g(t)F_n(t)dt\right)F_n(x), \tag{5.2.27}$$

the series being uniformly convergent on $[0,\pi]$. For g in $L[0,\pi]$, let us write $S_n(g)$ for the n-th partial sum of the series in (5.2.27), then (cf. [21], p. 313)

$$\int_o^{\pi} |S_n(g)(y)-g(y)|dy \leqslant B \sup_{0 \leqslant h \leqslant \frac{1}{n}} \int_o^{\pi-h} |g(x+h)-g(x)|dx, \tag{5.2.28}$$

where B is a constant independent of n. In what follows, we use the symbol B for such a constant and assume it to be not necessarily the same at each occurrence.

Let us write $K_n(x,y) = \sum_{i=0}^{n} F_i(x)F_i(y)$, the Dirichlet kernel of $\{F_n\}$. Then K_n is continuous in each variable x and y separately, $-\pi \leq x,\ y \leq \pi$ and it is known that, for $n \geq 2$,

$$|K_n(x,y)| \leq Bn \exp(-Bn\ |x-y|), \tag{5.2.29}$$

valid for $-\pi \leq x,\ y \leq \pi$; cf. [21], p. 301. Since $K_n(x,y)$ is linear in between the consecutive nodal points $\pm\ \pi\ell/2^{m+1}$, whenever $n \geq 2$ is given by $n = 2^m+p$, $m = 0,1,\ldots,$ and $1 \leq p \leq 2^m$, where $\ell = 0,1,\ldots,2p,\ 2p+2,\ 2p+4,\ldots,2^{m+1}$, we easily conclude that

$$\int_0^{1/n} \left|\frac{K_n(x+t,y)-K_n(x-t,y)}{t}\right| dt \leq Bn \int_{-7/n}^{7/n} |K_n(x+t,y)|dt. \tag{5.2.30}$$

[(5.2.30) also follows from Markov's inequality [242], p.40]. Hence, using (5.2.29) and (5.2.30), we have

$$\int_0^{\pi}\int_0^{1/n} \left|\ \frac{K_n(x+t,y)-K_n(x-t,y)}{t}\ \right| dt \leq B. \tag{5.2.31}$$

Now we have the main

<u>Theorem 5.2.32</u>: The sequence $\{G_n\}$ is an S.b. for A_c, where

$$G_n(z) = \frac{1}{2\pi}\int_{-\pi}^{\pi} \text{Rl}\left[\frac{e^{it} + z}{e^{it} - z}\right]G_n(t)dt,\ |z| < 1$$

and $G_n(t)$ is given by (5.2.25), $-\pi \leq t \leq \pi$.

<u>Proof</u>: Let $f \in A_c$ and decompose f as in (5.2.20) over $[-\pi,\pi]$. It is enough to prove that

$$\sum_{n=0}^{N} \alpha_n G_n(x) \to u(x) + iv(x) + i\ \text{Im}\ f(0) \tag{5.2.33}$$

uniformly in $-\pi \leq x \leq \pi$, where $\alpha_n (n \geq 0)$ is the Bočkarev-Fourier coefficient of f:

$$\alpha_n = \int_{-\pi}^{\pi} f(x)\ \overline{G_n(x)}dx.$$

Here

$$\alpha_o = (1+i)\sqrt{\pi}\ \text{Im } f(0) + \frac{1}{2\sqrt{\pi}}(1-i)\int_{-\pi}^{\pi} u(x)dx \tag{5.2.34}$$

and, using (5.2.24) through (5.2.26),

$$\alpha_n = \sqrt{2}\ (u,F_n) + i(v,F_n),\ n \geq 1. \tag{5.2.35}$$

Hence, using (5.2.34) and (5.2.35), we find

$$\sum_{n=0}^{N} \alpha_n G_n(x) = \frac{1}{2\pi}\int_{-\pi}^{\pi} u(x)dx + \sum_{n=1}^{N} (u,F_n)F_n(x) - \sum_{n=1}^{N} (v,F_n)\tilde{F}_n(x)$$

$$+ i\ [\text{Im } f(0) + \sum_{n=1}^{N} (v,F_n)F_n(x) + \sum_{n=1}^{N} (u,F_n)\tilde{F}_n(x)] \tag{5.2.36}$$

$$= \Sigma_N^{(1)} + i\Sigma_N^{(2)},\ \text{(say)}.$$

In order to prove (5.2.33), we proceed to show that $\{\Sigma_N^{(1)}\}$ and $\{\Sigma_N^{(2)}\}$ converge to $u(x)$ and $\text{Im } f(0) + v(x)$ respectively, uniformly over $[-\pi,\pi]$.

Estimation of $\Sigma_N^{(1)}$

For any function g defined over $[-\pi,\pi]$, let us write g_1 and g_2 respectively for even and odd components of g. Then

$$\Sigma_N^{(1)} = \frac{1}{2\pi}\int_{-\pi}^{\pi} u_1(x)dx + \sum_{n=1}^{N} (u_1,F_n)F_n(x) - \sum_{n=1}^{N} (v_1,F_n)\tilde{F}_n(x). \tag{5.2.37}$$

Since u_1 is even, the first two terms on the right of (5.2.37) converge to $u_1(x)$ uniformly on $[-\pi,\pi]$; cf. (5.2.27).

Define the operator $\tilde{S}_n : C[-\pi,\pi] \to C[-\pi,\pi]$ by

$$\tilde{S}_n(g) = \sum_{i=0}^{n} (g,F_i)\tilde{F}_i.$$

Let us prove that if g is continuous on $[-\pi,\pi]$, is even and is of period 2π, then

$$\|\tilde{S}_n(g)\| \leq B\ \|g\|,\quad \forall n \geq 0. \tag{5.2.38}$$

It suffices to prove (5.2.38) when $n \geq 2$. Observe that

$$-\pi\tilde{S}_n(g)(x) = \int_{-\pi}^{\pi} \left[\int_0^{\pi} \frac{K_n(x+t,y)-K_n(x-t,y)}{2\tan(t/2)}\, dt\right] g(y)dy. \tag{5.2.39}$$

Introduce $\Lambda_{n,x}(t)$ for $-\pi \leq t, x \leq \pi$ as follows:

$$\Lambda_{n,x}(t) = \begin{cases} 0, & |x-t| \leq 1/n, \\ \frac{1}{2}\cot\left(\frac{t-x}{2}\right), & |x-t| > 1/n. \end{cases}$$

Then (5.2.39) yields

$$-\pi\tilde{S}_n(g)(x) = \int_{-\pi}^{\pi} \left[\int_0^{1/n} \frac{K_n(x+t,y)-K_n(x-t,y)}{2\tan(t/2)}\, dt\right]\, g(y)dy$$

$$+ \int_{-\pi}^{\pi} \left[\int_{-\pi}^{\pi} K_n(t,y)\Lambda_{n,x}(t)dt - \Lambda_{n,x}(y)\right] g(y)dy$$

$$+ \int_{-\pi}^{\pi} \Lambda_{n,x}(y)g(y)dy$$

$$= I_1 + I_2 + I_3, \text{ (say)}, \tag{5.2.40}$$

since

$$\int_{-\pi}^{\pi} K_n(t,y)\Lambda_{n,x}(t)dt = \int_{1/n}^{\pi} \frac{K_n(x+t,y)-K_n(x-t,y)}{2\tan(t/2)}\, dt.$$

But $\cot(x/2) \leq 2/x$ for $0 \leq x \leq \pi$, and so

$$|I_1| \leq B\,\|g\|\,;\ \forall x,\ -\pi \leq x \leq \pi. \tag{5.2.41}$$

Also

$$I_3 = \int_{1/n}^{\pi} \frac{g(x+t) - g(x-t)}{2\tan(t/2)}\, dt,$$

and this yields

$$|I_3| \leq B\,\|g\|\,;\ \forall x,\ -\pi \leq x \leq \pi. \tag{5.4.42}$$

To estimate I_2, let us write $\Lambda_{n,x}^{(1)}$ for the even part of $\Lambda_{n,x}$, then, using (5.2.27), we get

$$I_2 = 2\int_0^{\pi} [S_n(\Lambda_{n,x}^{(1)})(y) - \Lambda_{n,x}^{(1)}(y)]g(y)dy$$

$$\Rightarrow \ |I_2| \leqslant B\,\|g\|\int_0^{\pi} |S_n(\Lambda_{n,x}^{(1)})(y) - \Lambda_{n,x}^{(1)}(y)|dy. \qquad (5.2.43)$$

But $\Lambda_{n,x}^{(1)}$ can be decomposed into four monotonic functions over $[0,\pi]$, say, u_1, u_2, u_3 and u_4 (for instance, if $x > 1/n$, then with $\alpha = \cot(1/2n)$, let

$$4u_1(t) = \begin{cases} \cot\frac{1}{2}(t-x), & 0 < t < x - \frac{1}{n}, \\ -\alpha, & t \geqslant x - \frac{1}{n}, \end{cases}$$

$$4u_2(t) = \begin{cases} \alpha, & t \leqslant x + \frac{1}{n}, \\ \cot\frac{1}{2}(t-x), & t > x + \frac{1}{n}, \end{cases}$$

$$4u_3(t) = \begin{cases} -\alpha-\cot\frac{1}{2}(t+x), & t < x - \frac{1}{n}, \\ -\cot\frac{1}{2}(t+x), & t \geqslant x - \frac{1}{n}, \end{cases}$$

$$4u_4(t) = \begin{cases} 0, & t \leqslant x + \frac{1}{n}, \\ \alpha, & t > x + \frac{1}{n}, \end{cases}$$

similarly, we may decompose $\Lambda_{n,x}^{(1)}$ in other cases as well) and consequently, using (5.2.28) and (5.2.43), we obtain

$$|I_2| \leqslant B\,\|g\|, \ \forall\, x, \ -\pi \leqslant x \leqslant \pi. \qquad (5.2.44)$$

Inequalities (5.2.41), (5.2.42) and (5.2.44) result in (5.2.38).

Next, if P is any polynomial of the form $P(x) = \sum_{i=0}^{N} \alpha_i F_i(x)$, $-\pi \leqslant x \leqslant \pi$, then $\tilde{S}_n(P)(x) = \tilde{P}(x)$, $-\pi \leqslant x \leqslant \pi$ for all $n \geqslant N$. Hence, using (5.2.38) we conclude that

$$\|\tilde{S}_n(v_1)-\tilde{v}_1\| \leqslant B\,\|v_1-P\| + \|\tilde{P}-\tilde{v}_1\|, \tag{5.2.45}$$

for all $n \geqslant N$, N being dependent on the polynomial considered above. By using Fejér's theorem on (C,1)-summability of Fourier series and (5.2.27), we can determine P of the type considered above such that $\|v_1-P\|$, $\|\tilde{v}_1-\tilde{P}\| \leqslant \varepsilon$. Hence, from (5.2.45),

$$\|\tilde{S}_n(v_1) - \tilde{v}_1\| \leqslant B\varepsilon, \tag{5.2.46}$$

for all large n. Therefore, from (5.2.21), (5.2.37) and (5.2.46), we conclude that

$$\Sigma_N^{(1)} \to u_1(x) + u_2(x) \tag{5.2.47}$$

uniformly in $-\pi \leqslant x \leqslant \pi$.

Estimation of $\Sigma_N^{(2)}$

By (5.2.21), we have $\tilde{u}_1(x) = v_2(x)$, $-\pi \leqslant x \leqslant \pi$. Making use of (5.2.22) and proceeding as before, we conclude that

$$\Sigma_N^{(2)} \to \operatorname{Im} f(0) + v_1(x) + v_2(x) \tag{5.2.48}$$

uniformly in $-\pi \leqslant x \leqslant \pi$.

The desired conclusion (5.2.33) now follows by combining (5.2.47) and (5.2.48). □

6 The weak basis theorem

6.1 BASES IN COMPATIBLE TOPOLOGIES

It is comparatively straightforward to ascertain the Schauder base character of a sequence in an l.c. TVS with respect to its weak topology rather than its original topology. Assuming this to be the case, a related question is: given an S.b. with respect to the weak topology of an l.c. TVS, when can we infer the S.b. character of this base with respect to the original topology of the space? More generally, we seek the solution of (cf. [190])

Problem 6.1.1: Let $\{x_n\}$ be a nonzero sequence in a vector space X equipped with two compatible l.c. topologies T_1 and T_2. If $\{x_n\}$ is a t.b. (resp. an S.b.) for (X,T_1), does this imply that $\{x_n\}$ is a t.b. (resp. an S.b.) for (X,T_2)?

In general, the solution to this problem is negative, cf. Examples 6.2.1, 6.2.2 and 6.2.3. Let us therefore pose a slightly simpler but more general problem in the form of

Problem 6.1.2: Let $\{x_n\}$ be a nonzero sequence in a TVS (X,T). If $\{x_n\}$ is a t.b. for $(X,\sigma(X,X^*))$, is $\{x_n\}$ a t.b. for (X,T)?

In the literature, the above problem, when particularized to a given class of TVS, is also known as the *weak basis theorem* (WBT) for that class.

The WBT for Banach spaces is the famous *Banach-Mazur theorem*, namely,

Theorem 6.1.3: A weak t.b. for a Banach space is a t.b. for this space in its initial topology.

Note: The above theorem is stated in [5], p. 238 without proof. Its proof was first outlined by Karlin [138] and later it was proved by Day [35] under certain restrictions.

Possible extensions

Essentially, possible extensions of Theorem 6.1.3 have appeared in the literature in one of two forms, namely, (a) the basis hypothesis is replaced by decomposition or (b) the Banach space structure is weakened to more general topological vector spaces. However, another possible extension of Theorem 6.1.3 takes the form (c) of a combination of (a) and (b).

Note: Concerning (a), it was possibly Ruckle [195] who first extended Theorem 6.1.3 for decompositions. For subsequent extensions of Theorem 6.1.3 in the direction of (c), one may refer to [41], [164] and [231], details of which will appear in [81].

Regarding direction (b), Bessaga and Pelczynski [13] proved

Theorem 6.1.4: A weak base for a Fréchet space is an S.b. in its original topology.

Note: There is an exhaustive proof of Theorem 6.1.4 in [54] on the lines of [5], p. 110 and a weak version of

Proposition 6.1.5: Each $\sigma(X,X^*)$-S.b.$\{x_n,f_n\}$ for a barrelled space (X,T) is a T-S.b. for X.

Proof: Since $f_n(x_m) = \delta_{mn}$ and $\{S_n\}$ is equicontinuous, (5.1.3) is satisfied. Also $[x_n]^T = X$. Now apply Theorem 5.1.2. □

Note: For the importance of barrelledness in Proposition 6.1.5, see Example 6.2.1.

Note: Proposition 6.1.5 is a special case of a result proved in [43], cf. also [3] or Chapter 7 and [76]. For further results in the direction of this proposition, one may look into [36] (cf. also [37] and [38]), [164] and [231].

An F-space analogue of Theorem 6.1.4 will be discussed in Chapter 10. There is a proof of Theorem 6.1.4 in [7], also on the lines of [164]. An extension of this theorem for LF-spaces is given in [7] as follows.

Theorem 6.1.6: Every $\sigma(X,X^*)$-t.b. $\{x_n;f_n\}$ for an LF-space (X,T) is an S.b. for (X,T).

We need two lemmas for the proof of Theorem 6.1.6.

Lemma 6.1.7: Let $\{x_n\}$ be a sequence in an l.c. TVS (X,T) and $f_n \in X'$ $(n \geq 1)$ so that $f_m(x_n) = \delta_{mn}$. Let Y be the subspace of X such that each y in Y is uniquely expressible as $\sum_{n\geq 1} f_n(y)x_n$, the convergence of the series being considered relative to $\sigma(X,X^*)$. Then there exists a unique l.c. topology $\bar{T}$ on Y such that $T_1 \equiv T|Y \subset \bar{T}$ and $\{R_n\}$ is $\bar{T}$-T_1 equicontinuous, where for y in Y, $R_n(y) = \sum_{i=1}^{n} f_i(y)x_i$. Further, if T is complete (resp. metrizable), then so is $\bar{T}$.

Proof: For p in D_T and y in Y, define $\bar{p}(y) = \sup \{p(R_n(y)):n \geq 1\}$. Starting with an arbitrary equicontinuous subset M of X^*, we find some p in D_T so that $|f(y)| \leq \bar{p}(y)$ for each y in Y and all f in M. Hence $T_1 \subset \bar{T}$. For the rest of the proof, make use of Lemma 1.2.20 and the proof of Theorem 2.2.5. □

Exercise 6.1.8: Prove Theorem 6.1.4. [Hint: Use Lemma 6.1.7.]

Lemma 6.1.9: Let (X,T) be the SIL of Fréchet spaces (X_n,T_n), $n \geq 1$ and $\{x_n;f_n\}$ be a t.b. for $(X,\sigma(X,X^*))$. Then, for each M in $\mathbb{N}$, there corresponds an N in $\mathbb{N}$ such that for every x in X_M, we have

$$x = \sigma(X_N,X_N^*) - \lim_{n\to\infty} \sum_{i=1}^{n} f_i(x)x_i, \tag{6.1.10}$$

where $f_i(x) = 0$ if $x_i \notin X_N$; and $X_n^* = (X_n,T_n)^*$, $n \geq 1$.

Proof: For $k \geq 1$, let

$$X_M^k = \{x \in X_M: x = \sigma(X_k,X_k^*)\text{-}\lim_{n\to\infty} \sum_{i=1}^{n} f_i(x)x_i,\ f_i(x) = 0 \text{ if } x_i \notin X_k\}.$$

From a well-known theorem of Dieudonné-Schwartz (cf. [93], p. 159 and 161), one can deduce (i) $\sigma(X,X^*)|X_n = \sigma(X_n,X_n^*)$, and (ii) a $\sigma(X,X^*)$-convergent sequence $\{x_n\}$ in X is contained in some X_N and $\sigma(X_N,X_N^*)$ converges. Hence, from (ii),

$$X_M = \bigcup_{k\geq 1} X_M^k.$$

Thus, by Baire's theorem, there exists N in $\mathbb{N}$ such that X_M^N is of second category in X_M. Regarding X_M^N as a subspace of X_N and applying Lemma 6.1.7, we get a Frechét topology T_M^N on X_M^N such that $T_N|X_M^N \approx T_M|X_M^N \subset T_M^N$ and so $X_M = X_M^N$; cf. [192], p. 125. □

Proof of Theorem 6.1.6: In view of Proposition 6.1.5, it suffices to show that each $f_n \in X^*$. For each M in $\mathbb{N}$, let $S_n^M = S_n|X_M$. Also, there exists N such that (6.1.10) holds. By using Lemmas 6.1.7 and 6.1.9, there exists a Fréchet topology $\bar{T}_M$ on X_M such that $\{S_n^M: n \geq 1\}$ is $\bar{T}_M$-T_M equicontinuous and $T_M \subset \bar{T}_M$. But then $T_M \approx \bar{T}_M$ and so $f_n|X_M$ is continuous for $M \geq 1$ and $n \geq 1$. □

6.2 Failure of the WBT

We now turn to the failure of the WBT on a certain class of spaces. Before we actually state the main result, let us consider the next three nontrivial examples exhibiting the failure of the WBT in some l.c. TVS. We respectively follow [48], [8] and [114] for these examples.

Example 6.2.1: The biorthogonal sequence $\{e^n;e^n\}$ is a $\sigma(k,\ell^1)$-S.b. for $(k,\tau(k,\ell^1))$. However, $e^{(n)} \not\to e$ in $\tau(k,\ell^1)$, cf. [132], p. 123, and so $\{e^n,e^n\}$ is not an S.b. for $(k,\tau(k,\ell^1))$ which is neither barrelled nor ω-complete.

Example 6.2.2: We consider here a complete Mackey space. Let λ be a BK-sequence space such that, for each x in λ, $||x|| \equiv |x_1| + \sup |x_n - x_{n+1}| < \infty$. Suppose further $W \equiv W_\lambda = \{x \in \lambda: ||x^{(n)}-x|| \to 0\}$. Now consider the space $(\ell_\lambda^\infty,\tau(\ell_\lambda^\infty,\ell^1))$, where $\ell_\lambda^\infty = W \cap \ell^\infty$. For brevity, let $\sigma = \sigma(\ell_\lambda^\infty,\ell^1)$ and $\tau = \tau(\ell_\lambda^\infty,\ell^1)$. It is clear that $\{e^n;e^n\}$ is an S.b. for $(\ell_\lambda^\infty,\sigma)$. We show that this sequence is not a t.b. for $(\ell_\lambda^\infty,\tau)$. If $x_i^n = (n-i)/n$, $1 \leq i \leq n$ and $x_i^n = 0$ for $i > n$, then $x^n \to e$ in λ and $||x^n||_\infty \leq ||e||_\infty$, $n \geq 1$. Hence $e \in \ell_\lambda^\infty$, cf. [132]; p. 214. If $e^{(n)} \to e$ in τ then $||e^{(n)}-e|| \to 0$ and $\sup ||e^{(n)}-e||_\infty < \infty$; cf. [132], p. 219. However, this is not true as $||e^{(n)}-e|| = 1$, $n \geq 1$. Also, this space is complete ([132], p. 220).

Example 6.2.3: The space $(C^*, \tau(C^*,C))$ does possess a $\sigma(C^*,C)$-S.b. by Example 5.2.11, where $C = C[0,1]$. For our purpose, it suffices to show that this space no longer possesses a t.b. Using the following result of Grothendieck ([54], p. 283 and 621; [72]; [74], p. 229): "a subset of C* is $\tau(C^*,C)$-relatively compact if and only if this set is $\sigma(C^*,C^{**})$-relatively compact", we conclude that any $\tau(C^*,C)$-t.b. is a $\sigma(C^*,C^{**})$-t.b. for C* and hence, from

Theorem 6.1.3 or 6.1.4, C* is norm separable. However, using an equivalent norm on C*, given by the total variation (cf. [156], p. 126), it is known that C* is not separable ([224], p. 92).

Failure in non-locally convex spaces

In this subsection we discuss a general result regarding the failure of the WBT for a certain class of TVS which are not l.c. TVS; the motivation for this is derived from the next example. There is no such general result for any subclass of l.c. TVS, although we do have occasional examples, as mentioned earlier, in different types of l.c. TVS, exhibiting the failure of the WBT.

Example 6.2.4: This is the space $(\ell^p, |\cdot|_p)$, $0 < p < 1$ considered in [222]; [223] and is already known to be a non-locally convex complete l.b. TVS with $(\ell^p)^* = \ell^\infty = (\ell^p)^\times$; cf. [132], p. 123 and 125. It is also known ([132], p. 125; [210] and [222]) that ℓ^p contains a proper closed subspace M such that $\bar{M}^\sigma = \ell^p$ and so M is dense in ℓ^1. This gives a sequence $\{h^n\}$ in M with $\|e^n-h^n\|_1 < 1/2^{n+1}$, $n \geq 1$. Since $\{h^n\}$ cannot be complete (for, otherwise, $M = \ell^p$), it is therefore not a t.b. for the space in question. Finally, we apply the following result on the stability of bases in Banach spaces, known as the Krein-Milman-Rutman theorem ([149]; cf. also [162], p. 63), namely, "if $\{x_n;f_n\}$ is an S.b. for a Banach space $(X,\|\cdot\|)$ with $\sum_{n\geq 1}\|x_n-y_n\|\,\|f_n\| < 1$, for some $\{y_n\} \subset X$, then $\{y_n\}$ is also an S.b. for X". We thus easily conclude that $\{h^n\}$ is an S.b. for $(\ell^p,\sigma(\ell^p,\ell^\infty))$.

Example 6.2.4 is a particular case of the following result (see [212]).

Theorem 6.2.5: Every non-locally convex l.b. F-space (X,T) with a $\sigma(X,X^*)$-t.b. $\{x_n;f_n\}$ has a $\sigma(X,X^*)$-t.b. which is not a t.b. for (X,T).

Proof: In view of Theorem 2.2.12, we may assume that $\{x_n;f_n\}$ is an S.b. for (X,T); otherwise there is nothing to prove.

By using Theorem 5.1.8 and the fact that $\tau(X,X^*)$ is given by a norm $\|\cdot\|$ (Proposition 1.2.15), one easily computes that $\{S_n\}$ is $\|\cdot\| - \|\cdot\|$ equicontinuous on X. Invoking both the notation and proof of Proposition 3.2.1, we find that $\{x_n;\hat{f}_n\}$ is an S.b. for the completion $(\hat{X}, \|\hat{\cdot}\|)$ of $(X, \|\cdot\|)$.

If possible, multiply $\{x_n\}$ by a nonzero sequence of scalars so that $x_n \to 0$

in T. As $\tau \subsetneq T$, we find a sequence $\{y_k\} \subset X$ so that $y_k \to 0$ in τ and $q(y_k) \geq \varepsilon$ for some $\varepsilon > 0$ (q is the F-norm, generating T) and

$$\|y_k\| < \frac{1}{2^{k+1}\|\hat{f}_k\|}, \quad k \geq 1.$$

Therefore, from the K-M-R theorem referred to in the preceding example, the sequence $\{u_n\} \equiv \{x_n + y_n\}$ is an S.b. for $\hat{X}$ and let $\{\phi_n\}$ be the corresponding s.a.c.f. If $\{u_n;\phi_n\}$ were a t.b. for (X,T), then $\{\phi_n\}$ is T-equicontinuous and therefore $\tau(X,X^*)$-equicontinuous. This contradicts that $u_n \to 0$ in $\tau(X,X^*)$. Hence $\{u_n;\phi_n\}$ is not a t.b. for (X,T), although it is an S.b. for $(X,\sigma(X,X^*))$.□

As an application of Theorem 6.2.5, we give another important example of a space exhibiting the failure of the WBT.

Example 6.2.6: Let H^p denote the Hardy class of order p, $0 < p < 1$, consisting of all analytic functions f in the open unit disc D_1 such that

$$\|f\|_p^p \equiv \sup_{0 \leq r < 1} \{\int_0^{2\pi} |f(re^{i\theta})|^p d\theta\} < \infty, \ \forall f \in H^p.$$

Also, let B^p $(0 < p < 1)$ denote the class of all functions analytic in D_1 and, for each f in B^p,

$$\|f\| \equiv \frac{1}{2\pi}\int_0^1 \int_0^{2\pi} (1-r)^{(1/p)-2} |f(re^{i\theta})| dr\, d\theta < \infty .$$

It is known (cf. [87]; [52], p. 36) that $\|\cdot\|$ is inferior to $\|\cdot\|_p^p$ on H^p and the completion of $(H^p, \|\cdot\|)$ is B^p; cf. [53]. Since $B^p \simeq \ell^1$ (cf. [153]; [213]), there exists a sequence $\{h^n\}$ in H^p, which can be made an S.b. for B^p by using the K-M-R theorem referred to in Example 6.2.4. Write H^{p*} for $(H^p, \|\cdot\|_p^p)^*$ and τ for $\tau(H^p,H^{p*})$. It is also known that $(H^p, \|\cdot\|)^* = H^{p*}$; cf. [53]. Hence $\{h^n\}$ is a $\sigma(H^p,H^{p*})$-S.b. for H^p. Finally, note that $(H^p, \|\cdot\|_p^p)$ is a non-locally convex l.b. F-space ([52], p. 37) and it remains to apply Theorem 6.2.5 to derive the desired conclusion.

The next result ([212]) is a variation of Theorem 6.2.5.

Theorem 6.2.7: Every non-locally convex F-space (X,T) admitting a continuous norm $\|\cdot\|$ and possessing a $\sigma(X,X^*)$ t.b. $\{x_n;f_n\}$, has a $\sigma(X,X^*)$-t.b. which is not a T-t.b.

The proof of this theorem requires (cf. [112])

<u>Proposition 6.2.8</u>: Let (X,T) be a complete l.c. TVS having an S.basic sequence $\{x_n;f_n\}$ such that, for some p_o in D_T, $|f_n(x)| \leq p_o(x)$ for all $n \geq 1$ and each x in $X_o \equiv [x_n]$. Consider any sequence $\{y_n\}$ in X satisfying

$$\sum_{n\geq 1} p_o(x_n-y_n) = K_o < 1;\ K_p \equiv \sum_{n\geq 1} p(x_n-y_n) < \infty\ ,\ \forall p \in D_T.$$

Then $\{y_n\}$ is an S.basic sequence in (X,T) and $[x_n] \simeq [y_n]$ under A with $Ax_n = y_n (n \geq 1)$.

<u>Proof</u>: It follows immediately that, for each x in X_o, the series $\sum_{n\geq 1} f_n(x)(y_n-x_n)$ converges in X (in fact, this series is absolutely convergent) and so we can define $A:X_o \to X_1 \equiv [y_n]$ by $Ax = \sum_{n\geq 1} f_n(x)y_n$. Then $Ax_n = y_n$, $n \geq 1$. Since $p(Ax) \leq p(x) + K_p p_o(x)$ and $p(x) \leq p(Ax) + K_p p_o(x) \leq p(Ax) + K_p(1-K_o)^{-1} p_o(Ax)$, the map A is a topological isomorphism from X_o onto $A[X_o] \subset X_1$. However, $A[X_o]$ is closed and so $A[X_o] = X_1$. □

<u>Proof of Theorem 6.2.7</u>: As before, assume that $\{x_n;f_n\}$ is an S.b. for (X,T). Here $\tau(X,X^*)$ is generated by a nondecreasing sequence $\{p_n:n \geq 1\}$ of seminorms on X. Observe that $\{x_n;\hat{f}_n\}$ is an S.b. for $(\hat{X},\hat{\tau})$ such that $\{S_n\}$ is equicontinuous, where $S_n:\hat{X} \to X$, $S_n(\hat{x}) = \sum_{i=1}^{n} \hat{f}_i(\hat{x})x_i$ and $\hat{X}$ is the completion of (X,τ), $\tau \equiv \tau(X,X^*)$ etc. Without loss of generality, we may assume that $\{x_n\}$ is regular in $\hat{X}$ and so $\{\hat{f}_n\}$ is equicontinuous on $(\hat{X},\hat{\tau})$; that is, for some N, $|\hat{f}_n(\hat{x})| \leq M\hat{p}_N(\hat{x})$ for all $\hat{x}$ in $\hat{X}$ and $n \geq 1$.

Choose $\{\alpha_n\}$ in $\mathbb{K}$ with $\alpha_n \neq 0$ $(n \geq 1)$ such that $q(\alpha_n x_n) \to 0$, q being the F-norm generating T. Since $\tau \subsetneq T$, there exist $\varepsilon > 0$ and a sequence $\{y_k\}$ in X such that $q(y_k) > \varepsilon$ and $p_{N+k}(y_k) < 2^{-k-1}|\alpha_k|$, for $k \geq 1$. Hence, using Proposition 6.2.8, the sequence $\{x_n + \alpha_n^{-1} y_n\}$ is an S.b. for $(\hat{X},\hat{\tau})$ and let $\{\phi_n\}$ be the corresponding s.a.c.f. Then $\{x_n + \alpha_n^{-1} y_n;\phi_n\}$ is a $\sigma(X,X^*)$-S.b. for (X,T) and, proceeding as in Theorem 6.2.5, we conclude that this sequence is not a t.b. for (X,T). □

<u>Remark</u>: The use of the continuous norm $\|\cdot\|$ in the foregoing theorem appears to force the weak S.b. $\{x_n\}$ to be $\tau(X,X^*)$-regular (shortly we will prove another result devoid of this restriction). This observation is

illustrated (cf. [212]) by

Example 6.2.9: Let $X^p_\omega = \ell^p \oplus \omega$, $0 < p < 1$, be equipped with the F-norm $\|\cdot\|^p_\omega$ with $\|\alpha \oplus \beta\|^p_\omega = |\alpha|_p + |\beta|_\omega$; $\alpha \in \ell^p$, $\beta \in \omega$. Since $(\omega, |\cdot|_\omega)$ does not admit a continuous norm (cf. [11]; [132], p. 106), neither does the space X^p_ω. If $\{h^n\}$ is the sequence constructed in Example 6.2.4, then $\{h^n \oplus e^n\}$ is a weak S.b. but not a t.b. for X^p_ω.

Earlier we have seen the failure of the WBT for certain subclasses of non-locally convex F-spaces. However, the problem as to whether the WBT fails in all non-locally convex F-spaces with weak bases was raised by Shapiro [212], p. 1299 5(b), and answered by Drewnowski [44] in the affirmative as follows:

Theorem 6.2.10: Every non-locally convex F-space (X,T) having a $\sigma(X,X^*)$-t.b. admits a $\sigma(X,X^*)$-t.b. which is not a T-t.b. for X.

Some intermediary discussion is needed before we pass on to the proof of Theorem 6.2.10. At this stage it is convenient to introduce

Definition 6.2.11: An S.b. $\{x_n;f_n\}$ for a TVS (X,T) is called an e-*Schauder base* (e-S.b. or T-e.S.b.) if $\{S_n\}$ is T-equicontinuous on X.

Note: We do not intend to deal with e-S.b. here, but rather postpone its general discussion to [133]. However, every S.b. in a TVS having a set of second category is an e-S.b.; on the other hand, $\{e^n;e^n\}$ is an e-S.b. for $(\lambda,\sigma(\lambda,\lambda^\times))$ if and only if $\lambda^\times = \phi$, λ being an arbitrary sequence space; cf. [132], p. 65.

The next result is reproduced from [44].

Proposition 6.2.12: Let (X,p) be an F*-space having an e-S.b. $\{x_n;f_n\}$ such that $\{x_{n_k}\}$ is regular in (X,p) for some infinite $\{n_k\}$ in $\mathbb{N}$. Consider a sequence $\{y_n\}$ in X with $y_n = 0$, $n \in \mathbb{N}\setminus\{n_k\}$ and $\sum_{n\geq 1} p(y_n) < \infty$, and let $u_n = x_n + y_n$, $n \geq 1$. Then there exists m such that $\{u_n:n \geq m\}$ is an S.b. for $Y \equiv [u_n:n \geq m]$, $[x_n:n \geq m] \simeq [u_n:n \geq m]$ and $\dim X/Y < \infty$.

Proof: By using Lemma 1.2.21, the equicontinuity of $\{f_{n_k}\}$ which is a consequence of the regularity of $\{x_{n_k}\}$ and the equicontinuity of $\{S_n\}$, the linear map $K:X \to X$, $Kx = \sum_{n \geq 1} f_n(x)y_n$ is compact. Let $A = I+K$, I being the identity operator on X. From the theory of Riesz compact operators ([192], p. 144; cf. also [188]), we find that $A[X]$ is closed and $\dim A^{-1}(0) < \infty$. Hence there exists $m \geq 1$ such that $A|[x_n:n \geq m]$ is 1-1. Consequently, $[x_n:n \geq m] \simeq Y$ under A. □

Next, we have (cf. [116])

Proposition 6.2.13: Let $\{x_n;f_n\}$ be an S.b. for a non-locally convex F-space (X,T). For a sequence $\{\alpha_n\}$ in $\mathbb{K}$, $\alpha_n x_n \to 0$ in T if and only if $\alpha_n x_n \to 0$ in $\tau \equiv \tau(X,X^*)$. Hence $\{x_n\}$ is T-bounded (resp. regular, irregular) if and only if it is τ-bounded (resp. regular, irregular).

Proof: Let $\alpha_n x_n \to 0$ in τ. If possible, suppose for some $\varepsilon > 0$ and an increasing subsequence $\{n_k\}$, $q(\alpha_{n_k} x_{n_k}) \geq \varepsilon$ for $k \geq 1$, q being the F-norm generating T. Then $\alpha_{n_k}^{-1} f_{n_k}(x) \to 0$ for each x in X. Therefore, by Lemma 1.2.20, $p(x) \equiv \sup |\alpha_{n_k}^{-1} f_{n_k}(x)| \leq Mq(x)$ for each x in X, M being some positive constant. But then p is also τ-continuous. Hence for some constant $K > 0$ and τ-continuous seminorm p_N, $p_N(\alpha_{n_k} x_{n_k}) \geq K$ for $k \geq 1$, a contradiction. □

The following result is stated without proof in [45]. We recall the notation R and δ from Section 2.3.

Proposition 6.2.14: Let (X,T) be an l.c. TVS having an S.b. $\{x_n;f_n\}$ such that $\{x_n\}$ is irregular. Then $(X,T) \simeq (\delta,\sigma(\delta,\phi))$ under R, $R(x) = \{f_n(x)\}$ and $T \approx \sigma(X,X^*)$.

Proof: It suffices to prove that R^{-1} is continuous. By the irregularity of $\{x_n\}$, it is easily seen that, for each p in D_T, the set $N_p = \{n \in \mathbb{N}:p(x_n) \neq 0\}$ is finite and let $n_p = \max\{i \in N_p\}$. Let $\alpha^\lambda \to 0$ in $(\delta,\sigma(\delta,\phi))$. Since $\alpha^\lambda = \{f_j(y^\lambda)\}$ and for any p in D_T, $p(x_j) = 0$ for $j \geq n_p+1$,

$$p(y^\lambda) \leq p\left(\sum_{j=1}^{n_p} f_j(y^\lambda)x_j\right),$$

$y^\lambda \to 0$ in D_T. □

Proof of Theorem 6.2.10: Let q (resp. p) denote the F-norm on X generating the topology T (resp. $\tau \equiv \tau(X,X^*)$) on X. Let $\{x_n;f_n\}$ be a $\sigma(X,X^*)$-t.b. for X, which we may assume as before an S.b. for (X,T). Then $\{x_n;f_n\}$ is an e-S.b. for (X,T) as well as for (X,τ).

Consider two mutually exclusive cases.

Case 1: Let $\{x_n\}$ be not irregular. Without loss of generality, this allows us to choose an $\varepsilon > 0$ and an increasing subsequence $\{n_k\}$ with $q(x_{n_k}) > \varepsilon$, $k \geq 1$. This yields the equicontinuity of $\{f_{n_k}\}$ on (X,T) as well as on (X,τ).

Choose $\{\alpha_n\}$ in $\mathbb{K}(\alpha_n \neq 0$, $|\alpha_n| \leq 1$; $n \geq 1$ with $\alpha_n = 1$ for n in $\mathbb{N}\setminus\{n_k\})$ and $q(\alpha_{n_k} x_{n_k}) \to 0$. Following the last paragraph of the proof of Theorem 6.2.7, we find a sequence $\{w_n\}$ in X and a positive δ so that $p(\alpha_n^{-1} w_n) < 1/2^n$ and $q(w_n) > \delta$ for $n \geq 1$. From here, we easily get a sequence $\{u_n\}$ in X such that $u_n = 0$ for n in $\mathbb{N}\setminus\{n_k\}$ and $\sum_{n \geq 1} p(\alpha_n^{-1} u_n) < \infty$. Therefore, by Proposition 6.2.12, there exists $m \geq 1$ such that $\{y_n\}$ is an S.b. for $Y \equiv [y_n; n \geq m]^\tau$, where $y_n = x_n + \alpha_n^{-1} u_n$, $n \geq 1$. Hence $\{\alpha_n y_n: n \geq m\}$ is a weak t.b. for (Y,q). Observe that $p(\alpha_{n_k} y_{n_k}) \to 0$ but $q(\alpha_{n_k} y_{n_k}) \not\to 0$ and hence $\{\alpha_n y_n : n \geq m\}$ cannot be a t.b. for the F-space (Y,q), for otherwise a subsequence of a.s.c.f. corresponding to $\{\alpha_n y_n : n \geq m\}$ is p-equicontinuous, giving $p(\alpha_{n_k} y_{n_k}) \not\to 0$. Thus the theorem is proved in this case.

Case 2: Let $\{x_n\}$ be irregular in (X,T). Then $\{x_n\}$ is irregular in (X,τ). By Propositions 1.2.16 and 6.2.14, there exists a proper T-closed and $\sigma(X,X^*)$-dense subspace Y of X such that $(Y,\sigma(X,X^*)|Y) \simeq (\omega_1, \sigma(\omega,\phi)|\omega_1)$, where $\bar{\omega}_1^{\sigma} = \omega$, $\sigma \equiv \sigma(X,X^*)$. Hence $(Y,\sigma|Y)$ has an e-S.b., say, $\{y_n\}$. Clearly $\{y_n\}$ is a $\sigma(X,X^*)$-S.b. for X but not a t.b. for (X,T). □

The next result ([212], p. 1299) shows that the failure of the WBT in Theorem 6.2.10 is sufficiently attributed to the F-structure of the space (X,T).

Proposition 6.2.15: In every Fréchet space (X,T) with $\sigma(X,X^*) \neq T$, there exists a nonconvex non-metrizable l. topology T_1 on X such that every weak t.b. $\{x_n;f_n\}$ for (X,T_1) is indeed a T_1-t.b. for X.

Proof: Apply Proposition 1.2.17 and Theorem 6.1.4. □

Note: Although the WBT fails in every non-locally convex F-space, yet a weaker version of the WBT holds good in every F-space X with a genuine dual X*. Indeed, we have after Kalton [115] the following

Theorem 6.2.16: For every weak t.b. $\{x_n;f_n\}$ in an F-space (X,T) with a separating dual X*, f_n is T-continuous for each $n \geq 1$.

Proof: We may clearly assume that T is nonconvex. By the Baire category theorem, the topology $\tau \equiv \tau(X,X^*)$ is barrelled; also $\tau \subsetneqq T$. Observe that τ is metrizable and is given by $\{p_n\}$, p_n being the Minkowski functional of V_n; cf. Proposition 1.2.15. If $\bar{p}_n(x) = \sup\{p_n(\sum_{i=1}^{m} f_i(x)x_i): m \geq 1\}$, then $\{\bar{p}_n\}$ generates a metrizable l.c. topology $\bar{\tau} \supset \tau$ such that $\{S_m\}$ is $\bar{\tau}$-τ equicontinuous and so each f_i is $\bar{\tau}$-continuous.

In order to prove the τ-continuity (and hence the T-continuity) of each f_i, it is enough to establish the continuity of the identity map $I:(X,\tau) \to (\hat{X},\hat{\bar{\tau}})$, the completion of $(X,\bar{\tau})$. Let $y_n \to y$ in τ and $Iy_n \to \bar{y}$ in $\bar{\tau}$. Then, to each $\varepsilon > 0$ and $n \geq 1$, there exists p_o such that $p_n(S_m(y_p - y_q)) \leq \varepsilon$, for all $p, q \geq p_o$ and $m \geq 1$. For each f in X*, there exists p_n with $|f(x)| \leq Kp_n(x)$, $x \in X$. Consequently,

$$\lim_{p\to\infty}\lim_{m\to\infty} \sum_{i=1}^{m} f_i(y_p)f(x_i) = \lim_{m\to\infty}\lim_{p\to\infty} \sum_{i=1}^{m} f_i(y_p)f(x_i). \qquad (*)$$

Also, we can define α_i in $\mathbb{K}$ by $\alpha_i = \lim f_i(y_p)$, $i \geq 1$. Then using (*), we get $\alpha_i = f_i(y)$, $i \geq 1$. Hence $\bar{p}_n(y_p - y) \leq \varepsilon$ for all $p > p_o$ and so the graph of I is closed. The continuity of I now follows from Theorem 1.2.7. □

6.3 Further Extensions and Remarks

We know that the solution to Problem 6.1.1 is, in general, negative. By restricting this problem slightly, let us seek the solution of the related problem [36]:

Problem 6.3.1: If $\{x_n;f_n\}$ is a $\sigma(X,X^*)$-S.b. for an l.c. TVS (X,T), what are those compatible l.c. topologies on X for which $\{x_n;f_n\}$ is an S.b.?

Proposition 2.3.8 provides an answer to the above problem in terms of the topology $\tilde{\sigma} \supset \sigma \equiv \sigma(X,X^*)$, σ being the weakest among all such topologies as required in Problem 6.3.1. Accordingly, let us find out the strongest l.c.

topology which answers the above problem in the affirmative. For this, let $\{x_n;f_n\}$ and (X,T) stand as in the above problem and then introduce $\mathfrak{P} = \{B \subset X^* : \tilde{B} = B, B \text{ is } \hat{\sigma}\text{-compact}\}$, where $\tilde{B}$ and $\hat{\sigma}$ are defined in Section 2.3. Let us write $\hat{T}$ for the l.c. topology on X, generated by the polars of B with B in $\mathfrak{P}$. Now, we have the desired

Theorem 6.3.2: Let $\{x_n;f_n\}$ be a $\sigma(X,X^*)$-S.b. for an S-space (X,T). Then $\hat{T}$ is the strongest compatible l.c. topology on X such that $\{x_n;f_n\}$ is an S.b. for $(X,\hat{T})$.

Proof: By Propositions 2.3.11 and 2.3.13(ii), $\hat{T}$ is compatible and so we first show that $S_n(x) \to x$ in $\hat{T}$ for each x in X. Since $S_n^*(f) \to f$ in $\hat{\sigma}$, B in $\mathfrak{P}$ is $\hat{\sigma}$-precompact and $\{S_n^*\}$ is $\hat{\sigma}$-$\hat{\sigma}$ equicontinuous, we have (e.g. [140], p. 76), for each x in X,

$$\sup_{f \in B} |\langle x, S_n^*(f) - f\rangle| \to 0.$$

Thus $S_n(x) \to x$ in $\hat{T}$. Finally, if S is any other compatible l.c. topology on X such that $\{x_n;f_n\}$ is an S.b. for (X,S), then for any S-equicontinuous subset B of X^*, $\tilde{B}$ is $\hat{\sigma}$-precompact; cf. Proposition 2.3.13. As $\tilde{\tilde{B}} = \tilde{B}$, we get $S \subset \hat{T}$. □

Exercise 6.3.3: Let $\{x_n;f_n\}$ be a $\sigma(X,X^*)$-S.b. for an S-space (X,T). Prove that $\{x_n;f_n\}$ is an S.b. for (X,T) if and only if, for each T-equicontinuous subset B, $\tilde{B}$ is $\hat{\sigma}$-relatively compact. Using this result, prove also Proposition 6.1.5; cf. [36], p. 283.

WBT in webbed spaces

We now come to the final result of this chapter, which further extends Theorem 6.1.6. This result, which is due to De Wilde [39], finds itself enveloped in another result of De Wilde [41], details of which will appear elsewhere in [81].

Theorem 6.3.4: Let $\{x_n;f_n\}$ be a $\sigma(X,X^*)$-t.b. for a bornological ω-complete space (X,T) having a web $\mathfrak{N}$. Then $\{x_n;f_n\}$ is an S.b. for (X,T).

Proof: Let $\bar{T}$ be the topology on X defined by $\bar{D} = \{\bar{p} : p \in D_T\}$, $\bar{p}(x) = \sup p(S_n(x))$.

As in Lemma 6.1.7, each f_n is $\bar{T}$-continuous, $T \subset \bar{T}$ and $\{S_n\}$ is $\bar{T}$-T equicontinuous. The desired conclusion will follow on the lines of the proof of Theorem 5.1.2 (cf. also Proposition 6.1.5), once we have proved $\bar{T} \approx T$. To accomplish the objective, we make use of Theorem 1.2.8. and for this we have to show that $(X,\bar{T})$ has a web $\mathfrak{m}$ of type $\mathfrak{C}$. We prove this fact by considering two cases: Case 1 when $\mathfrak{N}$ is a strict web, and Case 2 when $\mathfrak{N}$ is a web of type $\mathfrak{C}$.

<u>Case 1</u>: The desired construction of $\mathfrak{m}$ is obtained in several steps.

(i) Suppose $\mathfrak{N} = \{A_{n_1,\ldots,n_k}\}$ is a strict web. For indices n_k and m_k in $\mathbb{N}$, write $r_k = (m_k,n_k)$, $k \geq 1$ and let

$$A'_{r_1} = m_1 A_{n_1}\ ;\ A'_{r_1,\ldots,r_k} = A'_{r_1,\ldots,r_{k-1}} \cap m_k A_{n_1,\ldots,n_k},\ k > 1.$$

Then $\mathfrak{N}_1 = \{A'_{r_1,\ldots,r_k}\}$ is clearly a web, it being understood that $\{r_k\}$ is in one-to-one correspondence with $\mathbb{N}$.

To prove the strict web character of $\mathfrak{N}_1$, let us consider the sequence $\{\mu_k\}$ corresponding to $\{n_k\}$ in Definition 1.2.2 for the strict web $\mathfrak{N}$. For the given $\{r_k\}$, $r_k = (m_k,n_k)$, write $\nu_k = \mu_k/m_k$, $k \geq 1$. If $0 \leq \lambda_k \leq \nu_k$ and $v_k \in A'_{r_1,\ldots,r_k}$, then clearly $\sum_{k\geq 1} \lambda_k v_k$ converges in (X,T). Further

$$\sum_{k\geq k_o} \lambda_k v_k = \sum_{k\geq k_o} (m_k\lambda_k)(v_k/m_k) \in m_i A_{n_1,\ldots,n_i}\ ;\ \forall\ i \leq k_o,$$

and so $\mathfrak{N}_1$ is a strict web.

(ii) For each x in X, let $B_x = \Gamma(S_x)^T$ where $S_x = \{x,S_1(x),S_2(x),\ldots\}$ and hence, by Proposition 1.2.4, there exist sequences $\{n_k\}$ and $\{\alpha_k\}$, $\alpha_k > 0$ such that $B_x \subset \alpha_k A_{n_1,\ldots,n_k}$, $k \geq 1$. Choose $m_i \geq \alpha_i$, then $B_x \subset A'_{r_1,\ldots,r_k}$, $k \geq 1$. With this property of $\mathfrak{N}_1$ in mind, let us rewrite for convenience $\mathfrak{N}_1$ as $\{A_{n_1,\ldots,n_k}\}$ and define the subset $A^*_{n_1,\ldots,n_k} = \{x \in X: B_x \subset A_{n_1,\ldots,n_k}\}$. Suppose that $\mathfrak{m}$ is the collection of all such subsets. $\mathfrak{m}$ is clearly a web and we proceed to show that $\mathfrak{m}$ is of type $\mathfrak{C}$ in the space $(X,\bar{T})$.

(iii) Since the web $\mathfrak{N}_1$ is strict, for each $\{n_k\}$ there exists $\{\mu_k\}$ such that for all ν_k $(0 \leq \nu_k \leq \mu_k)$ and all y_k in $A_{n_1,\ldots,n_k}$, $\sum_{k\geq 1} \nu_k y_k$ converges in (X,T), $\sum_{i \geq k} \nu_i y_i \in A_{n_1,\ldots,n_k}$, $k \geq 1$ where we may assume without loss of generality that $0 < \mu_{k+1} < \mu_k < 1$, $k \geq 1$.

Let now $u_k \in A^*_{n_1,\ldots,n_k}$ and $0 \leq \lambda_k \leq \mu_k^2$. Suppose $v = \sum_{k \geq 1} \lambda_k u_k$ and $v_m = \sum_{k \geq 1} \lambda_k S_m(u_k)$, $m \geq 1$ (the existence of v and v_m's in X are guaranteed by the strict character of $\mathbb{N}_1$). Put $\alpha_m = \sum_{k \geq 1} \lambda_k f_m(u_k)$, $m \geq 1$. Then $v_m = \sum_{i=1}^{m} \alpha_i x_i$, $m \geq 1$. We now show that $v_m \to v$ in $\sigma(X,X^*)$ and this will yield $v_m = S_m(v)$, $m \geq 1$ which will, in turn, lead to the convergence of $\sum_{k \geq 1} \lambda_k u_k$ in the space $(X,\bar{T})$ as shown in (iv) below.

Choose an arbitrary f in X*. Then, for given $\varepsilon > 0$, there exists W in $\mathcal{B}_T$ so that $|f(x)| \leq \varepsilon/3$ for all x in W; and also there exists an L_o in $\mathbb{N}$ with $\mu_L A_{n_1,\ldots,n_L} \subset W$ for all $L \geq L_o$. Further, for $L \geq 1$,

$$|f(v_m - v)| \leq |f(v_m - \sum_{k=1}^{L} \lambda_k S_m(u_k)| + |f(\sum_{k=1}^{L} \lambda_k u_k - v)|$$

$$+ |f(\sum_{k=1}^{L} (\lambda_k S_m(u_k) - \lambda_k u_k)|$$

Since $\lambda_k/\mu_{k+1} \leq \mu_k$, $k \geq L+1$, we have, for $m \geq 1$,

$$v - \sum_{k=1}^{L} \lambda_k u_k,\ v_m - \sum_{k=1}^{L} \lambda_k S_m(u_k) \in \mu_{L+1} A_{n_1,\ldots,n_{L+1}} \subset W \qquad (+)$$

for $L \geq L_o$. Fix $L \geq L_o$; then there exists m_o such that

$$|f(\sum_{k=1}^{L} (\lambda_k S_m(u_k) - \lambda_k u_k))| \leq \frac{\varepsilon}{3},\ \forall m \geq m_o.$$

Consequently, $|f(v_m - v)| \leq \varepsilon$ for all $m \geq m_o$.

(iv) Finally, let $\varepsilon > 0$ and p in D_T be arbitrary. If $W = \{x \in X : p(x) \leq \varepsilon\}$, then, from (+) in (iii), we easily conclude that

$$p(v_m - \sum_{k=1}^{L} \lambda_k S_m(u_k)) \leq \varepsilon;\ \forall m \geq 1,\ L \geq L_o$$

and this immediately leads to $\bar{p}(v - \sum_{k=1}^{L} \lambda_k u_k) \leq \varepsilon$, for all $L \geq L_o$.

Case 2: (v). Here $\mathfrak{N} = \{A_{n_1,\ldots,n_k}\}$ is a web of type $\mathfrak{C}$ and we may assume without loss of generality that $\lambda A_{n_1,\ldots,n_k} \subset A_{n_1,\ldots,n_k}$ for each λ with $0 \leq \lambda \leq 1$.

For each x in X, consider the Banach space $Y_x = \cup \{nB_x : n \geq 1\}$ equipped with the norm $\|\cdot\|_x$ generated by B_x defined in Case 1. If $I_x : Y_x \to X$ is the identity map in place of R in Proposition 1.2.5, then (a) we find the existence of n_1 in $\mathbb{N}$ and $M_1 > 0$ such that $Y_x \cap A_{n_1}$ is of second category and $B_x \subset M_1 \; \overline{\langle A_{n_1} \rangle}$ and (b) if $A_{n_1,\ldots,n_k} \cap Y_x$ is of second category, there exist n_{k+1} in $\mathbb{N}$ and $M_{k+1} > 0$ such that $A_{n_1,\ldots,n_{k+1}} \cap Y_x$ is of second category and $B_x \subset M_{k+1} \; \overline{\langle A_{n_1,\ldots,n_{k+1}} \rangle}$. By the property of the members of $\mathfrak{N}$ mentioned in the first paragraph above, we may assume all M_k's to be in $\mathbb{N}$. If $r_k = (m_k, n_k)$, let $A^*_{r_1,\ldots,r_k} = \{x \in A^*_{r_1,\ldots,r_{k-1}} : A_{n_1,\ldots,n_k} \cap Y_x$ is of second category and $B_x \subset m_k \; \overline{\langle A_{n_1,\ldots,n_k} \rangle}\}$, $k > 1$ and $A^*_{r_1} = \{x \in X : A_{n_1} \cap Y_x$ is of second category and $B_x \subset m_1 \; \overline{\langle A_{n_1} \rangle}\}$. The collection $\mathfrak{M} = \{A^*_{r_1,\ldots,r_k}\}$ is clearly a web for X. We next show that $\mathfrak{M}$ is of type $\mathfrak{C}$.

(vi) Fix $\{r_k\}$, $r_k = (m_k, n_k)$. Let the sequence $\{\rho_k\}$ of positive reals correspond to $\{n_k\}$ for the purpose of the $\mathfrak{C}$ character of $\mathfrak{N}$. For each V in $\mathcal{B}_T$, there exists k_o such that (cf. [147], p. 55)

$$\rho_k \overline{\langle A_{n_1,\ldots,n_k} \rangle} \subset V, \quad \forall\, k \geq k_o. \tag{*}$$

Put $\rho'_k = \rho_k / m_k$ and $\alpha_k = \rho'_k / 2^k$, $k \geq 1$. Define $\mu_1 = \alpha_1$ and $\mu_k = \inf \{\mu_1 / 2^{k-1} m_1, \ldots, \mu_{k-1} / 2m_{k-1}, \alpha_k\}$. Then

$$\sum_{k \geq n+1} \mu_k \leq \frac{\mu_n}{M_n}, \quad \forall\, n \geq 1. \tag{**}$$

Let now $u_k \in A^*_{r_1,\ldots,r_k}$ and $0 \leq \lambda_k \leq \mu_k$. By (*), $\{\rho'_k u_k\}$ is bounded in (X,T). Since $\lambda_k u_k = (\lambda_k/\mu_k)(\mu_k/\alpha_k)(\rho'_k u_k / 2^k)$, the series $\sum_{k \geq 1} \lambda_k u_k$ converges to v in (X,T). Similarly, $\sum_{k \geq 1} \lambda_k S_m(u_k)$ converges to v_m in (X,T).

(vii) The rest of the arguments runs similar to (iii) and (iv) of Case 1. Indeed, in place of (+) in (iii) above, we here obtain, for $L \geq 1$,

$$v - \sum_{k=1}^{L} \lambda_k u_k = \mu_L \sum_{k \geq L+1} \left(\frac{m_L \lambda_k}{\mu_L}\right) \frac{u_k}{m_L} \in \mu_L \, \overline{\langle A_{n_1,\ldots,n_L} \rangle}$$

and

$$v_m - \sum_{k=1}^{L} \lambda_k S_m(u_k) \in \mu_L \, \overline{\langle A_{n_1,\ldots,n_L} \rangle}; \; \forall\, m,\, L \geq 1.$$

by using (**). Now proceed as in (iii) and (iv) above. □

7 Biorthogonal systems

7.1 INTRODUCTION

In this chapter we study the properties of some of the basic components comprising a topological base in a TVS. Such a study is helpful in two ways: (i) in ascertaining the t.b. character of a sequence in a TVS and (ii) in realizing the importance of different components of a t.b. in the structural study of a TVS.

If $\{x_n;f_n\}$ is a t.b. for a TVS, then $f_m(x_n) = \delta_{mn}$. We extend this concept in the form of

Definition 7.1.1: Let $\langle X,Y\rangle$ be a dual system and Λ a directed set. Write Φ_Λ for the family of all finite subsets of Λ. A system $\{x_\lambda;y_\lambda\} \equiv \{x_\lambda;y_\lambda:\lambda \in \Lambda\}$, with x_λ in X and y_λ in Y for λ in Λ, is called a *biorthogonal system* (b.sy.) if $\langle x_\lambda,y_\mu\rangle = \delta_{\lambda\mu}$, the generalized Kronecker delta; and if $F \in \Phi_\Lambda$, the operators $S_F:X \to X$ and $S_F^*:Y \to Y$ defined by

$$S_F(x) = \sum_{\lambda\in F} \langle x,y_\lambda\rangle x \; ; \quad S_F^*(y) = \sum_{\lambda\in F} \langle x_\lambda,y\rangle y_\lambda ,$$

where $x \in X$, $y \in Y$, are called the *expansion* and the *dual expansion* operators respectively.

Exercise 7.1.2: Show that each S_F (resp. S_F^*) is $\sigma(X,Y)$-continuous on X (resp. $\sigma(Y,X)$-continuous on Y); and, for all E, F in Φ_Λ, $S_F(S_E(x)) = S_{E\cap F}(x)$, $x \in X$ and $S_F^*(S_E^*(y)) = S_{E\cap F}^*(y)$, $y \in Y$.

Definition 7.1.3: If $\Lambda = \mathbb{N}$, the b.sy. $\{x_n;y_n\}$ is generally called a *biorthogonal sequence* (b.s.) for $\langle X,Y\rangle$.

7.2 DIFFERENT SYSTEMS

We consider several properties of a biorthogonal system relevant to topological bases and deal with each one of them separately with examples and counter-examples.

Minimality

This notion is essentially introduced in [105] in the form of

Definition 7.2.1: A family $\{x_\lambda : \lambda \in \Lambda\}$ in a TVS (X,T) is called T-*minimal* or T-*topologically free* if $x_\lambda \notin [x_\mu : \mu \neq \lambda, \mu \in \Lambda]$ for each λ in Λ.

Note: If no emphasis of the underlying topology T is required, we will merely write *minimal* for T-minimal.

Remark: Every S.b. in a TVS is minimal and every minimal sequence is ω-linearly independent and hence linearly independent. A t.b. may not be minimal. The converse of the above statements is not true in general and we offer the following counter-examples.

Example 7.2.2: The sequence $\{e^n\}$ is minimal but not a S.b. for $(\ell^\infty, \|\cdot\|_\infty)$.

Example 7.2.3: The t.b. $\{x^n\}$ of Example 2.2.3 is not minimal as $\|x^n - x^2\|_\infty = 1/n$, $n \geq 3$ and so $x^2 \in [x^3, x^4, \ldots]$. However, this sequence is ω-linearly independent.

Exercise 7.2.4: Prove that the sequence $\{x^n\}$ of Example 2.2.4 is not minimal but ω-linearly independent for $(\ell^1, \sigma(\ell^1, c_0))$. Also, show that the sequence $\{f_n\}$ of Example 5.2.6 is not minimal (cf. [176], p. 162) but ω-linearly independent. [Hint: For the last part, use term-by-term differentiation.]

Example 7.2.5: Consider the sequence $\{x^n\}$ in $(\delta, \eta(\delta, \delta^\times))$ where $x^1 = \sum_{n \geq 1} e^n/n!$ and $x^n = e^{n-1}$, $n \geq 2$. Since $x^1 = \sum_{n \geq 2} x^n/(n-1)!$, $\{x^n\}$ is not ω-linearly independent; however, it is linearly independent.

Exercise 7.2.6: Prove that the sequence $\{f_n\}$ in the space (E_∞, T_∞) of Example 5.2.17 is not ω-linearly independent but linearly independent, where $f_1(z) = e^z$; $f_n = \delta_{n-2}$, $n \geq 2$.

Notation: Throughout $\langle X,Y \rangle$ stands for an arbitrary dual system of vector spaces X and Y. The letters $T_1 \equiv T_{X,Y}$ (resp. $T_2 \equiv T_{Y,X}$) will mean hereafter, unless the contrary is specified, for an l.c. topology on X (resp. Y) compatible with $\langle X,Y\rangle$. Without further reference, X (resp. Y) is assumed to have the topology T_1 (resp. T_2).

We essentially follow [43] for the following few results and begin with a result which shows the utility of minimal families in constructing b.sy.; cf. also [17], p. 10; [61].

<u>Proposition 7.2.7</u>: A family $\{x_\lambda\} \equiv \{x_\lambda : \lambda \in \Lambda\}$ in (X,T_1) is minimal if and only if there exists $\{y_\lambda\}$ in Y such that $\{x_\lambda; y_\lambda\}$ is a b.sy. for $\langle X,Y\rangle$.

<u>Proof</u>: If $\{x_\lambda; y_\lambda\}$ is a b.sy. and if, for some λ, $x_\lambda \in [x_\mu : \mu \neq \lambda]$, then, for some $F \in \Phi_\Lambda \smallsetminus \{\lambda\}$, $|\langle x_\lambda - \sum_{\mu \in F} \alpha_\mu x_\mu, y_\lambda\rangle| < 1/2$. However, this is absurd and so $\{x_\lambda\}$ is minimal. The other part is a simple consequence of the Hahn-Banach theorem. □

<u>Exercise 7.2.8</u>: Let $\{x_\lambda; y_\lambda : \lambda \in \Lambda\}$ and $\{x_\lambda; g_\lambda : \lambda \in \Lambda\}$ be two b.sy. for $\langle X,Y\rangle$. Prove that $f_\lambda - g_\lambda \in [x_\mu : \mu \in \Lambda]^o$; and hence, in particular, $f_\lambda = g_\lambda$ for each λ if and only if $X = [x_\mu]$.

<u>Exercise 7.2.9</u>: If $\{x_\lambda\}$ is minimal in (X,T_1), prove that, for any x in $[x_\lambda]$ and each λ, there exists a unique scalar a_λ such that $x - a_\lambda x_\lambda \in [x_\mu : \mu \neq \lambda]$.

<u>Note</u>: For other results on minimal sequences in Banach spaces, see [219].

<u>Maximality</u>

Orthonormal bases in Hilbert spaces display certain pathological properties of themselves and one of these leads us to introduce

<u>Definition 7.2.10</u>: A b.sy. $\{f_\mu; g_\mu : \mu \in \Lambda_1\}$ for $\langle X,Y\rangle$ is said to be an *extension* of another b.sy. $\{x_\lambda; y_\lambda : \lambda \in \Lambda\}$ for $\langle X,Y\rangle$ provided that $\Lambda \subset \Lambda_1$ and $x_\lambda = f_\lambda; y_\lambda = g_\lambda$, $\lambda \in \Lambda$. The b.sy. $\{x_\lambda; y_\lambda\}$ is said to be *maximal* if there is no proper extension of this b.sy.

We have ([43]; cf. also [25])

<u>Proposition 7.2.11</u>: Every b.sy. $\{x_\lambda; y_\lambda\}$ for $\langle X,Y\rangle$ has a maximal extension.

Proof: Consider the family of all b.sy.'s which are extensions of $\{x_\lambda;y_\lambda\}$ and partially order it through the notion of the extension of a b.sy. Now apply Zorn's lemma. □

Proposition 7.2.12: Let $\{x_\lambda;y_\lambda:\lambda \in \Lambda\}$ be a b.sy. for $\langle X,Y\rangle$. Then (i) $\langle x_\lambda;y_\lambda\rangle$ is maximal if and only if (ii) $[y_\lambda]^o \subset [x_\lambda]$ if and only if (iii) $[x_\lambda]^o \subset [y_\lambda]$.

Proof: (ii) $\Longleftrightarrow$ (iii) Immediate.

(i) $\Rightarrow$ (iii) Suppose that there is y in $[x_\lambda]^o$ such that $y \notin [y_\lambda]$. Hence we find an x in X so that $\langle x,y\rangle = 1$ and $\langle x,u\rangle = 0$ for all u in $[y_\lambda]$. Since $\{x_\lambda,x;y_\lambda,y\}$ is a proper extension of $\{x_\lambda,y_\lambda\}$, (iii) is true.

(iii) $\Rightarrow$ (i) This easily follows by the contradiction method. □

Remark: Every separable Banach space is known to have a maximal b.sy. which is countable. Unfortunately, this pleasant situation is not generally found in an arbitrary separable l.c. TVS. For instance, consider

Example 7.2.13: Let $X = \mathbb{R}^{\aleph}$ equipped with its usual product topology and let $Y = X^* = \bigoplus_{\aleph} \mathbb{R}$. As pointed out in Example 4.2.4, X is separable. The system $\{e^a;e^a:a \in \aleph\}$ discussed in Example 3.2.5 is a maximal b.sy.

Definition 7.2.14: For a dual system $\langle X,Y\rangle$, let $A \subset X$ and $B \subset Y$. Then A is called *complete* or *fundamental* if $[A] = X$ and B is called *total on* X or *in* Y if $\langle x,y\rangle = 0$ for every y in B implies that $x = 0$. The *orthogonal complement* $A^\perp$ of A is defined as the set $\{y \in Y:\langle x,y\rangle = 0, \forall x \in A\}$. If $\{x_\lambda;y_\lambda:\lambda \in \Lambda\}$ is a b.sy. for $\langle X,Y\rangle$, then the map $\psi \equiv \psi_\Lambda : X \to \mathbb{K}^\Lambda$ defined by $\psi(x) = \{\langle x,y_\lambda\rangle: \lambda \in \Lambda\}$ is called the *coefficient map*.

Note: In particular, a b.sy. $\{x_\lambda;y_\lambda\}$ for $\langle X,Y\rangle$, is said to be *complete* (resp. *total*) if $\{x_\lambda\}$ is complete in X (resp. $\{y_\lambda\}$ is total on X).

Proposition 7.2.15: Let $\langle x_\lambda;y_\lambda\rangle$ be a b.sy. for $\langle X,Y\rangle$. Then
(a) $\{y_\lambda\}$ is total on X if and only if $\ker(\psi) = 0$; and
(b) $\{x_\lambda;y_\lambda\}$ is maximal if and only if $[x_\lambda]^\perp \subset [y_\lambda]$ if and only if $\cap\,\{\ker(y_\lambda): \lambda \in \Lambda\} \subset [x_\lambda]$.

Proof: Straightforward. □

Remark: If $\{x_n;f_n\}$ is an S.b. for an l.c. TVS X, then $\{x_n;f_n\}$ is a b.s. for $\langle X,X^*\rangle$ such that it is maximal, $\{x_n\}$ is complete and $\{f_n\}$ is total on X. The converse of this statement is not always true, for we have

Example 7.2.16: Let $\{x_n : n \geqslant 0\}$ be an arbitrary orthonormal base for a separable Hilbert space H. If $y_n = x_n - x_{n+1}$, and $g_n = \sum_{i=0}^{n} x_i$, $n \geqslant 0$, then $\{y_n;g_n : n \geqslant 0\}$ is a maximal b.s. for $\langle H,H^*\rangle$, where $H^* \simeq H$. Let S_n denote the expansion operator corresponding to $\{y_n;g_n : n \geqslant 0\}$. Choose x and y in H with $x = \sum_{i \geqslant 0} x_i/(1+i)$, $y = \sum_{i \geqslant 0} \alpha_i x_i$, where $\alpha_{i+1} = 1/\log\log i$, $i = 2^{2^n}$ $(n \geqslant 0)$ and $= 0$, otherwise. Then

$$\langle S_n(x),y\rangle = \sum_{i=0}^{n} \langle x_i,y\rangle\langle x,x_i\rangle - \alpha_{n+1} \sum_{i=0}^{n} \frac{1}{1+i}$$

and hence $\{y_n;g_n\}$ cannot be an S.b. for H.

Completeness and maximality

In order to infer the S.b. character of a b.s. $\{x_n;f_n\}$ for an l.c. TVS (X,T), the latter has to satisfy the conditions of (i) completeness, (ii) maximality and (iii) totality. However, (iii) always implies (ii). Since (iii) cannot be ignored in case a b.s. is to become an S.b., conditions (i) and (iii) are indeed essential for a b.s. to be an S.b. On the other hand, conditions (i) and (ii) are independent of each other. Julia [104] constructed maximal b.s. in a Hilbert space for which the respective orthonormal complements are of infinite dimension and consequently these sequences have no proximity to orthonormal bases. A similar study has also recently been undertaken by Cook [25]; based on several examples he has constructed in this direction, we can prove the following

Theorem 7.2.17: Let (X,T) be an l.c. TVS containing a $\sigma \equiv \sigma(X,X^*)$-S.b. $\{x_n;f_n\}$ with $x_n \to 0$ in σ and $f_n \to 0$ in $\sigma^* \equiv \sigma(X^*,X)$. Then there exists a maximal b.s. $\{y_n;g_n\}$ for $\langle X,X^*\rangle$ such that $\dim(X/[y_n]) = \infty$ and $\dim(X^*/[g_n]) = \infty$, where $[g_n] = \overline{\mathrm{sp}\,\{g_n\}}^{\sigma^*}$.

Proof: For convenience, the proof is divided into four parts, (i) through (iv).

(i) Let $u_n = x_1 + x_{n+1}$ and $v_n = f_{n+1}$, $n \geqslant 1$. Then $[u_n] = X$ and $\dim (X^*/[v_n]) = 1$, where $[v_n] = \overline{sp\{v_n\}}^{\sigma^*}$. It is easily seen by contradiction that $\{u_n;v_n\}$ is a maximal b.s. for $\langle X,X^*\rangle$.

(ii) Let now $z_n = u_{n+1}$ and $h_n = v_1 + v_{n+1}$, $n \geqslant 1$. Then $[z_n]$ and $[h_n]$ both have codimension 1. Again one can easily check that $\{z_n;h_n\}$ is a maximal b.s.

(iii) Since $\mathbb{N} \times \mathbb{N}$ is equivalent to $\mathbb{N}$, we may write $\{x_n\}$ and $\{f_n\}$ as the union of countably many disjoint countable sets:

$$\{x_n\} = \bigcup_{i \geqslant 1} \{x_j^i : j \geqslant 1\}; \quad \{f_n\} = \bigcup_{i \geqslant 1} \{f_j^i : j \geqslant 1\}.$$

Let $u_j^i = x_1^i + x_{j+1}^i$ and $v_j^i = f_{j+1}^i$, $i, j \geqslant 1$. As in (i), one can show that $\{u_j^i;v_j^i:(i,j) \in \mathbb{N} \times \mathbb{N}\}$ is a maximal b.s. Since $[f_1^i:i \geqslant 1] \simeq \omega$ and $[f_j^i:i,j \geqslant 1] = X^*$, the codimension of $[v_j^i:i,j \geqslant 1]$ is $\aleph$. Also, note that $[u_j^i:i,j \geqslant 1] = X$.

(iv) Put $z_j^i = u_{j+1}^i$ and $h_j^i = v_1^i + v_{j+1}^i$; $i,j \geqslant 1$. Then $\{z_j^i,h_j^i:(i,j) \in \mathbb{N} \times \mathbb{N}\}$ is the required maximal b.s. Indeed, following (iii), we find that the codimension of each of the spaces $[z_j^i:i,j \geqslant 1]$ and $[h_j^i:i,j \geqslant 1]$ is $\aleph$. □

Completeness and totality

As pointed out in the last subsection, a b.s. relevant to an S.b. is one which is both complete and total. Accordingly, we dissect the notion of an S.b. to yield the following

Definition 7.2.18: A b.sy. $\{x_\lambda;y_\lambda\}$ for $\langle X,Y\rangle$ is called (i) an M-*base system* if it is complete and total, (ii) an M-*semibase system* or an M-*dual base system* if it is complete, and (iii) an M-*generalized base system* if it is total.

Note: If $\{x_n;f_n\}$ is a b.s., the concepts introduced in (i) through (iii) are respectively referred to as (i)' an M-*base* or a *Markushevich base*, (ii)' an M-*semibase* or an M-*dual base* and (iii)' an M-*generalized base*.

Proposition 7.2.19: Let X be an l.c. TVS and $\{x_n;f_n\}$ a b.s. for $\langle X,X^*\rangle$. Then for the statements, namely, $\{x_n;f_n\}$ is (i) an S.b., (ii) an M-base, (iii) an M-semibase, (iv) an M-generalized base and (v) a maximal b.s., the following implications hold:

$$(i) \Longrightarrow (ii) \Longrightarrow (iii) \Longrightarrow (iv)$$

$$(ii) \searrow (iv) \nearrow (iv)$$

Proof: Straightforward. □

Remark: We offer several examples to show that the converse implications of the statements in Proposition 7.2.19 are not true in general.

Example 7.2.20: Consider the Fréchet space (E_1,T_1) of Example 5.2.17. Fix α in D_1, $\alpha \neq 0$ and set $f_n(z) = (z-\alpha)^n$, for z in D_1 and $n \geqslant 0$. Then $f_n \in E_1$, $n \geqslant 0$. Let $F_n(f) = f^{(n)}(\alpha)/n!$, for f in E_1 and $n \geqslant 0$. By using the classical complex variable theory, in particular the Taylor series at $z = \alpha$, one concludes that $\{f_n;F_n\}$ is an M-base for $\langle E_1,E_1^*\rangle$. The function f in E_1 with $f(z) = 1/(1-z)$ is, however, not representable as $\sum_{n\geqslant 1} F_n(f)f_n$ and so the M-base obtained above is not an S.b. for (E_1,T_1).

Example 7.2.21: The b.s. $\{e^n;e^n\}$ for $\langle \ell^\infty, (\ell^\infty)^* \rangle$ is clearly an M-generalized base and as $[e^n] = c_o$, it is not an M-semibase.

Next, we have (cf. [3]; [158])

Example 7.2.22: Consider the space of Example 7.2.20 and let $f_n = \delta_o+\ldots+\delta_n$, $n \geqslant 0$. Define F_n in E_1^* by $F_n(f) = f^{(n)}(0)/n! - f^{(n+1)}(0)/(n+1)!$; then $\{f_n;F_n\}$ is a b.s. for $\langle E_1,E_1^*\rangle$. If $F_n(f) = 0$ for some f in E_1 and each $n \geqslant 0$, then by the Taylor series $f(z) = K/(1-z)$ in D_1 for some constant K. Hence $\{f_n;F_n\}$ is not an M-generalized base. If $\{f_n;F_n\}$ were not maximal, then for some f in E_1 and F in E_1^*, $F(f_n) = 0$, $n \geqslant 0$ and $F(f) = 1$. Hence $F(\delta_n) = 0$ and so, from the Taylor expansion of f around $z = 0$, we deduce that $F(f) = 0$. Thus $\{f_n;F_n\}$ is maximal.

Exercise 7.2.23: Show that the b.s. of Example 7.2.22 is an M-semibase but not an M-base.

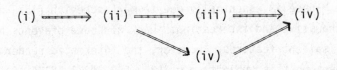

Exercise 7.2.24: Prove that a TVS admitting an M-semibase is necessarily

separable and hence ℓ^∞ cannot have an M-semibase.

Note: The notion of an M-base is essentially due to Markuschevich [157] and its history is quite exhaustive and interesting. Lack of space prevents us from recalling even its salient features. However, the interested reader may consult various papers on this subject, e.g. [3], [4], [27], [30], [31], [32], [36], [76], [100], [101], [106], [118], [142], [196], [219] and [237].

Let us now return to the discussion on the bases of types M: we begin with (cf. [3])

Proposition 7.2.25: Corresponding to a given family $\{x_\lambda : \lambda \in \Lambda\}$ in an l.c. TVS (X,T), an M-generalized base system $\{x_\lambda ; f_\lambda\}$ for $\langle X, X^* \rangle$ is an M-base system if and only if $\{f_\lambda : \lambda \in \Lambda\}$ is the only family of coefficient functionals associated with $\{x_\lambda\}$.

Proof: The necessity part is obvious. For the converse, let $[x_\lambda] \subsetneqq X$. Then we get an f in X^* so that $f \neq 0$ and $f | [x_\lambda] = 0$. Fix μ and define $g_\mu = f_\mu + f$ and $g_\lambda = f_\lambda$ for all $\lambda \neq \mu$. If $g_\lambda(x) = 0$ for each λ, and $y = -f(x) x_\mu$, then $f_\lambda(x-y) = 0$ for each λ and so $x = y$. Hence $f_\lambda(x) = 0$ for every λ, giving $x = 0$. Consequently, $f_\lambda = g_\lambda$ by hypothesis and so $f = 0$, a contradiction. □

The next result [30] is dual to Proposition 7.2.25.

Proposition 7.2.26: Let X be a TVS and $\{x_\lambda : f_\lambda\}$ an M-semibase for $\langle X, X^* \rangle$. The family $\{x_\lambda\}$ is uniquely determined if and only if $\{f_\lambda\}$ is total on X.

Proof: The 'if' part is obvious. For the other part, let $\{f_\lambda\}$ be not total. Then for some $x \neq 0$, $f_\lambda(x) = 0$ for every λ. Fix μ and let $y_\lambda = x_\lambda$ for all $\lambda \neq \mu$ and $y_\mu = x_\mu + x$. By Exercise 7.2.9, $y_\mu - a_\mu x_\mu \in [x_\lambda : \lambda \neq \mu]$ for some a_μ. Clearly $a_\mu = 1$ and so $x \in [x_\lambda : \lambda \neq \mu]$. Therefore $[x_\lambda] = [y_\lambda]$, giving $\{y_\lambda ; f_\lambda\}$ to be another M-semibase. Hence $x = 0$, a contradiction. □

In contrast to the basis problem, the existence of an M-base in every separable TVS is guaranteed in the form of the following result; cf. [100].

Theorem 7.2.27: Let (X,T) be a separable TVS and Y a $\beta(X^*,X)$-closed separable subspace of X^* such that Y is total on X. Then there exists $\{y_n\}$ in X and $\{g_n\}$ in X^* such that $\{y_n ; g_n\}$ is an M-base for $\langle X, X^* \rangle$ and $Y = [g_n]^\beta$,

$\beta \equiv \beta(X^*,X)$.

Proof: By the hypothesis, we can find the sequences $\{x_n\}$ in X and $\{f_n\}$ in Y such that $X = [x_n]$, $Y = [f_n]^\beta$, $\{f_n\}$ is total on X and $f_1(x_1) \neq 0$. We prove the result by the induction process.

So, let $y_1 = x_1$, $k_1 = 1$ and $g_1 = f_1/f_1(x_1)$ and suppose that we have already chosen $y_1,\ldots,y_{k_n}$ in X and $g_1,\ldots,g_{k_n}$ in Y such that (i) $\{y_i;g_i:1 \leqslant i \leqslant k_n\}$ is a b.s., (ii) $\text{sp}\,\{x_1,\ldots,x_n\} \subset \text{sp}\,\{y_1,\ldots,y_{k_n}\}$ and (iii) $\text{sp}\,\{f_1,\ldots,f_n\} \subset \text{sp}\,\{g_1,\ldots,g_{k_n}\} \subset Y$. Consider the cases (a) and (b) which finally yield the required result

Case (a) If x_{n+1} belongs to $\text{sp}\,\{y_i:1 \leqslant i \leqslant k_n\}$, we let $k = k_n$ and proceed hereafter to Case (b), otherwise we put $k = k_n+1$ and assume

$$y_k \equiv y_{k_n+1} = x_{n+1} - \sum_{i=1}^{k_n} g_i(x_{n+1})y_i.$$

By (i), $g_i(y_k) = 0$ for $1 \leqslant i \leqslant k_n$. By the Hahn-Banach theorem, we find g_k in Y with $g_k(y_k) = 1$ and $g_k(y_i) = 0$ for $1 \leqslant i \leqslant k_n$. Thus using (i), (ii) and (iii), we conclude that (i)' $\{y_i;g_i:1 \leqslant i \leqslant k\}$ is a b.s., (ii)' $\text{sp}\,\{x_i:1 \leqslant i \leqslant n+1\} \subset \text{sp}\,\{y_i:1 \leqslant i \leqslant k\}$ and (iii)' $\text{sp}\,\{g_i:1 \leqslant i \leqslant k\} \subset Y$. Let us now proceed to

Case (b) Having chosen the integer k through either of the alternatives described in Case (a), consider once again two situations: (1) $f_{n+1} \in \text{sp}\,\{g_i:1 \leqslant i \leqslant k\}$ and (2) $f_{n+1} \notin \text{sp}\,\{g_i:1 \leqslant i \leqslant k\}$. For (1), we put $k_{n+1} = k$ and we are done. If (2) is true, we put $k_{n+1} = k+1$ and let

$$g_{k+1} = f_{n+1} - \sum_{i=1}^{k} f_{n+1}(y_i)g_i.$$

Then $g_{k+1} \in Y \setminus \text{sp}\,\{g_i:1 \leqslant i \leqslant k\}$ and, applying the Hahn-Banach theorem to $(Y,\sigma(Y,X))$, we find y_{k+1} in X so that $g_i(y_j) = \delta_{ij}$; $1 \leqslant i, j \leqslant k+1$. Hence the b.s. $\{y_i;g_i:1 \leqslant i \leqslant k+1\}$ is an extension of $\{y_i;g_i:1 \leqslant i \leqslant k_n\}$ and satisfies relations of the types (i), (ii) and (iii) above. □

Corollary 7.2.28: Let (X,T) be a separable TVS and $\{f_n\}$ a total sequence in

X^*. Then there exist $\{y_n\}$ in X and $\{g_n\}$ in X^* such that $\{y_n;g_n\}$ is an M-base for $\langle X,X^*\rangle$.

The totality of $\{f_n\}$ in the above corollary is indispensable, for we have

Example 7.2.29: Recall the separable space X of Example 7.2.13 and let $\{f_n\}$ be any sequence in X^*. If $A_n = \{\lambda \in \aleph : f_n(\lambda) \neq 0\}$, then $\#(A_n) < \infty$ for each $n \geq 1$. Choose x in X so that $x(\lambda) = 0$ for $\lambda \in A_n$, $n \geq 1$ and $x(\lambda) \neq 0$ for λ in $\aleph \setminus \cup\{A_n : n \geq 1\}$. Hence $x \neq 0$ but $\langle x,f_n\rangle = 0$ for $n \geq 1$. Hence no b.s. can be an M-semibase for $\langle X,X^*\rangle$.

7.3 SIMILAR SYSTEMS

Given two b.sy. $\{x_\alpha;f_\alpha\}$ and $\{y_\alpha;g_\alpha\}$ over the same index set Λ in two different TVS (X_1,T_1) and (X_2,T_2) respectively, it is quite natural to enquire the manner in which they should appear to be similar. The intriguing part at this stage is to define suitably the notion of similarity between two biorthogonal systems, although the reasons for it can be traced back as early as 1925; cf. [5], p. 112. However, the concept of similarity between two biorthogonal systems that we use hereafter is essentially adapted to accommodate the analysis of cones (cf. Part Three and the stability of Schauder bases to be discussed in our subsequent work [133]).

Definition 7.3.1: A b.sy. $\{x_\alpha;f_\alpha:\alpha \in \Lambda\}$ in a TVS (X_1,T_1) is said to *dominate* another b.sy. $\{y_\alpha;g_\alpha:\alpha \in \Lambda\}$ in a TVS (X_2,T_2) if, for each x_1 in X_1, there exists x_2 in X_2 such that $f_\alpha(x_1) = g_\alpha(x_2)$ for each α in Λ. If both these systems dominate each other, they are said to be *equivalent* or *similar*.

Notation: Throughout this section, (X_1,T_1) and (X_2,T_2) stand for two arbitrary TVS containing respectively the b.sy. $\{x_\alpha;f_\alpha:\alpha \in \Lambda\}$ and $\{y_\alpha;g_\alpha:\alpha \in \Lambda\}$, while ω_Λ denotes the vector space over $\mathbb{K}$, consisting of all $\mathbb{K}$-valued families indexed on Λ; that is, $\omega_\Lambda = \mathbb{K}^\Lambda$. Define subspace μ_Λ and ν_Λ by $\mu_\Lambda = \{\{f_\alpha(x)\} \in \omega_\Lambda : x \in X_1\}$ and $\nu_\Lambda = \{\{g_\alpha(y)\} \in \omega_\Lambda : y \in X_2\}$.

Proposition 7.3.2: The b.sy. $\{x_\alpha;f_\alpha\}$ and $\{y_\alpha;g_\alpha\}$ are similar if and only if $\mu_\Lambda = \nu_\Lambda$.

Proof: Obvious. □

Proposition 7.3.3: Let $\{x_\alpha;f_\alpha\}$ and $\{y_\alpha;g_\alpha\}$ be equivalent and total. Then there exists an algebraic isomorphism R from X_1 onto X_2 such that $Rx_\alpha = y_\alpha$, $\alpha\in\Lambda$.

Proof: If the algebraic isomorphism F_1, F_2 and I are defined by $F_1x = \{f_\alpha(x)\}$, $F_2y = \{g_\alpha(y)\}$ and $I(\{f_\alpha(x)\}) = \{g_\alpha(y)\}$, then the required R is given by $R = F_2^{-1} \circ I \circ F_1$ as shown below

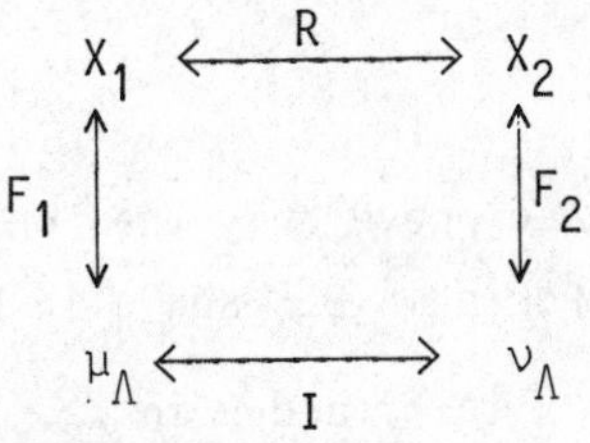

□

Conversely, we have

Proposition 7.3.4: Let $\{x_\alpha;f_\alpha\}$ and $\{y_\alpha;g_\alpha\}$ be complete and let R be a topological isomorphism from (X_1,T_1) onto (X_2,T_2) such that $Rx_\alpha = y_\alpha$ for α in Λ. Then these two systems are equivalent.

Proof: For x in X_1, there exists y in X_2 so that $Rx = y$. Fix α, then $R^*g_\alpha \in X_1^*$. Hence, for some q in D_{T_1}, $|f_\alpha(u)|$, $|R^*g_\alpha(u)| \leq q(u)$ for all u in X_1. Further, for $\varepsilon > 0$ one finds z in $sp\{x_\beta\}$ so that $q(x-z) \leq \varepsilon$. Since $\langle v,R^*g_\alpha\rangle = \langle v,f_\alpha\rangle$ for any v in $sp\{x_\beta\}$, $|\langle x,R^*g_\alpha - f_\alpha\rangle| \leq 2\varepsilon$. Thus $R^*g_\alpha = f_\alpha$, giving $\langle x,f_\alpha\rangle = \langle y,g_\alpha\rangle$ for every α. This shows the domination of $\{x_\alpha;f_\alpha\}$ over $\{y_\alpha;g_\alpha\}$. Similarly, $\{y_\alpha;g_\alpha\}$ dominates $\{x_\alpha;f_\alpha\}$. □

For the main result of this section, we require

Lemma 7.3.5: Let a vector space X be equipped with two (Hausdorff) linear topologies T_1 and T_2. Suppose $\{f_\alpha\}$ is a total family on X such that each $f_\alpha \in (X,T_1)^*, (X,T_2)^*$. Let T be the linear topology on X generated by $D = \{r = p+q : p \in D_{T_1}, q \in D_{T_2}\}$. If (X,T_1) and (X,T_2) are complete, then so is (X,T); in particular, if (X,T_1) and (X,T_2) are F-spaces, then $T_1 \approx T_2$.

Proof: For each α, we find p_α in D_{T_1} and q_α in D_{T_2}, so that $|f_\alpha(x-y)| \leq p_\alpha(x-z) + q_\alpha(y-z)$ for all x,y,z in X. This inequality along with the totality of $\{f_\alpha\}$ yields the first part. The last part is a consequence of the closed graph theorem. □

Now we have the desired result (cf. [3])

Theorem 7.3.6: Let (X_1,T_1) and (X_2,T_2) be two F-spaces having respectively M-base systems $\{x_\alpha;y_\alpha:\alpha \in \Lambda\}$ and $\{y_\alpha;g_\alpha:\alpha \in \Lambda\}$. Then these two systems are similar if and only if there exists a topological isomorphism $R:(X_1,T_1) \to (X_2,T_2)$ such that $Rx_\alpha = y_\alpha$ for each α in Λ.

Proof: In view of Proposition 7.3.4, we need prove only the "only if" part. Accordingly, we have $\mu_\Lambda = \nu_\Lambda = \lambda_\Lambda$ (say). For p in D_{T_1} and q in D_{T_2}, let $Q_p(\{f_\alpha(x)\}) = p(x)$, $Q_q^*(\{g_\alpha(y)\}) = q(y)$ for x in X_1 and y in X_2. If T and T* are the topologies on λ_Λ generated respectively by $\{Q_p:p \in D_{T_1}\}$ and $\{Q_q^*:q \in D_{T_2}\}$, then, by Lemma 7.3.5, $T \approx T^*$. Hence the result follows by an application of Proposition 7.3.3. □

7.4 QUASI-REGULAR SYSTEMS

Earlier in this chapter we discussed in some detail the properties of an important ingredient forming an S.b., namely, the b.s. Our attempt is now how to recover an S.b. from a given b.s. To begin with, let us recall (cf. [43])

Definition 7.4.1: A b.s. $\{x_n;f_n\}$ for <X,Y> is called *unordered quasi-regular* (resp. *quasi-regular*), abbreviated hereafter as u.q.r. (resp. q.r), if the family $\{S_F(x):F \in \Phi\}$ (resp. $\{S_n(x):n \geq 1\}$) is $\sigma(X,Y)$-bounded in X for each x in X.

We also need

Definition 7.4.2: A dual system <X,Y> is called an M-*system* if bounded sets in $(X,\sigma(X,Y))$ and $(X,\beta(X,Y))$ are the same.

We quote the following elementary result (cf. [76]; [235], p. 158) in the form of

Lemma 7.4.3: For an M-system <X,Y>, bounded sets in $(Y,\sigma(Y,X))$ and $(Y,\beta(Y,X))$ are the same.

The main result of this section is

Theorem 7.4.4: Let $\{x_n;f_n\}$ be a q.r.b.s. for an M-system $\langle X,Y\rangle$. Then $\{x_n;f_n\}$ is an S.b. for $(X_1,\beta_1|X_1)$ and $\{f_n,\psi x_n\}$ is an S.b. for $(Y_2,\beta_2|Y_2)$, where $X_1 = [x_n]^{\beta_1}$, $\beta_1 \equiv \beta(X,Y); Y_2 = [f_n]^{\beta_2}$, $\beta_2 \equiv \beta(Y,X)$ and $\psi: X \to (Y,\beta(Y,X))^*$, is the usual canonical injection.

Proof: For each x in X, the sequence $\{S_n(x)\}$ is β_1-bounded and so, from Proposition 1.2.6, $\{S_n\}$ is β_1-β_1 equicontinuous on X. Hence, to every $\sigma(Y,X)$-bounded subset B of Y, there exists a similar subset D of Y such that $p_B(S_n(x)) \leqslant p_D(x)$, for each x in X and all $n \geqslant 1$, where $p_B(x) = \sup\{|\langle x,y\rangle|: y \in B\}$ etc. Let $\varepsilon > 0$ and $E = B \cup D$. If $x \in X_1$, then there exists y in sp $\{x_n\}$ so that $p_E(x-y) \leqslant \varepsilon/2$. Since we may write

$$y = \sum_{i=1}^{N} \langle y,f_i\rangle\, x_i,$$

$p_B(S_n(x)-x) \leqslant p_B(S_n(x-y)) + p_B(x-y) \leqslant \varepsilon$ for all $n \geqslant N$, giving $S_n(x) \to x$ in β_1. This shows that $\{x_n;f_n\}$ is an S.b. for $(X_1,\beta_1|X_1)$. Similarly the other part follows. □

Note: When X is a Banach space and $Y = X^*$, Theorem 7.4.4 was first proved by Banach ([5], p. 107; cf. also [59], p. 1067) and its extension to barrelled spaces was given by Dieudonné in [43]. It was further sharpened in [237] for W-spaces. A still general form of Theorem 7.4.4 is given in [76] where one may find more information in this direction.

Exercise 7.4.5: Give an example of a b.s. which is not q.r. [Hint: Example 7.2.16.].

Remark: In terms of Definition 5.1.1, the foregoing theorem tells that every q.r.b.s. in W-space is S. basic in the strong topology. However, such a sequence in this space may fail to be its S.b. in the strong topology; for instance, consider the familiar b.s. $\{e^n;e^n\}$ for $\langle \ell^\infty, (\ell^\infty)^*\rangle$. Another non-trivial example in this direction is the following (cf. [43], [67]).

Example 7.4.6: Let $x^n = e^n - e^{n+1}$. $f^n = e^1 + \dots + e^n$, $n \geqslant 1$. Then $\{x^n;f^n\}$ is a total b.s. for $\langle \ell^1,\ell^\infty\rangle$. For y in ℓ^1 and g in ℓ^∞,

$$\sum_{i=1}^{N} \langle y,f^i\rangle\langle x^i,g\rangle = \sum_{i=1}^{N} y_i g_i - g_{N+1} \sum_{i=1}^{N} y_i,$$

and thus $\{x^n;f^n\}$ is q.r. If $y_n = 1/2^n$ and $g_n = (-1)^{n-1}$, $n \geq 1$, then $\{\langle S_n(y),g\rangle\}$ does not converge. Hence $\{x^n;f^n\}$ is not an S.b. for the W-space $(\ell^1,\tau(\ell^1,\ell^\infty))$.

Finally, we have

Proposition 7.4.7: Let $\{x_n;y_n\}$ be a q.r.b.s. for an M-dual system $\langle X,Y\rangle$ such that each $\beta \equiv \beta(Y,X)$-bounded subset of Y is $\sigma \equiv \sigma(Y,X)$-relatively compact. Then the following statements are true.

(i) For each sequence $\{\lambda_n\}$ of scalars with $M = \{\sum_{i=1}^{n} \lambda_i y_i\}$ being σ-bounded, there corresponds y in Y such that $\langle x_n,y\rangle = \lambda_n$, $n \geq 1$.

(ii) For each F in $Y^* = (Y,\beta)^*$, the sequence $\{\sum_{i=1}^{n} F(y_i)x_i\}$ is $\sigma(X,Y)$-bounded.

(iii) For each sequence $\{\mu_n\}$ of scalars with $\{\sum_{i=1}^{n} \mu_i x_i\}$ being $\sigma(X,Y)$-bounded, there exists an F in Y^* such that $F(y_n) = \mu_n$, $n \geq 1$.

Proof: (i) By the hypothesis, M has a σ-adherent point y in Y. Now the result follows by considering $\sigma(X,Y)$-neighbourhoods of y relative to each x_i.

(ii) For each y in Y, the sequence $\{S_n^*(y)\}$ is σ-bounded and so it is β-bounded. Since $\{F(S_n^*(y))\}$ is bounded, the result follows.

(iii) Recall the map ψ of Theorem 7.4.4. Since, for each $\sigma(X,Y)$-bounded subset A, $|\psi(x)(y)| \leq \sup\{|\langle x,y\rangle| : x \in A\}$ for y in Y, the set $\psi[A]$ is $\sigma(Y^*,Y)$-relatively compact by the Alaoglu-Bourbaki theorem. Now proceed as in (i). □

8 Types of Schauder bases

8.1 INTRODUCTION

To derive further benefit from the presence of an S.b. $\{x_n;f_n\}$ in an l.c.TVS (X,T), we have to examine carefully the behaviour of $\{x_n;f_n\}$ in terms of (a) the convergence of the underlying infinite series and (b) the dual properties of $\{x_n\}$ and $\{f_n\}$. The motivation for (a) comes from the fact that by now there are different modes of convergence of an infinite series in an arbitrary l.c. TVS, while a study of (b) is prompted by the desirability of knowing the S.b. character of $\{f_n\}$ for $(X^*,\beta(X^*,X))$. In this chapter, we examine both (a) and (b) in a limited manner; their detailed study is postponed to our next project [133]. We assume throughout this chapter, unless otherwise indicated, that (X,T) is an arbitrary l.c.TVS containing an S.b. $\{x_n;f_n\}$.

Definition 8.1.1: The S.b. $\{x_n;f_n\}$ is called (i) a *bounced* multiplier (resp. *subseries, unconditional*) according as $\sum_{n\geq 1} f_n(x)x_n$ is bounded multiplier (resp. subseries, unconditionally) convergent; (ii) *shrinking* if $\{f_n\}$ is an S.b. for $(X^*,\beta(X^*,X))$; and (iii) *boundedly complete* or γ*-complete* if $\sum_{n\geq 1} \alpha_n x_n$ converges, whenever $\{\sum_{i=1}^{n} \alpha_i x_i\}$ is bounded in (X,T) for α in ω.

Note: We will abbreviate hereafter a bounded multiplier (resp. a subseries, an unconditional) Schauder base as a b.m-S.b. (resp. an s-S.b., an u-S.b.).

We first discuss (i) and then pass on to a combined study of (ii) and (iii).

8.2 UNCONDITIONAL BASES

In general a bounded multiplier base is subseries and a subseries base is always unconditional. In an ω-complete l.c.TVS, these three notions are the same; cf. Theorem 1.4.5. The concept of an unconditional base is the weakest among the three notions described in Definition 8.1.1, (i) and is quite useful in analyzing the geometry of Schauder bases. Accordingly, we pay specific attention to unconditional bases in terms of their characterizations.

However, we do intend to touch upon the other two related notions as well.

Let us begin with a simple

Proposition 8.2.1: Let an s.s. λ be equipped with $\sigma(\lambda,\lambda^{\times})$. Then $\{e^n;e^n\}$ is an u-S.b. for $(\lambda,\sigma(\lambda,\lambda^{\times}))$.

Proof: The result follows from Example 2.2.2 and the observation that, if $x \in \lambda$, $y \in \lambda^{\times}$ and $\pi \in \Pi$, then

$$q_y(x - \sum_{i=1}^{n} \langle x,e^{\pi(i)}\rangle e^{\pi(i)}) \leqslant \sum_{i>n} |x_{\pi(i)} y_{\pi(i)}| \to 0$$

as $n \to \infty$. □

Proposition 8.2.2: For an s.s.λ, let μ be a subspace of λ^{β}. Then $\{e^n;e^n\}$ is a b.m-S.b. (resp. an s-S.b.) for $(\lambda,\sigma(\lambda,\mu))$ if and only if λ is normal (resp. monotone).

Proof: Let $\{e^n;e^n\}$ be a b.m-S.b. If $x \in \lambda$ and $u \in \omega$ with $|u_n| \leqslant |x_n|$, then $u_n = \alpha_n x_n$, where $|\alpha_n| \leqslant 1$ for $n \geqslant 1$. Here $\sum_{n\geqslant 1} \alpha_n x_n e^n$ converges in $(\lambda,\sigma(\lambda,\mu))$, say, to v and so $u_n = v_n$, $n \geqslant 1$; that is, $u \in \lambda$. On the other hand, if λ is normal, then for x in λ and α in ℓ^{∞}, $\alpha x = \{\alpha_n x_n\} \in \lambda$. Hence $\{e^n;e^n\}$ is a b.m.-S.b. Similarly the other part follows. □

Recall the notation of sequence spaces $\delta \equiv \delta_X$ and $\mu \equiv \mu_{X*}$ from Section 2.3. We have

Proposition 8.2.3: $\{x_n,f_n\}$ is a b.m-S.b. (resp. an s-S.b.) for (X,T) if and only if δ is normal (resp. monotone).

Proof: This is a direct consequence of Propositions 2.3.1 and 8.2.2. □

Proposition 8.2.4: Let $\{x_n;f_n\}$ be an s-S.b. for (X,T). Then $\mu = \delta^{\times}$ if and only if (X,T) is an S-space.

Proof: Let $\mu = \delta^{\times}$, where δ is monotone by Proposition 8.2.3. The S-character of (X,T) now follows with the help of Propositions 1.3.6 and 2.3.1.

Conversely, by Proposition 2.3.2, $\mu = \delta^{\beta}$. Since $\{e^n;e^n\}$ is an s-S.b. for $(\delta,\sigma(\delta,\mu))$, δ is monotone. Hence, by Theorem 1.3.2, $\mu = \delta^{\beta} = \delta^{\times}$. □

Characterizations: In order to obtain the first characterization of unconditional bases due to Weill (cf. [232]; [233]), we need a lemma which reflects an extension of certain techniques developed in [191] and where we do not suppose the existence of any S.b. for the space considered.

Lemma 8.2.5: Let (X,T) be an l.c. TVS and $\{g_\alpha : \alpha \in \Lambda\}$ a net of pointwise bounded linear operators on X to itself. For each v in $\mathcal{B} \equiv \mathcal{B}_T$, let

$$v^* = \{x \in X : g_\alpha(x) \in v, \ \forall \alpha \in \Lambda\} = \cap \{g_\alpha^{-1}[v] : \alpha \in \Lambda\}.$$

Then the following statements are true:

(i) The family $\mathcal{B}^* = \{v^* : v \in \mathcal{B}\}$ is a fundamental neighbourhood system at the origin of X for a unique l.c. topology T^* on X.

(ii) If there is a subnet $\{g_\beta : \beta \in \Lambda_1\}$ of the net $\{g_\alpha : \alpha \in \Lambda\}$ such that for each x in X, $g_\beta(x) \to x$ in T, then $T \subset T^*$.

(iii) If in addition to (ii), each g_α is continuous and (X,T) is barrelled, then $T \approx T^*$.

Proof: (i) Straightforward.

(ii) Here $v^* \subset v$ for each v in $\mathcal{B}$.

(iii) Observe that each v^* is T-closed. □

Notation: The following notation will be used in the rest of this chapter. First, let us introduce the subsets $\bar{a}$, $\bar{b}$ and $\bar{e}$ of ω as follows:

$$\bar{a} = \{\alpha \in \omega : \alpha_n = 1 \text{ or } -1, \ \forall\, n \geq 1\};$$

$$\bar{b} = \{\beta \in \omega : \beta_n = 0 \text{ or } 1, \ \forall\, n \geq 1\};$$

$$\bar{e} = \{\varepsilon \in \omega : |\varepsilon_n| \leq 1, \ \forall\, n \geq 1\}.$$

Further, define the operators $S_{n,\alpha} : X \to X$ by

$$S_{n,\alpha}(x) = \sum_{i=1}^{n} \alpha_i f_i(x) x_i; \quad x = \sum_{n \geq 1} f_n(x) x_n,$$

where $\alpha \in \omega$. If $\sigma \in \Phi$, the family of all finite subsets of $\mathbb{N}$, then recall the operator $S_\sigma : X \to X$ introduced in Chapter 7, namely,

$$S_\sigma(x) = \sum_{i\in\sigma} f_i(x)x_i .$$

Also, if $v \in \mathcal{B} \equiv \mathcal{B}_T$, let us write $v_{\bar{a}}$, $v_{\bar{b}}$ and $v_{\bar{e}}$ for the following sets:

$$v_{\bar{a}} = \{x \in X : S_{n,\alpha}(x) \in v;\ \forall\, n \geq 1,\ \alpha \in \bar{a}\};$$

$$v_{\bar{b}} = \{x \in X : S_\sigma(x) \in v,\ \forall \sigma \in \Phi\}$$

$$= \{x \in X : S_{n,\alpha}(x) \in v,\ \forall\, n \geq 1,\ \alpha \in \bar{b}\};$$

$$v_{\bar{e}} = \{x \in X : S_{n,\varepsilon}(x) \in v;\ \forall\, n \geq 1,\ \varepsilon \in \bar{e}\}.$$

Finally, put

$$\mathcal{B}_{\bar{a}} = \{v_{\bar{a}} : v \in \mathcal{B}\};\ \mathcal{B}_{\bar{b}} = \{v_{\bar{b}} : v \in \mathcal{B}\};\ \mathcal{B}_{\bar{e}} = \{v_{\bar{e}} : v \in \mathcal{B}\}.$$

Theorem 8.2.6: Let $\{x_n;f_n\}$ be an u-S.b. Then each of the families $\mathcal{B}_{\bar{a}}$, $\mathcal{B}_{\bar{b}}$ and $\mathcal{B}_{\bar{e}}$ respectively forms a fundamental system of neighbourhoods at the origin of X for the unique l.c. topologies $T_{\bar{a}}$, $T_{\bar{b}}$ and $T_{\bar{e}}$ on X such that $T \subset T_{\bar{a}}, T_{\bar{b}}, T_{\bar{e}}$. If (X,T) is also barrelled, then

$$T \approx T_{\bar{a}} \approx T_{\bar{b}} \approx T_{\bar{e}} .$$

Proof: In order to get the result by applying Lemma 8.2.5, let us first show that the families $\mathcal{F}_{\mathbb{N}\times I} = \{S_{n,r} \equiv S_{(n,r)} : (n,r) \in \mathbb{N} \times I\}$ of operators on X form nets and are pointwise bounded, where $I = \bar{a}$ or $\bar{b}$ or $\bar{e}$ and note that $\mathcal{F}_{\mathbb{N}\times\bar{b}}$ is the same as the family $\{S_\sigma : \sigma \in \Phi\}$. Indeed, the set $\mathbb{N} \times I$ is a directed set under the usual ordering $<$ defined by the relation: $(n,\alpha) < (m,\beta)$ if and only if $n \leq m$. Since $\sum_{n\geq 1} f_n(x)x_n$ converges unconditionally for each x in X, the family $\mathcal{F}_{\mathbb{N}\times I}$ is pointwise bounded by Proposition 1.4.4 for each value of $I = \bar{a}$, $\bar{b}$ and $\bar{e}$. Recalling the unit vector e of ω, we find that $\{S_{n,e} : n \geq 1\}$ is a subnet for each of the nets $\mathcal{F}_{\mathbb{N}\times I}$, $I = \bar{a}$, $\bar{b}$ and $\bar{e}$ and observe that $S_{n,e}(x) \to x$ as $n \to \infty$. Hence the first part of the theorem follows from Lemma 8.2.5, (i) and (ii). Since each of the maps $S_{n,\alpha}$ ($\alpha \in I$; $I = \bar{a}, \bar{b}, \bar{e}$) is clearly continuous on (X,T), the last part of the theorem is proved by Lemma 8.2.5, (iii). □

It is often convenient to express the topologies $T_{\bar{a}}$, $T_{\bar{b}}$ and $T_{\bar{e}}$ in terms of the seminorms obtained with the help of members of D_T. Accordingly, we have

Proposition 8.2.7: Let $\{x_n;f_n\}$ be an u-S.b. Then the topology $T_I(I = \bar{a}$ or $\bar{b}$ or $\bar{e})$ of Theorem 8.2.6 is generated by the family $\{p_{v_I} : v \in \mathcal{B}_T\}$ of seminorms, where p_{v_I} is the Minkowski functional of v_I and for x in X

$$p_{v_I}(x) = \sup\{p_v(\sum_{i=1}^{n} \alpha_i f_i(x)x_i) : n \geq 1, \alpha \in I\}, \tag{8.2.8}$$

p_v being the Minkowski functional corresponding to v in $\mathcal{B}_T$.

Proof: Straightforward. □

The following result will be found quite useful. Let us recall Definition 7.4.1, then we have

Proposition 8.2.9: Let $\{x_n;f_n\}$ be an u.q.r. b.s. for $\langle X,X^*\rangle$, $X \equiv (X,T)$ being a W-space. Then $\{x_n;f_n\}$ is an u-S.b. for $[x_n] \equiv [x_n]^\beta$, $\beta \equiv \beta(X,X^*)$ and $\{f_n;\Psi x_n\}$ is an u-S.b. for $[f_n] \equiv [f_n]^\beta$, $\beta^* \equiv \beta(X^*,X)$, where $\Psi:X \to X^{**}$ the canonical map.

Proof: Since each S_σ is $\sigma(X,X^*)$-$\sigma(X,X^*)$ continuous, the family $\{S_\sigma:\sigma \in \Phi\}$ is β-β equicontinuous on X by Proposition 1.2.6. Hence to every $\sigma(X^*,X)$-bounded subset B of X*, there exists a similar subset D of X* such that $p_B(S_\sigma(x)) \leq p_D(x)$ for each x in X and every σ in Φ. Now proceed as in Theorem 7.4.4 to get the u-S.b. character of $\{x_n;f_n\}$ for $[x_n]$. The other part follows by analogy. □

The next main theorem, its variations and the resulting corollary are due to Weill [233]; cf. also [83].

Theorem 8.2.10: Let there be an arbitrary l.c. TVS (X,T) having a sequence $\{x_n\}$ with $x_n \neq 0$ for $n \geq 1$. Consider the condition:

> Given a member u in $\mathcal{B} \equiv \mathcal{B}_T$, there exists v in $\mathcal{B}$ such that for arbitrary a in ω, m and n in $\mathbb{N}$ with $m \leqslant n$ (m and n may be ∞) and γ in I (I is either $\bar{a}$, $\bar{b}$ or $\bar{e}$; or, $\bar{a} \subset I \subset \bar{e}$, $\bar{b} \subset I \subset \bar{e}$), it follows that $\sum_{i=1}^{m} \gamma_i a_i x_i \in u$, whenever $\sum_{i=1}^{n} a_i x_i \in v$. (*)

(i) If (X,T) is ω-complete, then (*) implies that $\{x_n\}$ is an u-S.b. for $[x_n] \equiv [x_n]^T$; (ii) if (X,T) is barrelled and $\{x_n\}$ is an u-S.b. for (X,T), then (*) holds; and (iii) if (X,T) is ω-complete, barrelled and $\{x_n\}$ is an u-S.b. for (X,T), then the topology T is generated by the family $\{P_p : p \in D_T\}$ of seminorms on X, where for x in X,

$$P_p(x) = \sup_{|\alpha_n| \leqslant 1} p\Big(\sum_{n \geqslant 1} \alpha_n f_n(x) x_n\Big). \tag{8.2.11}$$

Before we pass on to the proof of Theorem 8.2.10, it will be convenient to mention two variations of the condition (*).

<u>Variation 8.2.12</u>: The condition (*) in Theorem 8.2.10 is equivalent to the following condition $(*_1)$:

> Given an equicontinuous subset A of X^*, there exists a similar subset B of X^* such that for an arbitrary constant $M > 0$; arbitrary a in ω, m and n in $\mathbb{N}$ with $p \leqslant q$ (p and q can be ∞) and γ in I (I as in (*)), whenever $|\sum_{i=1}^{n} a_i f_i(x)| \leqslant M$ for all f in B, it follows that $|\sum_{i=1}^{m} \gamma_i a_i f_i(x)| \leqslant M$ for all f in A. $(*_1)$

<u>Variation 8.2.13</u>: The condition (*) in Theorem 8.2.10 can be replaced by the following equivalent condition $(*_2)$:

> Given p in D_T, there exists q in D_T such that for arbitrary a in ω, m and n in $\mathbb{N}$ with $m \leqslant n$ (m and n can be ∞) and γ in I (I as in (*)) such that $p(\sum_{i=1}^{m} \gamma_i a_i x_i) \leqslant q(\sum_{i=1}^{n} a_i x_i)$. $(*_2)$

<u>Note</u>: It is a routine exercise to verify the equivalence of (*), $(*_1)$ and $(*_2)$.

<u>Proof</u>: (i) Assume the truth of (*) and hence that of $(*_2)$. Choosing each

$\gamma_i = 1$, we find from Theorem 5.1.2 that $\{x_n\}$ is an S.b. for $[x_n]$. Let $\{f_n\}$ denote the s. a.c.f. corresponding to $\{x_n\}$. Let u and v in $\mathcal{B}$ be as in (*) and fix x in X. Since $\{\sum_{i=1}^{n} f_i(x)x_i\}$ is Cauchy in (X,T), $\{\sum_{i=m}^{n} \gamma_i f_i(x)x_i\} \in u$ for all large m and n and arbitrary γ in I. Hence $\sum_{n\geqslant 1} f_n(x)x_n$ is unconditionally convergent in $[x_n]$.

(ii) By Theorem 8.2.6, for each u in $\mathcal{B}$ there exist v in $\mathcal{B}$ so that $v \subset u_I$ and this proves (*).

(iii) By the ω-completeness of (X,T), the series $\sum_{n\geqslant 1} f_n(x)x_n$ is bounded multiplier convergent for each x in X. Consequently by Variation 8.2.13, to every p in D_T there exists q in D_T such that $P_p(x) \leqslant q(x)$. Since $p(x) \leqslant P_p(x)$, the family $\{P_p : p \in D_T\}$ generates T. □

<u>Corollary 8.2.14</u>: Suppose (X,T) is barrelled and $\{x_n;f_n\}$ is an u-S.b. For a sequence $\{y_n\}$ in X and $\{\sigma_n\}$ in Φ with $\sigma_p \cap \sigma_q = \emptyset$ if $p \neq q$, set

$$z_q = \sum_{i\in\sigma_q} f_i(y_q)x_i.$$

Then corresponding to each u in $\mathcal{B}$, there exists v in $\mathcal{B}$ such that for arbitrary a in ω, γ in I, p and q in $\mathbb{N}$ with $p \leqslant q$ (p and q may be ∞) we have

$$\sum_{i=1}^{q} a_i z_i \in v \Rightarrow \sum_{i=1}^{p} \gamma_i a_i z_i \in u. \qquad (*)$$

In addition, if (X,T) is ω-complete and $z_n \neq 0$ for $n \geqslant 1$, then $\{z_n\}$ is an u-S.b. for $[z_n] \equiv [z_n]^T$.

Indeed, if $b_{jk} = a_k f_j(y_k)$, then

$$\sum_{k=1}^{q} a_k z_k = \sum_{k=1}^{q} \sum_{j\in\sigma_k} b_{jk}x_j,$$

and since $\sigma_1,\dots,\sigma_q$ are all distinct, the last sum can be arranged as $\sum_{i=1}^{n} c_i x_i$. Now apply Theorem 8.2.10, (ii) to get (*). For the last part, make use of Theorem 8.2.10, (i).

Next, we pass on to a more analytical characterization of unconditional bases in the form of (cf. [78])

Theorem 8.2.15: Let (X,T) be ω-complete and barrelled. If $\{x_n;f_n\}$ is an M-semibase for $\langle X,X^*\rangle$, then the following statements are equivalent:

(i) $\{x_n;f_n\}$ is an u-S.b. for (X,T).

(ii) For every x in X, the series

$$\sum_{n\geq 1} |f_n(x)|x_n \tag{8.2.16}$$

converges in (X,T) and for each p in D_T, there exists q in D_T such that

$$p(x) \leq q(\sum_{n\geq 1} |f_n(x)|x_n), \ \forall x \in X. \tag{8.2.17}$$

Proof: (i) ⇒ (ii). For x in X, let $\alpha_n(x) = \text{sgn } f_n(x)$. Hence by Theorem 1.4.5, there exists y in X such that $f_n(y) = |f_n(x)|$, $n \geq 1$ and this yields (8.2.16). But $\sum_{n\geq 1} \overline{\alpha_n(x)}f_n(y)x_n$ also converges; hence for p in D_T,

$$p(x) = p(\sum_{n\geq 1} \overline{\alpha_n(x)}\ |f_n(x)|x_n) \leq \sup_{n\geq 1} p(\sum_{i=1}^{n} \overline{\alpha_n(x)}f_n(y)x_n). \tag{+}$$

Since the condition $(*_2)$ in Variation 8.2.13 is satisfied, there exists q in D_T such that

$$p(\sum_{i=1}^{n} \overline{\alpha_n(x)}f_n(y)x_n) \leq q(\sum_{i=1}^{r} f_i(y)x_i), \ \forall r \geq n. \tag{++}$$

The required inequality (8.2.17) is now a consequence of (+) and (++).

(ii) ⇒ (i). Choose arbitrary $\varepsilon > 0$ and an equicontinuous subset M of X^*. Then for some p in D_T, $|f(y)| \leq p(y)$ for all f in M, y in Y. Also there exists q in D_T so that (8.2.17) is satisfied. Further, for x in X there exists N in $\mathbb{N}$ such that $q(\sum_{i=m}^{n} |f_i(x)|x_i) < \varepsilon$ for all $m,n \geq N$. If $f \in X^*$, let $\alpha_n(f) = \text{sgn}(f_n(x)f(x_n))$, $x \in X$. Put

$$z = \sum_{i=m}^{n} \alpha_i^x f_i(x)\ x_i = \sum_{i=m}^{n} f_i(z)x_i.$$

Thus for f in M

$$\sum_{i=m}^{n} |f_i(x)f(x_i)| = |f(z)| \leq q(\sum_{i=m}^{n} |f_i(x)|x_i) < \varepsilon,$$

for all m, n $\geq$ N. Therefore, for each x in X, the series $\sum_{n\geq 1} f_n(x)x_n$ converges unconditionally by Theorem 1.4.7. □

Examples of conditional bases

Propositions 8.2.1 through 8.2.3 provide a number of examples of S.b. which are either u-S.b. or b.m-S.b. or s-S.b. or non-b.m-S.b. or non-s-S.b. Therefore, to complete the discussion of this section, let us now pass on to a set of examples exhibiting a number of different S.b.'s which are not unconditional; that is, these bases are conditional.

In the course of the present discussion, we will need the following result ([124]):

Proposition 8.2.18: As before, let (X,T) contain an S.b. $\{x_n;f_n\}$. Assume that $\{x_n\}$ is regular and $\alpha \in \omega$ with $\alpha_n \neq 0$, $n \geq 1$. Define

$$y_n = \sum_{i=1}^{n} \alpha_i x_i \,;\; g_n = \frac{1}{\alpha_n} f_n - \frac{1}{\alpha_{n+1}} f_{n+1},\; n \geq 1.$$

If $\{y_n/\alpha_{n+1}\}$ is T-bounded, then $\{y_n;g_n\}$ is an S.b. for (X,T).

Proof: It is clear that $\{y_n;g_n\}$ is a b.s. for $\langle X,X^*\rangle$. For x in X and $n \geq 1$,

$$\sum_{i=1}^{n} g_i(x)y_i = \sum_{i=1}^{n} f_i(x)x_i - \frac{1}{\alpha_{n+1}} y_n f_{n+1}(x).$$

Observe that $f_n \to 0$ in $\sigma(X^*,X)$ and this proves the result. □

Now we have (cf. [67])

Example 8.2.19: Consider the S.b. $\{e^n;e^n\}$ for $(c_0,\|\cdot\|_\infty)$. Let $x^n = e^{(n)}$ and $f^n = e^n - e^{n+1}$. Then, from Proposition 8.2.18, $\{x^n;f^n\}$ is an S.b. for $(c_0,\|\cdot\|_\infty)$. If $\alpha \in \omega$, then

$$\sum_{i=m}^{n} \alpha_i x^i = \{\underbrace{\sum_{i=m}^{n} \alpha_i,\ldots,\sum_{i=m}^{n} \alpha_i}_{\text{m-times}},\; \sum_{i=m+1}^{n} \alpha_i,\ldots,\alpha_n,0,0,\ldots\}.$$

Consequently, $\sum_{n\geq 1} \alpha_n x^n$ converges in c_0 if and only if $\sum_{n\geq 1} \alpha_n$ converges, or equivalently $\alpha \in cs$. If $\alpha_n = (-1)^n/n$, then $\sum_{n\geq 1} \alpha_n x^n$ converges in c_0. But for $\varepsilon_n = (-1)^n$, $n \geq 1$

$$\|\sum_{i=m}^{n} \varepsilon_i \alpha_i x^i\|_\infty = \sup_{m \leq k \leq n} |\sum_{i=k}^{n} \frac{1}{i}| .$$

Hence $\sum_{n \geq 1} \varepsilon_n \alpha_n x^n$ does not converge in c_o: therefore, by Theorem 1.4.5, $\{x^n;f^n\}$ is not an u-S.b. for $(c_o, \|\cdot\|_\infty)$.

The next two examples are reproduced from [84] (cf. also [218]) and [215] respectively, and will be found quite useful in Part Three.

Example 8.2.20: Let $x^n = \sum_{i=1}^{n} (-1)^{n+i} e^i$ and $f^n = e^n + e^{n+1}$, $n \geq 1$. By using Proposition 8.2.18, $\{x^n;f^n\}$ is an S.b. for $(c_o, \|\cdot\|_\infty)$. Following Example 8.2.19, we easily conclude that $\{x^n;f^n\}$ is not an u-S.b. for $(c_o, \|\cdot\|_\infty)$.

Example 8.2.21: Consider the sequence $\{y^n\}$ with $y^1 = e^1$; $y^n = e^{n-1} - e^n$, $n \geq 2$. On account of Theorem 7.4.4 and the S.b. character of $\{x^n;f^n\}$ of Example 8.2.19 for the space c_o, we find that $\{y^2,y^3,\ldots\}$ is an S.b. for $[y^2,y^3,\ldots]$, the $\|\cdot\|_1$-closure of sp $\{y^2,y^3,\ldots\}$. Hence $\{y^n\}$ is an S.b. for $(\ell^1, \|\cdot\|_1)$. If $\alpha \in \omega$, then for $n > m \geq 2$,

$$\|\sum_{i=m}^{n} \alpha_i y^i\|_1 = |\alpha_m| + |\alpha_n| + \sum_{i=m}^{n-1} |\alpha_{i+1} - \alpha_i| .$$

The above relation shows that $\{y^n\}$ cannot be an u-S.b. for $(\ell^1, \|\cdot\|_1)$; for example, $\sum_{i \geq 1} (y^i/i)$ converges whereas $\sum_{i \geq 1} ((-1)^i y^i/i)$ diverges.

Remark: In the next example, we construct another S.b. which is not unconditional. Incidentally, this example also shows that condition (8.2.17) in Theorem 8.2.15 cannot be dropped in order to conclude the u-S.b. character of $\{x_n;f_n\}$. Indeed, we have (cf. [173])

Example 8.2.22: Let $X = \{\alpha \in \ell^1 : \sum_{n \geq 1} \alpha_n = 0\}$. Since $e \in (\ell^1)^*$ and $X = \ker(e)$, $(X, \|\cdot\|_1)$ is a Banach space. Put $x^n = e^n - e^{n+1}$; $f^n = e^{(n)}$, $n \geq 1$. For α in ω,

$$\sum_{i=1}^{n} \langle\alpha,f^i\rangle x^i = \sum_{i=1}^{n} \alpha_i e^i - (\sum_{j=1}^{n} \alpha_j)e^{n+1}.$$

The preceding relation shows that $\{x^n;f^n\}$ is an S.b. for $(X, \|\cdot\|_1)$. However, this base is not a b.m-S.b. and so it cannot be a u-S.b. Indeed, for $\alpha_1 = 1$, $\alpha_n = 1/n - 1/(n-1)$, $n \geq 2$, and, for $n > m$,

$$\|\sum_{i=m}^{n} (-1)^i\langle\alpha,f^i\rangle\, x^i\|_1 = \frac{1}{m} + \frac{1}{n} + \sum_{i=m+1}^{n} (\frac{1}{i-1} + \frac{1}{i}) \to \infty\ ,$$

as $m, n \to \infty$.

Next, we verify that (8.2.16) is satisfied but not (8.2.17). For x in X, let $R_n = \sum_{i=1}^{n} |\langle x,f^i\rangle| x^i$. Then

$$\|R_n - R_m\|_1 \leq |\sum_{i=1}^{m+1} x_i| + |\sum_{i=1}^{n} x_i| + \sum_{i=m+2}^{n} |x_i|,$$

and so $\sum_{n\geq 1} |\langle x,f^n\rangle| x^n$ converges to $|x|$ (say) in $(X, \|\cdot\|_1)$ for every x in X, proving thereby (8.2.16). Further, for $n \geq 1$, let

$$y^n = \frac{1}{n} \sum_{i=1}^{2n} (-1)^{i+1} x^i$$

$$= \frac{1}{n} \{e^1 - 2 \sum_{i=2}^{2n} (-1)^i e^i + e^{2n+1}\}.$$

Then $\|y^n\|_1 = 4$. Observe that

$$\langle y^n,f^i\rangle = \begin{cases} (-1)^{i+1}/n, & i = 1,\ldots,2n; \\ 0, & i > 2n. \end{cases}$$

Hence

$$|y^n| = \sum_{i\geq 1} |\langle y^n,f^i\rangle| x^i$$

$$= |\langle y^n,f^1\rangle| e^1 + \sum_{i\geq 2} \{|\langle y^n,f^i\rangle| - |\langle y^n,f^{i-1}\rangle|\} e^i.$$

This yields $\|\,|y^n|\,\|_1 = 2/n$. Consequently $n\|\,|y^n|\,\|_1 < \|y^n\|_1$ and this disproves (8.2.17).

8.3 SHRINKING AND γ-COMPLETE BASES

In this section, we take up the problem (b) mentioned in the introduction and introduce to the reader certain basic results on shrinking and γ-complete bases, including a few of their applications in the general setting of locally convex spaces.

Shrinking and γ-complete bases were essentially introduced by James [98] in connection with the characterization of reflexivity of Banach spaces; however, James preferred to use the term "boundedly complete" in place of "γ-complete" introduced in [111]. Tumarkin [228] calls shrinking bases *stretching bases*. In [129] there is a notion of $\mathfrak{S}$-uniform bases which is more general than that of shrinking bases; in particular, weakly uniform bases are the same as shrinking bases.

Our notation (X,T) and $\{x_n;f_n\}$ remains as given in the introductory section of this chapter.

Shrinking bases

Recall the canonical map $\Psi:X \to X^{**} \equiv (X^*,\beta)^*$, $\beta \equiv \beta(X^*,X)$ with $\Psi x(f) = \langle x,f\rangle \equiv f(x)$; $x \in X$, $f \in X^*$. Thus to find the shrinking character of $\{x_n;f_n\}$ is equivalent to ascertaining the S.b. character of $\{f_n;\Psi x_n\}$ for (X^*,β). Observe that $\{f_n;\Psi x_n\}$ is always an S.b. for $(X^*,\sigma(X^*,X))$. Besides, the introduction of shrinking bases is further justified by the following two examples.

Example 8.3.1: The S.b. $\{e^n;e^n\}$ for $(\ell^1,\sigma(\ell^1,\ell^\infty))$ is not shrinking. Indeed $e^{(n)} \not\to e$ in $\beta(\ell^\infty,\ell^1)$; cf. [132], p. 110.

Example 8.3.2: The S.b. $\{e^n;e^n\}$ for $(\phi,\eta(\phi,\omega))$ is shrinking since $\phi^* = \omega$ and $\beta(\omega,\phi) \approx \sigma(\omega,\phi)$.

Example 8.3.2 is included in the more general (cf. [49])

Proposition 8.3.3: The S.b. $\{e^n;e^n\}$ for any l.c.s.s. $(\lambda,\sigma(\lambda,\lambda^\times))$ or $(\lambda,\eta(\lambda,\lambda^\times))$ is shrinking if and only if $x^{(n)} \to x$ in $\beta(\lambda^\times,\lambda)$ for each x in $\lambda^\times$.

Proof: A direct consequence of the definition. □

To obtain a characterization of shrinking bases, let us introduce

Definition 8.3.4: An arbitrary l.c. TVS (X,T) not necessarily having an S.b. $\{x_n;f_n\}$, is said to be a *non-ℓ^1-space* if it has no subspace Y with $(Y,T|Y) \cong (\ell^1, \|\cdot\|_1)$.

Following [233], we have the next two results.

Theorem 8.3.5: Let (X,T) be an ω-complete barrelled space containing an u-S.b. $\{x_n;f_n\}$. If (X,T) is a non-ℓ^1-space, then $\{x_n;f_n\}$ is shrinking.

Proof: Let $\{x_n;f_n\}$be not shrinking. Hence there exist g in X^*, a $\sigma(X,X^*)$-bounded subset B of X, an increasing subsequence $\{n_i\}$ of $\mathbb{N}$ and a sequence $\{z_n\}$ in B such that (cf. [140], p. 71)

$$\left|\sum_{k=n_{i-1}+1}^{n_i} g(x_k)f_k(z_i)\right| > 1; \; \forall i \geqslant 1, \quad (n_o = 0).$$

Let $y_i = S_{n_i}(z_i)-S_{n_{i-1}}(z_i)$. Since $\{S_n\}$ is equicontinuous, $\{y_n\}$ is bounded with $|g(y_n)| > 1$ for $n \geqslant 1$. By Corollary 8.2.14, $\{y_n\}$ is an u-S.b. for $Y = [y_n]$. Here $Y = \{y = \sum_{n\geqslant 1} \alpha_n y_n : \alpha \in \ell^1\}$ and so we may define a natural map $H:Y \to \ell^1$, $Hy = \alpha$ with $y = \sum_{n\geqslant 1} \alpha_n y_n$. Also for such y in Y, choose $\{\varepsilon_n\}$ with $|\varepsilon_n| = 1$, $n \geqslant 1$ so that $|\alpha_n g(y_n)| = \varepsilon_n \alpha_n g(y_n)$ for $n \geqslant 1$. Define z_ε in Y by $z_\varepsilon = \sum_{n\geqslant 1} \varepsilon_n \alpha_n y_n$. Since $|f_n(y)| = |f_n(z_\varepsilon)|$, $n \geqslant 1$, we find that

$$\|Hy\|_1 \leqslant |g(z_\varepsilon)| \leqslant \sum_{n\geqslant 1} |f_n(y)g(x_n)| = \left|g\left(\sum_{n\geqslant 1} \delta_n f_n(y)x_n\right)\right|,$$

where $|\delta_n| = 1$ for $n \geqslant 1$. But $|g(x)| \leqslant p(x)$ for some p in D_T. Therefore, $\|Hy\|_1 \leqslant P_p(y)$, where P_p is defined by (8.2.11). The continuity of H now follows by Theorem 8.2.10, (iii). The map H^{-1} is clearly continuous, giving thereby $(Y,T|Y) \cong (\ell^1, \|\cdot\|_1)$. This contradicts the nature of (X,T). □

Theorem 8.3.6: Let (X,T) be an ω-complete and barrelled space with an u-S.b. $\{x_n;f_n\}$. Then the following statements are equivalent:

(i) $\{x_n;f_n\}$ is shrinking.

(ii) X^* is β-separable, $\beta \equiv \beta(X^*,X)$.

(iii) (X,T) is a non-ℓ^1-space.

The proof requires a

Lemma 8.3.7: Let Y be a subspace of an arbitrary l.c. TVS (X,T) not necessarily containing an S.b. If X^* is $\beta \equiv \beta(X^*,X)$-separable, then Y^* is $\beta(Y^*,Y)$-separable.

Proof: Let us consider the β-closed subspace $Y^\perp = \{f \in X^*: f(y) = 0, \forall y \in Y\}$. Then $X^*/Y^\perp$ is $T^\perp$-separable, where $T^\perp$ is generated by $\{\hat{p}_B: B$ is bounded in $(X,T)\}$ with

$$\hat{p}_B(\hat{f}) = \inf\{\sup\{|(f-g)(x)|: x \in B\}: g \in Y^\perp\},\ \hat{f} \in X^*/Y^\perp.$$

The map $F: X^*/Y^\perp \to Y^*$, $F(\hat{f}) = f|Y$ is a well defined linear map which is also 1-1. By the Hahn-Banach theorem, it is onto as well. Since $\hat{p}_B(\hat{f}) = \sup\{|f(x) : x \in B\}$ whenever B is a $\sigma(Y,Y^*)$-bounded subset of Y, F is $T^\perp$-$\beta(Y^*,Y)$ continuous. Consequently, the map $\Psi_Y \circ F: (X^*,\beta) \to (Y^*,\beta(Y^*,Y))$ is also continuous, Ψ_Y being the quotient map from X^* onto $X^*/Y^\perp$. Hence $(Y^*,\beta(Y^*,Y))$ is separable. □

Proof of Theorem 8.3.6: By virtue of Theorem 8.3.5, we have to show only (ii) ⇒ (iii). Assume therefore the truth of (ii). If (iii) were not true, then for some subspace Y of X, $(Y^*, \beta(Y^*,Y)) \simeq (\ell^\infty, \beta(\ell^\infty, \ell^1))$. However, by Lemma 8.3.7, this contradicts (ii). □

Exercise 8.3.8: Let (X,T) be a W-space and $\{x_n; f_n\}$ an u-S.b. If X^* is $\sigma(X^*,X^{**})$-ω-complete, prove that $\{f_n; \Psi x_n\}$ is an u-S.b. for (X^*,β). [Hint: Apply Theorem 1.4.6, (iii) ⇒ (ii).]

γ-complete bases

At the outset, let us prove a result offering several examples of Schauder bases which are γ-complete as well as non-γ-complete. Indeed, we have (cf. [49])

Proposition 8.3.9: The S.b. $\{e^n; e^n\}$ for an arbitrary l.c.s.s. $(\lambda, \sigma(\lambda,\lambda^\times))$ or $(\lambda, \eta(\lambda,\lambda^\times))$ is γ-complete if and only if λ is perfect.

Proof: We consider the space $(\lambda, \sigma(\lambda,\lambda^\times))$ and let $\lambda = \lambda^{\times\times}$. Suppose $\{\sum_{i=1}^{n} \alpha_i e^i\}$ is $\sigma(\lambda,\lambda^\times)$-bounded for α in ω. Let $\beta \in \lambda^\times$. Choose θ_n with

$|\theta_n| = 1$ so that $|\alpha_n\beta_n| = \theta_n\alpha_n\beta_n$, $n \geq 1$. Define γ in $\lambda^\times$ by $\gamma_n = \theta_n\beta_n$. Since for some constant $M > 0$

$$\sum_{i=1}^{n} |\alpha_i\beta_i| = |\langle \sum_{i=1}^{n} \alpha_i e^i, \gamma\rangle| < M; \ \forall\ n \geq 1,$$

we find $\alpha \in \lambda^{\times\times} = \lambda$. Therefore $\sum_{n\geq 1} \alpha_n e^n$ converges in $(\lambda, \sigma(\lambda,\lambda^\times))$. The converse is much easier to prove. Similarly the other part follows. □

Another result similar to Proposition 8.3.9 is the following:

Proposition 8.3.10: Let (λ,T) be a monotone l.c.s.s. with $\sigma(\lambda,\lambda^\times) \subset T \subset \tau(\lambda,\lambda^\times)$. Then $\{e^n;e^n\}$ is a γ-complete S.b. for (λ,T) if and only if λ is perfect.

Proof: Make use of Theorem 1.3.3 and Proposition 8.3.9. □

The following proposition [49] extends a similar result in [98].

Proposition 8.3.11: Let (X,T) be an S-space and $\{x_n;f_n\}$ a b.m - S.b. Then $\{x_n;f_n\}$ is γ-complete if and only if $(X,\sigma(X,X^*))$ is ω-complete.

Proof: By Propositions 8.2.3 and 2.3.2, the s.s. $\delta \equiv \delta_X$ is normal and $\delta^\beta = \mu \equiv \mu_{X^*}$.

If $\{x_n;f_n\}$ is a γ-complete S.b., then $\{e^n;e^n\}$ is also a γ-complete S.b. for $(\delta,\sigma(\delta,\delta^\times))$; cf. Proposition 2.3.1. Hence δ is perfect by Proposition 8.3.9. Therefore, using Theorem 1.3.2 and Proposition 2.3.1, $(X,\sigma(X,X^*))$ is ω-complete. Similarly the other part follows by reversing the steps. □

Definition 8.3.12: An arbitrary l.c.TVS (X,T) is said to be a *non-*c_0*-space* if it has no subspace Y with $(Y,T|Y) \simeq (c_0, \|\cdot\|_\infty)$.

Then we have (cf. [110]; [233])

Proposition 8.3.13: Let (X,T) be ω-complete, barrelled and non-c_0, and $\{x_n;f_n\}$ an u-S.b. Then $\{x_n;f_n\}$ is γ-complete.

Proof: At the outset, let us observe, in view of Theorem 8.2.10, (iii) that T is also generated by $\{P_p:p \in D_T\}$ where P_p is a seminorm defined by (8.2.11). If the required result were not true, there would exist α in ω such that

$\{\sum_{i=1}^{n} \alpha_i x_i\}$ is bounded but does not converge in (X,T). Hence there exist an increasing sequence $\{n_j\}$ and some p in D_T such that if

$$y_j = \sum_{i=n_{j-1}+1}^{n_j} \alpha_i x_i, \quad (n_o = 0)$$

then $P_p(y_j) \geqslant 1$ and $\{\sum_{i=1}^{n} y_i\}$ is bounded. If $\beta \in c_o$ and $\varepsilon > 0$ are arbitrary, then $|\beta_j| \leqslant \varepsilon$ for all $j \geqslant N$ (say). Hence for q in D_T and $r > m > N$,

$$q(\sum_{i=m}^{r} \beta_i y_i) \leqslant \varepsilon P_q(\sum_{i=m}^{r} y_i).$$

This shows that $\sum_{n \geqslant 1} \beta_n y_n$ converges in (X,T) for each β in c_o. Further, $\{y_n\}$ is an u-S.b. for $Y \equiv [y_n]$; cf. Corollary 8.2.14. Therefore, the natural map $H:Y \to c_o$, $Hy = \{\phi_n(y)\}$ is a 1-1 and onto linear map, where $\{\phi_n\}$ is the s.a.c.f. corresponding to $\{y_n\}$.

We next show that $(Y,T|Y)$ is barrelled. Put $X_j = sp\,\{x_{n_{j-1}+1},\ldots,x_{n_j}\}$. For each $j \geqslant 1$, we find h_j in X_j' such that $h_j(y_j) = p(y_j)$ and $|h_j(x)| \leqslant p(x)$ for all x in X_j. Define g_j in X' by

$$g_j(x) = h_j(\sum_{i=n_{j-1}+1}^{n_j} f_i(x)x_i).$$

Then $|g_j(x)| \leqslant P_p(x)$ for each $j \geqslant 1$ and all x in X. Consequently, $\{g_n\}$ has a $\sigma(X^*,X)$-cluster point g in X^* and we easily find that $g(x_i) = 0$, $i \geqslant 1$; that is, $g = 0$. Thus $\{g_n(x)\} \in c_o$ for each x in X. Define $R_k:X \to Y$ by $R_k x = \sum_{i=1}^{k} g_i(x)y_i$, $k \geqslant 1$. Since $\sum_{n \geqslant 1} g_n(x)y_n$ converges in Y for each x in X and each R_k is continuous, the operator $R:X \to Y$, $Rx = \lim R_k x$ is a continuous, projection from X onto Y. This shows that $(Y,T|Y)$ is barrelled.

For γ in ℓ^1, define f in Y' by $f(y) = \sum_{n \geqslant 1} \gamma_n \phi_n(y)$. By the barrelled character of Y, we conclude that $f \in Y^*$. Since $|\langle Hy,\gamma \rangle| = |f(y)|$, H is $\sigma(Y,Y^*)$-$\sigma(c_o,\ell^1)$ continuous and hence, by Proposition 1.2.19, H is $T|Y - \|\cdot\|_\infty$ continuous. Similarly the continuity of H^{-1} follows. Therefore, $(Y,T|Y) \simeq (c_o, \|\cdot\|_\infty)$ and we arrive at a contradiction. □

The following proposition is an improvement of a similar result in [233],

Theorem 2.11, (1) $\Rightarrow$ (2). Our proof differs and is shorter.

Proposition 8.3.14: Let (X,T) be an S-space. If $\{x_n;f_n\}$ is γ-complete and $\sigma(X,X^*)$-u-S.b., then $(X,\sigma(X,X^*))$ is ω-complete.

Proof: If $\alpha \in \omega$ with $|\alpha_n| < 1$ and $n \geq 1$ and $x \in X$, then $\{\sum_{i=1}^{n} \alpha_i f_i(x)x_i\}$ is weakly bounded and hence it converges. Therefore, $\{x_n;f_n\}$ is a b.m.-S.b. and so δ is normal and $\mu = \delta^\times$. Thus $\{e^n;e^n\}$ is γ-complete for $(\delta,\sigma(\delta,\delta^\times))$ and consequently δ is perfect. Now make use of Theorem 1.3.2 and Proposition 2.3.1 to get the required ω-completeness of $(X,\sigma(X,X^*))$. □

We need the following lemma [233] for the main result of this subsection.

Lemma 8.3.15: If an arbitrary l.c. TVS (X,T) is $\sigma(X,X^*)$-ω-complete, then it is a non-c_0-space.

Proof: On the other hand, let there be a subspace Y of X with $(Y,T|Y) \simeq (c_0, \|\cdot\|_\infty)$. Hence Y is $\sigma(X,X^*)$-closed and so we easily conclude that $(Y,\sigma(Y,Y^*))$ is ω-complete. This shows that $(c_0,\sigma(c_0,\ell^1))$ is ω-complete which is, however, not true; cf. [132], p. 118. □

Now we come to the main result [233], namely,

Theorem 8.3.16: Let (X,T) be barrelled and $\{x_n;f_n\}$ an u-S.b. Then the following statements are equivalent:

(i) $\{x_n;f_n\}$ is γ-complete.

(ii) $(X,\sigma(X,X^*))$ is ω-complete.

(iii) (X,T) is ω-complete and a non-c_0-space.

Proof: (i) $\Rightarrow$ (ii) Apply Proposition 8.3.14.

(ii) $\Rightarrow$ (iii) Use Lemma 8.3.15.

(iii) $\Rightarrow$ (i) Same as Proposition 8.3.13. □

Applications to reflexivity

The vast importance of the Schauder basis theory began to be recognized after James [98] had successfully applied it to the characterization of reflexive Banach spaces having Schauder bases. This result of James has since undergone several extensions under more general conditions; one such result in this direction which suits our present purpose is the following, essentially due to Cook [23].

<u>Theorem 8.3.17</u>: (X,T) is semireflexive if and only if $\{x_n;f_n\}$ is both shrinking and γ-complete.

<u>Proof</u>: Let (X,T) be semireflexive. Therefore $\{f_n;\Psi x_n\}$ is an S.b. for (X^*,σ_*) and as $[f_n]^\beta = [f_n]^{\sigma*} = X^*$, where $\sigma_* \equiv \sigma(X^*,X^{**})$ and $\beta \equiv \beta(X^*,X)$, $\{x_n;f_n\}$ is shrinking by Theorem 7.4.4. Thus, to every f in X^*, bounded subset B of (X,T) and $\varepsilon > 0$, there exists N in $\mathbb{N}$ so that

$$|f(x-S_n(x))| < \varepsilon; \ \forall n \geq N, \ x \in B. \qquad (*)$$

Let $\{\sum_{i=1}^{n} \alpha_i x_i\}$ be T-bounded in X for α in ω. It easily follows from $(*)$ that $\{\sum_{i=1}^{n} \alpha_i x_i\}$ is $\sigma(X,X^*)$-Cauchy and, since $(X,\sigma(X,X^*))$ is ω-complete (cf. [93], p. 228), $\{x_n;f_n\}$ turns out to be γ-complete.

Conversely, let $\{x_n;f_n\}$ be both shrinking and γ-complete. If $x^{**} \in X^{**}$, then $\{\sum_{i=1}^{n} <f_i,x^{**}> x_i\}$ is T-bounded and hence for x in X,

$$x = \sum_{n\geq 1} \langle f_n,x^{**}\rangle x_n.$$

It follows that $\langle x,f\rangle = \langle f,x^{**}\rangle$ for every f in X^*, giving $\Psi x = x^{**}$. □

<u>Exercise 8.3.18</u>: Construct an example of a non-semireflexive l.c. TVS (X,T) having an S.b. $\{x_n;f_n\}$ such that it is (i) shrinking but not γ-complete and (ii) γ-complete but not shrinking. [Hint: For instance, consider (i) $(c_0,\tau(c_0,\ell^1))$ and (ii) $(\ell^1,\sigma(\ell^1,\ell^\infty))$.]

For reflexivity, we have (cf. [49])

<u>Proposition 8.3.19</u>: Suppose $\mu = \delta^\times$ where $\mu \equiv \mu_{X^*}$ and $\delta \equiv \delta_X$. Then (X,T) is reflexive if and only if $\{x_n;f_n\}$ is both shrinking and γ-complete, $(X,\beta(X,X^*))$ is separable and $T \approx \tau(X,X^*)$.

<u>Proof</u>: The necessary part is a consequence of Theorem 8.3.17 and the reflexity of (X,T).

For the converse, we again apply Theorem 8.3.17 to infer the semireflexivity of (X,T). Since $\mu = \delta^{\times}$, $(X^*,\sigma(X^*,X))$ is ω-complete by Propositions 8.3.9, 1.3.6 and 2.3.1. Hence, by Proposition 1.2.14, (X,T) is barrelled. Thus (X,T) is reflexive. □

Remark: The above proposition is no longer true for spaces having conditional bases. There are several examples to support this statement. The next proposition is quite a general result in this direction.

Proposition 8.3.20: If $\{x_n,f_n\}$ is γ-complete and conditional for (X,T), then $\mu \neq \delta^{\times}$ and $\{e^n;e^n\}$ is not γ-complete for $(\delta,\sigma(\delta,\delta^{\times}))$.

Proof: If $\mu = \delta^{\times}$, then, by Proposition 8.3.9, $\delta = \delta^{\times\times}$ and so $\{x_n;f_n\}$ is an u-S.b., a contradiction. The other part can similarly be disposed of. □

Duality relationship

Here we briefly touch upon the duality relationship that exists between shrinking and γ-complete bases and recall (cf. [49])

Proposition 8.3.21: Let (X,T) be σ-infrabarrelled and $\{x_n;f_n\}$ a shrinking base. Then $\{f_n;\Psi x_n\}$ is γ-complete for (X^*,β), $\beta \equiv \beta(X,X^*)$.

Proof: Let $\{\sum_{i=1}^{n} \alpha_i f_i\}$ be β-bounded in X^* for α in ω. Put

$$A = \{\sum_{i=k}^{k+p} \alpha_i f_i ; k,p \geqslant 1\}.$$

Then A is equicontinuous. If $x \in X$ and $\varepsilon > 0$, then there exists N in $\mathbb{N}$ so that

$$|\langle \sum_{i \geqslant N} f_i(x)x_i, \sum_{i=k}^{k+p} \alpha_i f_i\rangle| \leqslant \varepsilon; \ \forall\, k,p \geqslant 1$$

$$\Rightarrow |\langle x, \sum_{i=N}^{N+p} \alpha_i f_i\rangle| \leqslant \varepsilon, \ \forall\, p \geqslant 1,$$

since $f_i(x_j) = 0$ for $1 \leqslant j \leqslant N-1$, $N \leqslant i \leqslant N+p$. Since A is $\sigma(X^*,X)$-relatively compact and $\{\sum_{i=1}^{n} \alpha_i f_i\} \subset A$ is $\sigma(X^*,X)$-Cauchy, $\sum_{i=1}^{n} \alpha_i f_i \to f$ in $\sigma(X^*,X)$ for some f in X*. □

Remark: The condition of σ-infrabarrelledness on (X,T) cannot be dropped in the above result, for we have

Example 8.3.22: Recall the space $(\ell^1,\tau(\ell^1,c_o))$ of Example 3.3.9, for which $\{e^n;e^n\}$ is an S.b. Since $\{e^n;e^n\}$ is an S.b. for $(c_o, \|\cdot\|_\infty)$, $\{e^n;e^n\}$ is shrinking for $(\ell^1,\tau(\ell^1,c_o))$. On the other hand, $\{e^n;e^n\}$ is not γ-complete for $(c_o, \|\cdot\|_\infty)$. Finally, $(\ell^1,\tau(\ell^1,c_o))$ is not σ-infrabarrelled; cf. [132], p. 120.

Besides Proposition 3.2.1 there are some other results exhibiting the duality relationship between shrinking and γ-complete bases, but rigorous discussion of these is postponed to [133]. All these results are indeed extensions or refinements of the following theorem of Singer [215]:

Theorem 8.3.23: Let (X,T) be a Banach space, $H = [f_n]^\beta$, $\beta \equiv \beta(X^*,X)$ and $\beta_H \equiv \beta|H$. Then $\{x_n;f_n\}$ is (i) shrinking if and only if $\{f_n;\Psi x_n\}$ is a γ-complete S.b. for (X^*,β), and (ii) γ-complete if and only if $\{f_n;\Psi x_n\}$ is a shrinking base for (H,β_H).

9 Basic sequences in Fréchet spaces

9.1 INTRODUCTION

The problem that we discuss in this chapter and in Chapter 10 is closely related to Chapter 4 in which we discussed at length the existence of bases in several different topological vector spaces. Although we have not settled the problem of finding the class of spaces devoid of bases, yet we know that the same spaces have generally subspaces which possess bases. Accordingly, we seek the solution of

Problem 9.1.1: Let (X,T) be a TVS having no t.b. and dim $X > \aleph_0$. Does there exist a subspace Y of X with dim $Y > \aleph_0$ such that Y has a t.b.?

Note: Obviously any subspace Y of X in the above problem having the desired property, has to be separable irrespective of the separability or inseparability of X. Let us mention here that there do exist inseparable subspaces of a separable TVS X (see, for instance, [155]) and clearly such a subspace Y cannot have a t.b. Hence it does not seem advantageous to solve Problem 9.1.1 in the affirmative by restricting (X,T) to be separable.

Originally, Problem 9.1.1 was considered by Banach (cf. [5], p. 238) for Banach spaces, who stated the following result without its proof:

Theorem 9.1.2: Every infinite dimensional Banach space contains an infinite dimensional closed subspace which contains an S.b.

The proof of this result was in fact known to S. Mazur who presented it at a seminar held at Warsaw University in 1955.

Since 1958, several extensions and proofs of Theorem 9.1.2 have appeared in the literature, e.g. [12], [13], [34], [68], [102], [149], [154], [179], [180] and [181]; cf. also [169] for the historical development of this result. Let us single out the following result [179]:

Theorem 9.1.3: Let an infinite dimensional Banach space $(X, \|\cdot\|)$ contain a sequence $\{x_n\}$ with $\inf \|x_n\| \geqslant \delta > 0$ and $f(x_n) \to 0$ for each f in X*.

Then there exists a subsequence $\{n_k\}$ of $\mathbb{N}$ such that $\{x_{n_k}\}$ is an S.b. for $[x_{n_k}]$.

The discussion above reveals the existence of S.basic sequences (Definition 5.1.1) of one or the other kind in every infinite dimensional Banach space. Another notably important recent development in this direction is the construction of bibasic sequences $\{x_n;f_n\}$ in an arbitrary Banach space X; that is, $\{x_n\}$ is S.basic in X and $\{f_n\}$ is S.basic in X*; see [31] for detail.

Our purpose in this chapter is to study the extensions of Theorems 9.1.2 and 9.1.3 to Fréchet spaces. Their further extensions to F-spaces will be taken up in the next chapter.

Some general results

Let us collect the necessary background that will be useful in the main results of this and the subsequent chapter.

Suppose $\{x_n\}$ is a linearly independent (l.i.) sequence in a vector space X. Let $Y = sp\{x_n\}$ and $Y_i = sp\{x_n : n \neq i\}$. Denote by $\{f_n\}$ the unique sequence in Y' such that $f_m(x_n) = \delta_{mn}$. The sequence $\{f_n\}$ is called the *sequence of associated functionals* (s.a.f.) corresponding to the l.i. sequence $\{x_n\}$. Define $S_n : Y \to Y$ by $S_n(y) = \sum_{i=1}^{n} f_i(y)x_i$; $y \in Y$, $n \geq 1$.

Definition 9.1.4: A sequence $\{x_n\}$ in a TVS (X,T) is called *semibasic* if $x_n \notin [x_i : i > n]$, $n \geq 1$.

Then we have (cf. [115])

Proposition 9.1.5: An l.i. sequence $\{x_n\}$ in a TVS (X,T) is semibasic if and only if the corresponding s.a.f. $\{f_n\}$ is contained in Y* or, equivalently, each S_n is continuous on $(Y, T|Y)$.

Proof: Let $\{x_n\}$ be semibasic and $N_k = \{y \in Y : f_k(y) = 0\}$. Each N_k is a maximal subspace of Y. Here $N_k = sp\{x_i : 1 \leq i < k\} + sp\{x_i : i > k\}$ and so $\bar{N}_k = sp\{x_i : 1 \leq i < k\} + [x_i : i > k]$ where the closures of spaces are taken in Y. If $x_k \in \bar{N}_k$, then

$$x_k = \sum_{i=1}^{k-1} \alpha_i x_i + y,$$

where $y \in [x_i : i > k]$. Since $x_k \notin [x_i : i > k]$, there is a first index i_o such that $\alpha_{i_o} \neq 0$. Consequently $x_{i_o} \in [x_i : i > i_o]$, a contradiction. Hence $N_k = \bar{N}_k$ and so f_k is continuous on Y; cf. [140], p. 37. The converse is trivially true. □

Exercise 9.1.6: Prove that a sequence $\{x_n\}$ in a TVS (X,T) is minimal if and only if it is l.i. and semibasic.

Proposition 9.1.7: Let $\{x_n\}$ be an l.i. sequence in a metrizable TVS (X,T) and let $[x_n]$ denote the closure of sp $\{x_n\}$ in the completion of (X,T). Then $\{x_n\}$ is an S.b. for $[x_n]$ if and only if $\{S_n\}$ is equicontinuous on $(Y, T|Y)$.

Proof: This result follows from Theorems 5.1.2 and 5.1.7. □

Definition 9.1.8: Corresponding to a sequence $\{x_n\}$ in a vector space X with $x_n \neq 0$, $n \geq 1$, let

$$y_n = \sum_{i=m_{n-1}+1}^{m_n} \alpha_i x_i,$$

where $0 = m_o < m_1 < \ldots < m_n < m_{n+1} \ldots$ and $\alpha \in \omega$ such that $y_n \neq 0$ for $n \geq 1$. Then $\{y_n\}$ is called a *block sequence* (bl.s) corresponding to $\{x_n\}$.

Theorem 9.1.9: Each block sequence corresponding to an S.b. in a TVS (X,T) is an S.basic sequence in (X,T).

Proof: Let $\{x_n\} \equiv \{x_n; f_n\}$ be an S.b. in (X,T) and consider an arbitrary bl.s. $\{y_n\}$ defined as above. It is easily seen that $\{y_n\}$ is l.i. and semibasic. If $Y = \text{sp}\,\{y_n\}$, then, from Proposition 9.1.5, there exists a unique sequence $\{h_n\}$ in Y* such that $h_m(y_n) = \delta_{mn}$. Let us write g_n for the continuous unique extension of h_n to $\bar{Y}$.

Further, if $y \in Y$ and $i \in \mathbb{N}$, then

$$f_i(y) = \begin{cases} g_j(y) f_i(y_j), & \text{if } m_{j-1}+1 \leq i \leq m_j; \\ 0, & \text{otherwise.} \end{cases} \tag{9.1.10}$$

Consequently (9.1.10) also holds for y in $\bar{Y}$. Thus for y in $\bar{Y}$,

$$y = \sum_{j \geq 1} \sum_{i=m_{j-1}+1}^{m_j} f_i(y)x_i = \sum_{j \geq 1} \sum_{i=m_{j-1}+1}^{m_j} g_j(y)f_i(y_j)x_i$$

$$= \sum_{j \geq 1} g_j(y) \sum_{i=m_{j-1}+1}^{m_j} \alpha_i x_i = \sum_{j \geq 1} g_j(y)y_j.$$

Therefore $\{y_n;g_n\}$ is an S.b. for $[y_n]$. □

9.2 CONSTRUCTION OF BASIC SEQUENCES

The fundamental results on the extraction of basic sequences in Fréchet spaces are due to Bessaga and Pelczynski [13] and Kadec and Pelczynski [107]; let us quote the following (cf. [13], Twierdzenie 2):

Theorem 9.2.1: Every infinite dimensional Fréchet space (X,T) contains an S.basic sequence.

In order to prove this result, we require

Lemma 9.2.2: Consider a finite dimensional vector space X equipped with a seminorm p. Let $U_p = \{x \in X:p(x) \leq 1\}$ and $S_p = \{x \in X:p(x) = 1\}$. Then both U_p and S_p are precompact.

Proof: If p is a norm, there is nothing to prove. Let $X_p = X/\ker p$ and $K_p:X \to X_p$ be the quotient map. Note that $\hat{S}_p = K_p[S_p]$ is compact in X_p and hence S_p is precompact in (X,p). Similarly, the other part follows. □

Proof of Theorem 9.2.1: At the outset, we may write $D_T = \{p_n\}$, where $p_1 \leq p_2 \leq \ldots \leq p_n \leq \ldots$. Put $X_i = X/\ker p_i$. Then X_i is a Banach space with respect to the norm $\hat{p}_i(\hat{x}) = \inf \{p_i(x+y): p_i(y) = 0\}$, where $\hat{x} \in X_i$. Consider two cases: Case (a), when dim $X_i < \infty$ for each $i \geq 1$, and Case (b), when dim $X_{i_0} = \infty$ for some i_0.

Case (a): Let $r_i = \dim X_i$ and we may suppose that $1 \leq r_1 < r_2 < \ldots$. By induction we get a sequence $\{x_n\}$ in X satisfying

(i) The set $\{x_1,\ldots,x_{r_i}\}$ is a maximal l.i. set in X relative to p_i; that is, whenever $p_i(\sum_{j=1}^{r_i} \alpha_j x_j) = 0$, then $\alpha_1 = \alpha_2 = \ldots = \alpha_{r_i} = 0$, $i \geq 1$; and

(ii) $p_i(x_j) = 0$ for $j > r_i$, $i \geq 1$.

Property (i) is obtained by the r_i-dimensionality of X_i, whereas (ii) is established by extending the pre-images of the maximal l.i. sequence in X_i to a maximal l.i. sequence in X_{i+1}.

Let $x \in X$ and write $x(i)$ for the corresponding element in X_i. There exist unique scalars $\alpha_j \equiv \alpha_j(x(i))$, $1 \leq j \leq r_i$, $i \geq 1$ and hence a unique sequence $\{\alpha_j\}$ such that

$$x(i) = \sum_{j=1}^{r_i} \alpha_j x_j(i);\ i = 1,2,\ldots,$$

and from this expression we find that $x = \sum_{j \geq 1} \alpha_j x_j$. Therefore $\{x_n\}$ is an S.b. for (X,T).

<u>Case (b)</u>: Here D_T may be taken as $\{q_i : i \geq 1\}$ with $q_i = p_{i_0+i}$ and let $Y_i = X_{i_0+i}$, $i \geq 1$. Let us note that each Y_i is infinite dimensional. Choose $\{\varepsilon_m\}$ with $0 < \varepsilon_m < 1$, $m \geq 1$ so that

$$\inf_{1 \leq p \leq q < \infty} \prod_{m=p}^{q} (1-\varepsilon_m) = \delta,\ 0 < \delta < 1. \tag{$*$}$$

The major step for the proof is to construct a sequence $\{x_n\}$ in X satisfying

$$q_i\left(\sum_{j=1}^{m+1} \alpha_j x_j\right) \geq (1-\varepsilon_m) q_i\left(\sum_{j=1}^{m} \alpha_j x_j\right), \tag{9.2.3}$$

valid for all i with $1 \leq i \leq m$, $m \geq 1$ and arbitrary scalars $\alpha_1,\ldots,\alpha_{m+1}$, and $q_1(x_n) \neq 0$, $n \geq 1$.

Assume temporarily the construction of $\{x_n\}$ as above. Then $\{x_n\}$ is l.i. with respect to q and hence relative to any q_i. For, let $q_1(\sum_{i=1}^{m} \alpha_i x_i) = 0$. Then, using (9.2.3), we conclude that

$$q_1(\alpha_k x_k) \leq \frac{2}{\delta} q_1\left(\sum_{i=1}^{m} \alpha_i x_i\right) = 0,\quad 1 \leq k \leq m.$$

Hence $\alpha_1 = \alpha_2 = \ldots = \alpha_m = 0$.

Next, we establish an inequality of the type (5.1.3). So, let $p,q, \in \mathbb{N}$ with $1 \leq p < q < \infty$ and choose arbitrary scalars $\alpha_1,\ldots,\alpha_q$. Further, pick up m in $\mathbb{N}$ arbitrarily. First, let $m \leq p$; then from (9.2.3),

$$q_m\left(\sum_{i=1}^{q} \alpha_i x_i\right) \geq \delta q_m\left(\sum_{i=1}^{p} \alpha_i x_i\right). \tag{9.2.4}$$

Next, suppose $m > p$. Also, let $Y = \mathrm{sp}\,\{x_n\}$ and $\{\phi_n\} \subset Y'$ denote the s.a.f. corresponding to $\{x_n\}$. Let us observe that $|\phi_i(z)| \leq (2/\delta q_1(x_i))q_1(z)$ for any z in Y and $i \geq 1$. If $z = \sum_{i=1}^{q} \alpha_i x_i$, then

$$\begin{aligned} q_m\left(\sum_{i=1}^{p} \alpha_i x_i\right) &\leq \sum_{i=1}^{m-1} |\phi_i(z)|q_m(x_i) \\ &\leq \left[\frac{2}{\delta}\sum_{i=1}^{m-1} \frac{q_m(x_i)}{q_1(x_i)}\right]q_m\left(\sum_{i=1}^{q} \alpha_i x_i\right). \end{aligned} \tag{9.2.5}$$

From (9.2.4) and (9.2.5), we get a constant $K_m > 0$ with

$$q_m\left(\sum_{i=1}^{p} \alpha_i x_i\right) \leq K_m q_m\left(\sum_{i=1}^{q} \alpha_i x_i\right). \tag{9.2.6}$$

Inequality (9.2.6) yields the required S.basic character of $\{x_n\}$; cf. Theorem 5.1.2.

We now return to the construction of $\{x_n\}$ promised earlier. Choose x_1 in X with $q_1(x_1) = 1$. Suppose that we have already got $x_1,\ldots,x_n$ so that $q_1(x_i) \neq 0$, $1 \leq i \leq n$ and (9.2.3) is satisfied for $m = 1,\ldots,n-1$.

Consider $R_n = \mathrm{sp}\,\{x_1,\ldots,x_n\}$ equipped with the seminorms $q_1,\ldots,q_n$. If $S_n^i = \{x \in R_n : q_i(x) = 1\}$, $1 \leq i \leq n$; then, by Lemma 9.2.2, there exists an ε_n-net $\{y_j^i : 1 \leq j \leq k_i\}$ for S_n^i such that $q_i(y_j^i) = 1$ for $1 \leq j \leq k_i$, $1 \leq i \leq n$.

Let $X_i = X/\ker q_i$ be the quotient space endowed with the norm $\hat{q}_i$ and denote by $x(i)$ an arbitrary element of X_i, $x \in X$. Then $\hat{q}_i(y_j^i(i)) = 1$; $1 \leq j \leq k_i$, $1 \leq i \leq n$. Hence, by the Hahn-Banach theorem, there exist F_j^i in X_i^* so that $\sup\{|F_j^i(x(i))| : \hat{q}_i(x(i)) = 1\} = 1$ and $|F_j^i(y_j^i(i))| = 1$; $1 \leq j \leq k_i$, $1 \leq i \leq n$.

If Ψ_i denotes the quotient map, $\Psi_i : X \to X_i$, we may define f_j^i in X^* by $f_j^i = F_j^i \circ \Psi_i$ for $1 \leq j \leq k_i$, $1 \leq i \leq n$. Let $A = \{f_j^i : 1 \leq j \leq k_i, 1 \leq i \leq n\}$ and let H be the intersection of the kernels of members in A. Then $\dim (X/H) < \infty$; cf. [234], p. 39.

Since X_1 is infinite dimensional, there exists an l.i. sequence $\{u_n(1)\}$ in X_1 such that $u_i \neq u_j$ for $i \neq j$. The sequence $\{u_n^H\}$ with $u_n^H = u_n + H$ cannot be l.i. in X/H and so, for some $i_1 \neq i_2$, $u_{i_1}^H = \alpha u_{i_2}^H$. If $x_{n+1} = u_{i_1} - \alpha u_{i_2}$, then $x_{n+1} \in H$ and consequently $f(x_{n+1}) = 0$ for every f in A; also $q_1(x_{n+1}) \neq 0$, for otherwise $u_{i_1}(1) = \alpha u_{i_2}(1)$.

For any index i with $1 \leqslant i \leqslant n$ and then any j satisfying $1 \leqslant j \leqslant k_i$, we have

$$1 = |f_j^i(y_j^i)| = |f_j^i(y_j^i + \alpha x_{n+1})| \leqslant q_i(y_j^i + \alpha x_{n+1})$$

for every α in $\mathbb{K}$. If $x \in S_n^i$, then, for some y_j^i, $q_i(x-y_j^i) \leqslant \varepsilon_n$. Therefore, for any α in $\mathbb{K}$ and x in S_n^i,

$$q_i(x + \alpha x_{n+1}) \geqslant q_i(y_j^i + \alpha x_{n+1}) - q_i(x-y_j^i)$$

$$\geqslant 1-\varepsilon_n.$$

Hence, for any x in R_n and α in $\mathbb{K}$,

$$q_i(x + \alpha x_{n+1}) \geqslant (1-\varepsilon_n)q_i(x); \; i = 1,2,\ldots,n.$$

The preceding inequality, along with $q_1(x_{n+1}) \neq 0$, completes the required induction process. □

Note: Case (a) in the preceding proof is also dealt with in [10], p. 376.

Our next result is on the selection of an S.basic sequence from a certain sequence in a Fréchet space. First we require the following definition [107]; cf. also [179]; [216]:

Definition 9.2.7: For an arbitrary l.c. TVS (X,T), a set $\Gamma \subset X^*$ is said to be *norming* provided that, for every or some fundamental system $\{B_\alpha : \alpha \in \Lambda\}$ of $\beta(X^*,X)$-bounded subsets of X^*, the topology T is generated by the family $\{q_\alpha : \alpha \in \Lambda\}$ of seminorms on X, where for x in X,

$$q_\alpha(x) = \begin{cases} \sup\{|f(x)| : f \in B_\alpha \cap \Gamma\}, & \text{if } B_\alpha \cap \Gamma \neq \emptyset; \\ 0, & \text{if } B_\alpha \cap \Gamma = \emptyset. \end{cases}$$

Notation: For a net $\{x_\alpha\}$ and an element x in an l.c. TVS (X,T) and a subset Γ of X*, we use the notation $x_\alpha \xrightarrow{\Gamma} x$ to mean $f(x_\alpha) \to f(x)$ for each f in Γ.

We present a nontrivial norming set in the form of

Example 9.2.8: The set $\Gamma = \{f \in c_0 : \|f\|_\infty = 1\}$ is norming for $(\ell^1, \|\cdot\|_1)$. Let $x \in \ell^1$ and choose β in ω with $\beta_i x_i = |x_i|$. Then $\beta^{(n)} \in c_0$ and $\|\beta^{(n)}\|_\infty = 1$, $n \geqslant 1$. Hence $\sup\{|f(x)| : f \in \Gamma\} \geqslant \|x\|_1$ and therefore $\sup\{|f(x)| : f \in \Gamma\} = \|x\|_1$.

Exercise 9.2.9: Prove that X* is a norming set for every infrabarrelled space (X,T).

Definition 9.2.10: If (X,T) is an l.c. TVS, then a seminorm p on X is called *ω-consistent* (resp. *consistent*) with a set $\Gamma \subset X^*$ if $x_n \xrightarrow{\Gamma} x$ (resp. $x_\alpha \xrightarrow{\Gamma} x$) implies that $\liminf_{n\to\infty} p(x_n) \geqslant p(x)$ (resp. $\liminf_{\alpha} p(x_\alpha) \geqslant p(x)$).

Proposition 9.2.11: Each seminorm q on an l.c. TVS (X,T) corresponding to a norming set $\Gamma \subset X^*$ is consistent with Γ.

Proof: Straightforward. □

Proposition 9.2.12: Let p be an ω-consistent seminorm on an l.c. TVS (X,T) with respect to a set $\Gamma \subset X^*$. Let a sequence $\{x_n\}$ in X satisfy either of the two conditions: (a) $x_n \xrightarrow{\Gamma} 0$ and $\inf p(x_n) \geqslant \delta > 0$; or (b) $\alpha_n x_n \xrightarrow{\Gamma} 0$ for each α in ω. Then

$$\lim_{n\to\infty} \inf_{\alpha} p(x + \alpha x_n) = p(x), \quad \forall x \in X. \tag{9.2.13}$$

Proof: (a) Fix x in X and define α_n by $p(x + \alpha_n x_n) = \inf\{p(x + \alpha x_n) : \alpha \in \mathbb{K}\}$. Then $|\alpha_n| \leqslant 2p(x)/\delta$. Consequently $x + \alpha_n x_n \xrightarrow{\Gamma} x$. Thus

$$\begin{aligned} p(x) &= \lim_{n\to\infty} p(x + 0.x_n) \\ &\geqslant \overline{\lim_{n\to\infty}} \inf_{\alpha} p(x + \alpha x_n) \\ &\geqslant \underline{\lim_{n\to\infty}} \inf_{\alpha} p(x + \alpha x_n) \geqslant p(x) \end{aligned}$$

and this proves (9.2.13).

(b) Let $J = \{n \in \mathbb{N}: p(x_n) = 0\}$ and $I = \mathbb{N} \setminus J$. Then, for x in X,

(i) $\lim_{n \in J} \inf_{\alpha} p(x + \alpha x_n) = p(x)$.

Put $y_m = x_{n_m}/p(x_{n_m})$ for n_m in I. Then, from (a),

(ii) $\lim_{m\to\infty} \inf_{\alpha} p(x + \alpha y_m) = p(x)$.

By (i) and (ii), we again get (9.2.13). □

Proposition 9.2.14: Let Γ be a norming subset of X^* for an l.c. TVS (X,T) and let $\{q_\alpha : \alpha \in \Lambda\}$ be the family of seminorms relative to Γ. Consider a sequence $\{x_n\}$ in X satisfying (i) $x_n \xrightarrow{\Gamma} 0$ and (ii) $q_\alpha(x_n) \geqslant A > 0$ for all n and α. Then, for any $\varepsilon(0 < \varepsilon < 1)$, finite subset Φ of Λ and finite dimensional subspace Y of X, there exists N such that

$$q_i(x + \alpha x_n) > (1-\varepsilon)q_i(x), \quad \forall n \geqslant N$$

valid for all x in Y, α in $\mathbb{K}$ and i in Φ.

Proof: Let $S_y^i = \{x \in Y : q_i(x) = 1\}$ for i in Φ. Then, for each $\varepsilon > 0$, there exists an $\varepsilon/2$-net $\{y_j^i : 1 \leqslant j \leqslant n_i\}$ for S_y^i such that $q_i(y_j^i) = 1$ for $1 \leqslant j \leqslant n_i$, i in Φ; cf. Lemma 9.2.2. Therefore, by Propositions 9.2.11 and 9.2.12, there exists N so that

$$\inf_{\alpha} q_i(y_j^i + \alpha x_n) > (1 - \tfrac{\varepsilon}{2})q_i(y_j^i)$$

for all $n \geqslant N$ and uniformly in $i \in \Phi$ and $1 \leqslant j \leqslant n_i$.

Choose x in Y and i in Φ with $q_i(x) = 1$. There exists $y_j^i (1 \leqslant j \leqslant n_i)$ such that $q_i(x - y_j^i) < \varepsilon/2$, Thus, for α in $\mathbb{K}$,

$$q_i(x + \alpha x_n) > (1 - \tfrac{\varepsilon}{2})q_i(y_j^i) - \tfrac{\varepsilon}{2} = (1-\varepsilon), \quad \forall n \geqslant N.$$

Hence, for α in $\mathbb{K}$ and $n \geqslant N$,

$$q_i(x + \alpha x_n) > (1-\varepsilon)q_i(x), \tag{*}$$

for any x in Y and i in Φ with $q_i(x) \neq 0$. (*) is trivially satisfied if $q_i(x) = 0$. □

Proposition 9.2.15: Let $q_1 \leq q_2 \leq \ldots \leq q_n \leq \ldots$ denote the sequence of seminorms on an l.c. TVS (X,T), obtained corresponding to a norming set $\Gamma \subset X^*$. If $\{x_n\}$ is a sequence in X with $x_n \xrightarrow{\Gamma} 0$ and for some $A > 0$, $q_1(x_n) \geq A$, $n \geq 1$; then there exists a subsequence $\{x_{n_i}\}$ such that $\{x_{n_i}\}$ is S.basic in $(X,\{q_n\})$.

Proof: Choose $\{\varepsilon_m\}$ with $0 < \varepsilon_m < 1$ as in Case (b) of the proof of Theorem 9.2.1; cf. $(*)$. Put $n_1 = 1$ and $Y_1 = sp\{x_{n_1}\}$. By letting $\Lambda = \mathbb{N}$ and $\Phi = \{1\}$, we find from Proposition 9.2.14, the existence of an $n_2 > n_1$, $n_2 \equiv n_2(Y_1,\varepsilon_1,\{1$ such that

$$q_1(\alpha_1 x_{n_1} + \alpha_2 x_{n_2}) > (1-\varepsilon_1)q_1(\alpha_1 x_{n_1});\ \forall\ \alpha_1,\alpha_2 \in \mathbb{K}.$$

Proceeding inductively, we find a sequence $\{n_j\}$ with $n_1 < n_2 < \ldots < n_m < \ldots$ such that

$$q_i\left(\sum_{j=1}^{m} \alpha_j x_{n_j}\right) > (1-\varepsilon_{m-1})q_i\left(\sum_{j=1}^{m-1} \alpha_j x_{n_j}\right);\ \forall\ \alpha_j \in \mathbb{K},$$

where $1 \leq i \leq m-1$, $n_m \equiv n_m(Y_{m-1},\varepsilon_{m-1},\{1,\ldots,m-1\})$ and $Y_{m-1} = sp\{x_1,\ldots,x_{n_{m-1}}$ $m = 2,3,\ldots$.

We now follow the proof of Theorem 9.2.1, especially after (9.2.3), to get the required S.basic character of $\{x_{n_i}\}$ for $(X, \{q_n\})$. □

It is now possible to give the second main result of this chapter. Indeed, we have (cf. [107])

Theorem 9.2.16: Let (X,T) be a Fréchet space and Γ a norming subset of X^*. Suppose $\{x_n\}$ is a sequence in X satisfying one of the conditions (a) or (b), where (a) $x_n \xrightarrow{\Gamma} 0$ but $x_n \not\to 0$ in T, and (b) $\alpha_n x_n \xrightarrow{\Gamma} 0$ for every α in ω but $x_n \neq 0$ for infinitely many n. Then $\{x_n\}$ contains a subsequence which is S.basic in (X,T).

Proof: At the outset, we may assume that T is generated by a nondecreasing sequence $\{q_n\}$ of seminorms generated by Γ; that is, $D_T = \{q_n\}$.

(a) Since deletion of a finite number of seminorms from D_T does not alter the topology, and, if necessary, writing a subsequence of $\{x_n\}$ by itself, we may assume that $x_n \xrightarrow{\Gamma} 0$ and $q_1(x_n) \geq A > 0$ for all $n \geq 1$. Then the required result follows from Proposition 9.2.15.

(b) Here we confine to two cases: (i) when $N_i = \{n \in \mathbb{N}: q_i(x_n) \neq 0\}$ is finite for each $i \geq 1$; and (ii) when N_{i_o} is countably infinite for some i_o.

If (i) is true, then (X,T) cannot have a continuous norm and $\alpha_n x_n \to 0$ for every α in ω. Let $n_i = \max \{j : j \in N_i\}$. Then we may write $n_1 < n_2 < \ldots$ and $q_i(x_{n_j}) \neq 0$ for $j = i$ and $q_i(x_{n_j}) = 0$ for $j > i$. Put $y_i = x_{n_i}$. Since $\{\sum_{i=1}^{n} \alpha_i y_i\}$ is Cauchy for every α in ω and so converges in (X,T), we can define a map $\Psi: \omega \to [y_i]$ by

$$\Psi(\alpha) = \sum_{i \geq 1} \alpha_i y_i.$$

The map Ψ is easily seen to be 1-1 with $\Psi(e^i) = y_i$, $i \geq 1$. Also Ψ is continuous and onto; hence $(\omega, \sigma(\omega,\phi)) \simeq ([y_i], T|[y_i])$ and the result is proved in this case.

Concerning (ii), let $y_k = x_{n_k}/q_{i_o}(x_{n_k})$, where $\{n_k\}$ is the infinite sequence for which $q_{i_o}(x_{n_k}) \neq 0$. By the hypothesis, $y_k \xrightarrow{\Gamma} 0$ and $q_i(y_k) \geq 1$ for $i \geq i_o$ and so (ii) reduces to (a) and we are done again. □

<u>Remark</u>: If $\sigma(X,X^*)$- and T-sequential convergence in an l.c. TVS (X,T) are the same, e.g. (X,T) is a Montel or a Fréchet nuclear space, then condition (a) of Theorem 9.2.16 is never satisfied for $\Gamma = X^*$. To overcome this difficulty, Mil'man and Tumarkin ([240], p. 64) have proposed a remedial measure in the form of

<u>Theorem 9.2.17</u>: Let $\{x_n\}$ be a sequence in a Fréchet space (X,T). If, for each α in ω with $\alpha_n > 0$, $\{\alpha_n x_n\}$ contains no subsequence converging weakly to $x_o \neq 0$, then $\{x_n\}$ contains a S.basic subsequence.

10 Basic sequences in F-spaces

10.1 INTRODUCTION

As mentioned in Chapter 9, we now discuss in detail the existence and construction of S.basic sequences in F-spaces. This is comparatively arduous in comparison to what we have done in Chapter 9. Hereafter, we follow Kalton [115] who has largely succeeded in extending most of the results of the last chapter to the setting of F-spaces; cf. also [118] for a similar study of the existence of M-basic sequences in F-spaces. Besides, we will also discuss some applications of the existence theorem of basic sequences in F-spaces and briefly touch upon the topic on block extensions of sequences.

Throughout this chapter, we consider all linear topologies (1.topologies) as having been generated by F-seminorms.

10.2 COMPATIBILITY AND POLARITY

In the absence of the local convexity structure of F-spaces, the usual notion of compatibility of two l.c. topologies can be modified in the form of

Definition 10.2.1: Two 1.topologies T_1 and T_2 on X are said to be *compatible* if they define the same closed subspaces of X.

Proposition 10.2.2: Let T_1 and T_2 be two compatible 1.topologies on X. An f in X' is T_1-continuous if and only if f is T_2-continuous.

Proof: Routine; e.g. see [140], p. 37. □

The concepts of norming sets and consistency of seminorms introduced in Chapter 9 play a useful role in extracting S.basic sequences from certain well behaved sequences. In place of these two notions, Kalton [115] has given the following definition and obtained the corresponding results.

Definition 10.2.3: Let a vector space X be endowed with two 1.topologies T_1 and T_2. T_2 is said to be T_1-*polar* if there exists a fundamental neighbourhood system for T_2, say, $\mathcal{B}_2 \subset \mathcal{B}_{T_2}$ such that $\mathcal{B}_2$ consists of T_1-closed sets.

The proof of the following result is omitted; one may see [28], p. 62; cf. also [75], [79] for further detail.

Proposition 10.2.4: Let X be equipped with two 1. topologies T_1 and T_2. Then T_2 is T_1-polar if and only if there exists a family D_2 of T_1-lower semicontinuous F-seminorms generating T_2.

Proposition 10.2.5: Let a vector space X be equipped with two 1. topologies T_1 and T_2. (i) If T_2 is T_1-polar, then T_2 is generated by a collection $\{q_\alpha : \alpha \in A\}$ of F-seminorms with $q_\alpha(x) = \sup\{p(x) : p \in \Lambda_\alpha\}$, where each Λ_α is a collection of T_1-continuous F-seminorms; further, if T_2 is metrizable, then T_2 is given by a single such F-norm. (ii) Let (X,T_2) be an F-space, $T_1 \subsetneqq T_2$ and $\{x_\alpha\}$ a net in X with $x_\alpha \to 0$ in T_1 and $x_\alpha \not\to 0$ in T_2; then there exist 1. topologies T_3 and T_4 on X such that (a) $T_1 \subset T_3 \subsetneqq T_4 \subset T_2$, (b) T_4 is metrizable and T_3-polar and (c) $x_\alpha \to 0$ in T_3 but $x_\alpha \not\to 0$ in T_4. (iii) Let T_1 and T_2 be the same as in (ii). If $v \in \mathcal{B}_{T_2}$, $\notin \mathcal{B}_{T_1}$, then there are 1. topologies T_3 and T_4 satisfying (a), (b) and (d) $v \in \mathcal{B}_{T_4}$, $\notin \mathcal{B}_{T_3}$.

Proof: (i) Let $D \equiv D_{T_1}$ and $\Gamma = \{r_\alpha : \alpha \in \Lambda\} \subset D_{T_2}$ be such that Γ generates T_2 and contains maximum of its every finite subfamily. For α in Λ and δ in D, define

$$p_\delta^\alpha(x) = \inf\{\delta(y) + r_\alpha(z) : x = y+z\};\ q_\alpha(x) = \sup\{p_\delta^\alpha(x) : \delta \in D\}.$$

Clearly each p_δ^α is a T_1- and T_2-continuous F-seminorm on X; indeed, $p_\delta^\alpha \prec \delta$, r_α for each δ in D. Also, $q_\alpha \prec r_\alpha$ for each α in Λ and so each q_α is a T_2-continuous F-seminorm on X. Therefore, if T is the 1.topology generated by the family $\{q_\alpha : \alpha \in \Lambda\}$, then $T \subset T_2$. For the other inclusion, consider $u \in \mathcal{B}_{T_2}$. Then there exist $\varepsilon > 0$ and $\alpha \in \Lambda$ such that $\overline{\{x : r_\alpha(x) < \varepsilon\}}^{T_1} \subset u$. For this choice of α and ε, one can easily show that $\{x : q_\alpha(x) < \varepsilon\} \subset u$. Hence $T \approx T_2$.

For the last statement, Γ is given by a single F-norm and this disposes of (i).

(ii) Let T_3 be the strongest l. topology on X with $T_1 \subset T_3 \subset T_2$ and $x_\alpha \to 0$ in T_3. Let T_4 be the l. topology with a 0-neighbourhood system consisting of T_3-closures of the members of $\mathcal{B}_{T_2}$. Using the closed graph theorem (cf. [139], p. 213), we infer $T_3 \subsetneq T_4$; for otherwise the identity map from (X,T_3) to (X,T_2) is almost continuous and so $T_3 = T_4$. The other parts are obvious.

(iii) Applying Zorn's lemma, we find a maximal l. topology T_3 on X such that $T_1 \subset T_3 \subset T_2$ and $v \notin \mathcal{B}_{T_3}$. Now proceed as in (ii) above. □

10.3 EXTRACTION OF BASIC SEQUENCES

The main result of this section depends upon

Lemma 10.3.1: Let B be a balanced closed subset of a finite dimensional space X. If $B \cap Y$ is bounded in X for every 1-dimensional subspace Y of X, then B is bounded in X.

Proof: We may assume that the topology of X is given by a norm $\|\cdot\|$. Suppose B is not bounded. Hence there exists a sequence $\{z_n\} \subset B$ with $\|z_n\| \geq n$, $n \geq 1$ and consequently a subsequence $\{y_n\}$ of $\{z_n/\|z_n\|\}$ such that $y_n \to y$, where $\|y\| = 1$. For N is $\mathbb{N}$ observe that $\|z_n\|^{-1} z_n \in N^{-1}B$ for all $n \geq N$ and so $\{y_n\}$ is eventually contained in $N^{-1}B$. Hence $y \in N^{-1}B$ for all N in $\mathbb{N}$. Consequently sp $\{y\} \subset B$, a contradiction. □

Theorem 10.3.2: Let (X,T_2) be a metrizable TVS and T_1 an l.topology on X such that T_2 is T_1-polar. Assume the existence of a net $\{x_\alpha\}$ in X such that $x_\alpha \to 0$ in T_1 and $x_\alpha \not\to 0$ in T_2. Further, let $z_1 \neq 0$, $z_1 \in X$. Then there exists a sequence $\{\alpha_k : k \geq 2\}$ so that $\alpha_{k+1} > \alpha_k$, $k \geq 2$ and $\{z_n\}$ is S. basic in (X,T_2), where $z_n = x_{\alpha_n}$, $n \geq 2$.

Proof: At the outset, let us observe that T_2 is given by an F-norm $\|\cdot\|$ with $\|x\| = \sup\{p(x) : p \in \Lambda\}$, where Λ generates T_1; cf. Proposition 10.2.5(i). Next, there exists $\theta > 0$ so that $\|z_1\| \geq 4\theta$, and, for every index β, there exists $\alpha \geq \beta$ satisfying $\|x_\alpha\| \geq 4\theta$. Put $B = \{x \in X : \|x\| \leq \theta\}$. Clearly, $B \cap$ sp $\{z_1\}$ is sequentially compact and so compact in (X,T_2).

Now there are two major steps in the proof: (i) construction of $\{\alpha_n\}$ as required such that $B \cap X_n$ is compact in (X,T_2), where $X_n = \text{sp}\,\{z_1, x_{\alpha_2}, \ldots, x_{\alpha_n}\}$

and (ii) proving the S.basic character of $\{z_1; x_{\alpha_n} : n \geq 2\}$ for (X, T_2) by using Theorem 5.1.2.

(i) Suppose that we have already constructed $\{\alpha_2, \ldots, \alpha_n\}$ having the required property.

For n,k in $\mathbb{N}$ with $1 \leq k \leq 2^{n+3}$, let $W_k^n = \{x \in X: \|x\| = k\theta/2^{n+3}\}$. Then each W_k^n has $\theta/2^{n+3}$-net $A(n;k) \subset W_k^n$ and let A_n be the union of A(n;k) for $1 \leq k \leq 2^{n+3}$. For each y in A_n, we can find p_y in Λ so that $p_y(y) \geq \|y\| - \theta/2^{n+3}$. Since A_n is finite, there exists $\beta > \alpha_n$ such that

$$p_y(x_\gamma) \leq 2^{-(n+3)}\theta, \ \forall \gamma \geq \beta, \ y \in A_n. \qquad (*)$$

Let $\{x_a : a \in \Sigma\}$ be a subnet of the net $\{x_\gamma : \gamma \geq \beta\}$ such that $\|x_a\| \geq 4\theta$ for each a in Σ.

Suppose $B \cap Y_a^n$ is T_2-unbounded for each a in Σ, where $Y_a^n = \text{sp}\,\{z_1, x_{\alpha_2}, \ldots, x_{\alpha_n}, x_a\}$. Hence, by Lemma 10.3.1, for each a in Σ there exists $y_a^n + t_a x_a \neq 0$ with y_a^n in X_n such that $\text{sp}\,\{y_a^n + t_a x_a\} \subset B_a^n = B \cap Y_a^n$. We may choose y_a^n so that $\|y_a^n\| \leq \theta$ and as $y_a^n \neq 0$, $\|y_a^n\| > 0$. It is clear that $|t_a| \leq 1$ and therefore $t_a x_a \to 0$ in T_1. Since $B \cap X_n$ is compact, there exists a subnet $\{u_a^n\}$ of $\{y_a^n\}$ with $u_a^n \to u^n$ in $B \cap X_n$ relative to the topology T_2. Consequently, for any t in $\mathbb{K}$ and p in Λ, we conclude that $p(t\,u^n) \leq \underline{\lim}_a \|t(u_a^n + t_a x_a)\| \leq \theta$. Thus $\text{sp}\,\{u^n\} \subset B \cap X_n$, a contradiction.

Hence we may choose $\alpha_{n+1} \geq \beta > \alpha_n$ such that $\|x_{\alpha_{n+1}}\| \geq 4\theta$ and $B \cap X_{n+1}$ is T_2-compact. This completes the required step (i).

(ii) To begin with, let $z_n = x_{\alpha_n}$, $n \geq 2$ and introduce an F-norm $\|\cdot\|^*$ on X equivalent to $\|\cdot\|$ by $\|x\|^* = \min(\|x\|, \theta)$. Our next step is to show that if $a_1, \ldots, a_{n+1}$ are any (n+1) scalars, then

$$\left\| \sum_{i=1}^{n+1} a_i z_i \right\|^* \geq \left\| \sum_{i=1}^{n} a_i z_i \right\|^* - 2^{-(n+1)}\theta. \qquad (10.3.3)$$

Assuming temporarily the truth of (10.3.3), it follows that $\{z_n\}$ is an l.i. sequence; indeed, if $\sum_{i=1}^{n+1} a_i z_i = 0$, then $\|s \sum_{i=1}^{n} a_i z_i\| \leq \theta$ for each s in $\mathbb{K}$, implying $\sum_{i=1}^{n} a_i z_i = 0$ and so $a_{n+1} = 0$; proceed similarly to get $a_1 = a_2 = \ldots = a_n = 0$. Let $Y = \text{sp}\,\{z_n\} = \cup\,\{X_n : n \geq 1\}$. Introduce the maps $P_{nm} : X_m \to X_n$ and $P_n : Y \to X_n$ with $P_n(x) = P_{nm}(x)$ as in the proof of Theorem 5.1.2. Then P_{nm} is clearly continuous in T_2 for $m \leq n$. Let $n < m$, $x_i \to 0$ in X_m and $P_{nm}(x_i) \not\to 0$

in $(X_n, T_2|X_n)$. Hence there exist $\varepsilon > 0$ and a subsequence of $\{P_{nm}(x_i)\}$ which we denote by itself such that $\|P_{nm}(x_i)\| > \varepsilon$ for $i \geq 1$. Also we may take $p_{nm}(x_i) \in B \cap X_n$ for all large i; therefore there exist z in $B \cap X_n$ and a subsequence of $\{P_{nm}(x_i)\}$ which we again denote by itself so that $P_{nm}(x_i) \to z$ in T_2. Observe that $\|z\| \geq \varepsilon > 0$. Choose k with $\|kz\| - \varepsilon \geq \theta$. Then, for all large $i \geq i_0$ (say), $\|kP_{nm}(x_i)\| \geq \theta$. But a repeated application of (3.3) yields

$$\|P_{nm}(x)\|^* \leq \|x\|^* + \theta \sum_{i=n+1}^{m} 2^{-i}, \; x \in X_m, \tag{10.3.4}$$

Therefore $\|kx_i\| \geq \theta/2$ for $i \geq i_0$ and this contradicts the choice of $\{x_i\}$. Consequently each P_{nm} is continuous and so is each P_n. By applying (10.3.4), we also conclude the equicontinuity of $\{P_n\}$ on Y and this yields an inequality of the type (5.1.3) for the sequence $\{z_n\}$, giving thereby the S.b. character of $\{z_n\}$ for $[z_n]^{T_2}$; alternatively, we may follow the arguments as given in the proof of Theorem 5.1.2 to infer the S.basic character of $\{z_n\}$ for (X, T_2).

Finally, it remains to prove (10.3.3). Choose largest k in $\mathbb{N}_0 = \mathbb{N} \cup \{0\}$ so that

$$\|\sum_{i=1}^{n} a_i z_i\|^* \geq k.2^{-(n+3)}\theta. \tag{+}$$

From the definition of $\|\cdot\|^*$, $0 \leq k \leq 2^{n+3}$. If $k = 0$, (10.3.3) easily follows from (+). So, fix $k \geq 1$ satisfying (+) and choose s with $0 \leq |s| \leq 1$ such that

$$\|s \sum_{i=1}^{n} a_i z_i\| = k.2^{-(n+3)}\theta.$$

Recall now the second paragraph of (i). Then we find y in A(n;k) such that

$$\|y - s \sum_{i=1}^{n} a_i z_i\| \leq 2^{-(n+3)}\theta,$$

and $p_y(z_{n+1}) \leq 2^{-(n+3)}\theta$; cf. (*). Considering two cases: (i) $|sa_{n+1}| \geq 1$ and (ii) $|sa_{n+1}| \leq 1$ and accordingly the inequalities:

$$\|y + sa_{n+1}z_{n+1}\| \geq \|z_{n+1}\| - \|y\|, \; p_y(y) - p_y(z_{n+1}),$$

we conclude that $\|y + sa_{n+1}z_{n+1}\| \geq (k-2)\theta/2^{n+3}$. Therefore

$$\|s \sum_{i=1}^{n+1} a_i z_i\| \geq (k+1)2^{-(n+3)}\theta - 2^2 . 2^{-(n+3)}\theta$$

$$\geq \|\sum_{i=1}^{n} a_i z_i\|^* - 2^{-(n+1)}\theta,$$

and since the right-hand side is at most equal to θ, we get (10.3.3). □

Exercise 10.3.5: Let (X,T_2) be an F-space and T_1 an l.topology on X with $T_1 \subset T_2$. If there is a net $\{x_\alpha\}$ in X such that $x_\alpha \to 0$ in T_1, $x_\alpha \not\to 0$ in T_2 and $0 \neq z_1 \in X$, then show that there exists a sequence $\{\alpha_k : k \geq 2\}$, $\alpha_{k+1} > \alpha_k$ so that $\{z_n\}$, $z_n = x_{\alpha_n}$, $n \geq 2$, is semibasic in (X,T_2).

10.4 THE EXISTENCE THEOREM

The existence theorem of basic sequences in every Fréchet space differs slightly from the corresponding situation in F-spaces. Indeed, the class of F-spaces where the existence theorem appears not to be valid is very small and is introduced in

Definition 10.4.1: A TVS (X,T) is called *minimal* if, there is no strictly weaker (Hausdorff) l.topology on X.

There are plenty of spaces which are non-minimal. On the other hand, when the l.c. topology is taken into consideration, there appears to be the only one known example of a minimal space contained in

Example 10.4.2: We prove that $(\omega,\sigma(\omega,\phi))$ is minimal. If this were not so, there would exist a net $\{x^\alpha\}$ in ω and l.topologies T, T_1 and T_2 by Proposition 10.2.5(ii) so that $T \subset T_1 \subsetneq T_2 \subset \sigma$, where T_2 is given by an F-norm $\|\cdot\|$ and is T_1-polar, $x^\alpha \to 0$ in T_1 but $x^\alpha \not\to 0$ in T_2. By Theorem 10.3.2, we can select a sequence $\{z^n\}$ from $\{x^\alpha\}$ such that z^n is an S.b. for $X_1 = [z^n]^{T_2}$ and $\|z^n\| \geq \theta > 0$. Let $\{h^n\}$ be its s.a.c.f. Put $X = [z^n]^\sigma$; then $(\omega,\sigma) \simeq (X,\sigma)$, say, under R. Since $\{h^n \circ R\}$ is total on ω, $\sigma(\omega,\phi) \approx \sigma(\omega, \{h^n \circ R\})$; cf. [235], p. 198 or use the Hahn-Banach theorem and a result in [234], p. 19. Observe that $h^i(z^n) \to 0$ for each $i \geq 1$ and hence $z^n \to 0$ in T_2, a contradiction.

Theorem 10.4.3: Every non-minimal F-space (X,T_2) contains an S.basic sequence.

Proof: Let T_1 be an l.topology on X such that $T_1 \subsetneq T_2$ and choose a fundamental T_2-neighbourhood system $\{v_n\}$ at the origin so that $v_1 \notin \mathcal{B}_{T_1}$. We may find l.topologies T_3 and T_4 on X, satisfying (a), (b) and (d) of Proposition 10.2.5.

By Theorem 10.3.2, we can find a sequence $\{y_k^1\}$ in X such that it is a T_4-S.b. for E_1, the T_2-closure of $F_1 = \text{sp}\,\{y_k^1\}$. Put $S_1 = T_4$. By the induction process, we construct sequences $\{y_k^n\}$, $\{E_n\}$, $\{F_n\}$ and $\{S_n\}$ such that E_n is the T_2-closure of $F_n = \text{sp}\,\{y_k^n : k \geq 1\}$, S_n is a metrizable l.topology on E_n and $\{y_k^n\}$ is an S.b. for (E_n, S_n); moreover, (i) $\{y_k^n\}$ is a block sequence relative to $\{y_k^{n-1}\}$, thus $F_n \subset F_{n-1} : E_n \subset E_{n-1}$, $n \geq 2$; (ii) $S_{n-1} | E_n \subset S_n \subset T_2 | E_n$, $n \geq 2$ and (iii) $v_n \cap E_n$ is a member of $\mathcal{B}_{S_n}$.

For n = 1, the above construction is obviously true.

If $v_{n+1} \cap E_n \in \mathcal{B}_{S_n}$, we let $S_{n+1} = S_n$ and $y_k^{n+1} = y_k^n$, $k \geq 1$; otherwise, we find l.topologies T and S_{n+1} on E_n such that $S_n \subset T \subsetneq S_{n+1} \subset T_2 | E_n$, S_{n+1} is T-polar and metrizable with $v_{n+1} \cap E_n \in \mathcal{B}_{S_{n+1}}$, $\notin \mathcal{B}_T$. Since $\bar{F}_n^{T_2} = E_n$, $E_n = \bar{F}_n^{S_{n+1}}$ and hence $T|F_n \subsetneq S_{n+1}|F_n$. Following the proof of Theorem 10.3.2, we find a sequence $\{z_k\}$ in F_n such that $\{z_k\}$ is S_{n+1}-regular, S.basic in (F_n, S_{n+1}) and $z_k \to 0$ in T.

Note that $z_k = \sum_{i=1}^{m(k)} \alpha_i^k y_i^n$, where $\alpha_j^k \to 0$ for each $j \geq 1$. Write $\|\cdot\|_{n+1}$ for the F-norm on F_n induced by S_{n+1}. Then we get sequences $1 = n_1 < n_2 < \ldots < n_j < \ldots$ and $m(n_1) < m(n_2) < \ldots < m(n_j) < \ldots$ such that

$$\Big\| \sum_{i=1}^{m(n_j - 1)} \alpha_i^{n_j} y_i^n \Big\|_{n+1} < \frac{1}{2^{j+1}}, \; j \geq 2; \; m(n_o) = 0.$$

If we put $y_j = z_{n_j}$ and

$$y_j^{n+1} = \sum_{i=m(n_{j-1})+1}^{m(n_j)} \alpha_i^{n_j} y_i^n, \; j \geq 1,$$

then $\sum_{j \geq 1} \|y_j - y_j^{n+1}\|_{n+1} < \infty$. If necessary, we consider the completion of (F_n, S_{n+1}) to infer the e-S.b. character (Definition 6.2.11) of $\{z_k\}$ and hence, from Proposition 6.2.12, we conclude that $\{y_j^{n+1} : j \geq 1\}$ is an S.basic sequence in (F_n, S_{n+1}). In order to complete the induction process, let us now put $F_{n+1} = \text{sp}\,\{y_k^{n+1}\}$ and $E_{n+1} = \bar{F}_{n+1}^{T_2}$.

Finally, we pass on to the construction of a basic sequence $\{x_n\}$ in (X,T_2). Put $x_n = y_n^n$, $n \geq 1$. It follows from what has preceded above that $\{x_n\}$ is S.basic in (E_1,S_1) and is a block sequence corresponding to $\{y_1^1, y_2^2, \ldots\}$. Denote by $\{f_n\}$ the S_1-continuous s.a.c.f. corresponding to $\{x_n\}$. Suppose $H = [x_n]^{T_2}$ and g_n denotes the unique T_2-continuous extension of f_n from sp $\{x_n\}$ to H.

Let $x \in H$ and for $n \geq 1$, put $R_n(x) = x - \sum_{i=1}^{n-1} g_i(x)x_i$. Since

$$[x_k : k \geq n]^{T_2} = \bigcap_{i=1}^{n-1} g_i^{-1}(0),$$

$R_n(x) = S_n - \lim_{m\to\infty} \sum_{i=n}^{m} g_i(x)x_i$. Hence there exists N such that

$$R_n(x) - \sum_{i=n}^{m} g_i(x)\, x_i \in V_n \cap E_n, \ \forall\, m \geq N,$$

and this proves the S.b. character of $\{x_n; g_n\}$ for $(H, T_2|H)$. □

10.5 APPLICATIONS

Thanks to the development of basic sequences in the general setting of TVS, it is now possible to answer, positively and negatively respectively, the analogues of the well known Mackey and Hahn-Banach theorems for a class of non-locally convex TVS. We essentially follow [115] for all the results of this section. We begin with

Proposition 10.5.1: Let a vector space X be equipped with two compatible l. topologies T_1 and T_2 with $T_1 \subset T_2$. Then T_1- and T_2-bounded sets in X are the same provided that (i) (X,T_2) is an F-space, or (ii) T_2 is T_1-polar.

Proof: (i) Let $A \subset X$ be T_1-bounded but T_2-unbounded. Then there exist sequences $\{x_n\}$ in A and a null-sequence $\{\varepsilon_n\}$ in $\mathbb{K}$ with $x_1, \varepsilon_n \neq 0$, $n \geq 1$ such that $y_n \to 0$ in T_1 but $y_n \not\to 0$ in T_2, where $y_n = \sqrt{\varepsilon_n}\,(\sqrt{\varepsilon_n}\, x_n + x_1)$. By Exercise 10.3.5, we can find a sequence $\{z_n\}$ which is semibasic in (X,T_2), where $z_1 = x_1$, $z_n = y_{m_n}$, $n \geq 2$. Thus $\varepsilon_{m_n}^{-1/2} z_n \to x_1$ in T_1, giving $x_1 \in [z_n : n \geq 2]^{T_2}$, and this contradicts the T_2-semibasic nature of $\{z_n\}$.

(ii) If T_2 is metrizable, the result follows as in (i), replacing therein the use of Exercise 10.3.5 by Theorem 10.3.2.

For the general case, assume that the result is not true and so, as in (i), we get $\{x_n\}$ in A, $\{\varepsilon_n\}$ in $\mathbb{K}$ and q in D_{T_2} so that $\varepsilon_n x_n \to 0$ in T_1 and $q(\varepsilon_n x_n) \not\to 0$. On account of Proposition 10.2.4, we can choose q to be T_1-lower semicontinuous. Since $q^{-1}(0)$ is T_1-closed, the quotient space $\hat{X} = X/q^{-1}(0)$ has two topologies $\hat{T}_1$ and $\hat{T}$ (generated by the quotient F-norm $\hat{q}$) such that $\hat{T}$ is $\hat{T}_1$-polar. Further, each $\hat{q}$-closed subspace of $\hat{X}$ is $\hat{T}_1$-closed (cf. [139], p. 94). Now observe that $\varepsilon_n \hat{x}_n \to 0$ in $\hat{T}_1$ while $\varepsilon_n \hat{x}_n \not\to 0$ in $\hat{T}$ and proceed as in the first paragraph. □

Let us take up the second problem of this section and introduce

Definition 10.5.2: A TVS (X,T) is said to have the *Hahn-Banach extension property* (HBEP) if every continuous linear functional on each closed subspace of X has a continuous linear extension to X.

The following two examples justify this definition.

Example 10.5.3: The space $L^p[0,1]$ with its usual F-norm has a trivial dual and so it does not possess the HBEP.

Example 10.5.4: Recall the space H^p, $0 < p < 1$, given in Example 6.2.6. Although this space has enough continuous linear functions ([52], p. 118), it fails to have the HBEP ([52], p. 123).

The next theorem of Kalton [115] solves a conjecture of Shapiro [211]; proposed earlier in [53], p. 59 in a different form.

Theorem 10.5.3: Every non-locally convex F-space (X,T) is devoid of the HBEP.

Remark: A weaker form of Theorem 10.5.3 was earlier proved by Shapiro [209], p. 18, who in fact established this result with an additional assumption of the presence of an S.b. in (X,T); in this case, one avoids the use of Theorem 10.3.2 and hence the proof is simpler. In any case, the next proposition is crucial in the proof of Theorem 10.5.3 or its weaker form.

Proposition 10.5.4: Let $\{y_n;g_n\}$ be a regular S.basic sequence in an F-space

(X,T) such that $y_n \to 0$ in $\tau \equiv \tau(X,X^*)$. If $Y = [y_n]^T$, then there exists an f in $Y^* \equiv (Y,T|Y)^*$ such that f cannot be extended to (X,T).

Proof: At the outset, let us observe that $g_n(y) \to 0$ for each y in Y. Recall the system $\{v_n\}$ associated with τ from Proposition 1.2.15 and let p_n be the Minkowski functional of v_n. For $\varepsilon > 0$, choose $\{y_{n_k}\}$ from $\{y_n\}$ so that $p_k(y_{n_k}) \leq 1/k^{\varepsilon+1}$, $k \geq 1$. Let $0 < \varepsilon_1 < \varepsilon$ and define α in ℓ^1 with $\alpha_{n_k} = 1/k^{\varepsilon_1+1}$, $k \geq 1$ and zero otherwise. By the Banach-Steinhaus theorem, the functional f defined by $f(y) = \sum_{n \geq 1} \alpha_n g_n(y)$, is a member of Y^*. The functional f defined in this way cannot be extended to (X,T), for if $\hat{f}$ is the assumed extension, then $\hat{f}$ is τ-continuous and therefore for some m in $\mathbb{N}$ and $A > 0$, $|\alpha_{n_k}| = |\hat{f}(y_{n_k})| \leq A\, p_k(y_{n_k})$, for each $k \geq m$. However, this gives a contradiction. □

Proof of Theorem 10.5.3: By Proposition 1.2.15, there exists $\{x_n\}$ in X with $x_n \to 0$ in τ but $x_n \not\to 0$ in T. Following the proof of Theorem 10.3.2, we can extract a subsequence $\{z_k\}$ from $\{x_n\}$ such that $\{z_k\}$ is regular, S.basic in (X,T) and $z_k \to 0$ in τ. It now remains to apply Proposition 10.5.4 to get the required result. □

The next result presents a partial solution to Problem 6.1.1.

Proposition 10.5.5: Let a vector space X be equipped with compatible l. topologies T_1 and T_2 such that $T_1 \subset T_2$ and (X,T_2) is an F-space. If $\{x_n;f_n\}$ is a t.b. for (X,T_1), then it is an S.b. for (X,T_2).

Proof: If T_3 is the topology generated by the T_1-closures of u in $\mathcal{B}_{T_2}$, then $T_3 \approx T_2$ (cf. Proposition 5.1 (i) and [146], p. 168). Hence T_2 is T_1-polar and so it is defined by a T_1-lower semicontinuous T_2-continuous F-norm p. Since bounded sets in (X,T_1) and (X,p) are the same, we may define a F-norm $\bar{p}$ exactly as in Theorem 2.2.5 and, using the T_1-lower semicontinuity of p, we conclude that $p(x) \leq \bar{p}(x)$, for x in X. But $(X,\bar{p})$ is complete and hence by the closed graph theorem, $p \approx \bar{p}$. Thus $\{S_n\}$ is p-p equicontinuous and this shows that $\{x_n\}$ is S.basic in (X,T_2). As $[x_n]^{T_2} = [x_n]^{T_1} = X$, we find that $\{x_n;f_n\}$ is an S.b. for (X,T_2). □

The final result of this section is an extension of a well known theorem of Eberlein-Šmulian on the characterization of compact sets in Banach spaces ([146], [180], [192] and [235]). Indeed, we have (cf. [115])

<u>Theorem 10.5.6</u>: Let a vector space X be equipped with compatible l.topologies T_1 and T_2 such that (X,T_2) is an F-space and $T_1 \subset T_2$. Then for a subset M of X, the following statements are equivalent:

(i) M is T_1-compact.

(ii) M is T_1-sequentially compact.

(iii) M is T_1-countably compact.

<u>Proof</u>: (i) ⇒ (iii) and (ii) ⇒ (iii) are known, cf. [146]. Before we prove other implications, let us observe that T_2 is T_1-polar and is given by a T_1-lower semicontinuous F-norm p.

(iii) ⇒ (i) Here it is enough to show that M is T_1-precompact and complete, the former being easily proved by contradiction and (iii).

To proceed further to prove completeness, let $(\hat{X},\hat{T}_1)$ be the completion of (X,T_1), $Y = \{x \in \hat{X}$: there exists a T_1-bounded net $\{y_\delta\}$ in X with $y_\delta \to x$ in $\hat{T}_1\}$, and $\hat{B}_k$ denote the $\hat{T}_1$-closure of $B_k = \{x \in X: p(x) \leq k\}$. Then for y in Y, the function $p^*(y) = \inf \{k > 0: y \in \hat{B}_k\}$ is well defined (cf. Proposition 10.5.1(i)). Since for x in Y and $t_n \to 0$, $\lim p^*(t_n x) \leq \lim \sup_\delta p(t_n y_\delta) = 0$, where $\{y_\delta\}$ is T_1-bounded net converging to x (cf. Remark 2.2.7(a)); and also if $p^*(x) = 0$, there is a net $\{x_{\varepsilon,u}\}$ with $p(x_{\varepsilon,u}) \leq \varepsilon$, $x_{\varepsilon,u} - x \in u$, $\varepsilon > 0$, $u \in \mathcal{B}_{\hat{T}_1}$, yielding thereby $x = 0$, we infer that p^* is an F-norm on Y. Moreover, it is easily seen that p^* is $\hat{T}_1$-lower semicontinuous.

Let us now consider a T_1-Cauchy net $\{x_\alpha\}$ in M, which is T_1-bounded and so converges to some y in Y relative to T_1. We deal with two cases: (a) $p^*(x_\alpha - y) \to 0$ and (b) $p^*(x_\alpha - y) \not\to 0$.

(a) Since $p(x) = p^*(x)$ for x in X and X is T_2-complete, $y \in X$. Therefore we can select a sequence $\{x_{\alpha_n}\}$ from $\{x_\alpha\}$ such that $x_{\alpha_n} \to y$ in T_1. Hence $y \in M$ by (iii).

(b) First, let $y \notin X$. Since $y \neq 0$, we may suppose that $x_\alpha \notin v$ for all α, for some v in $\mathcal{B}_{T_1}$. If T^* denotes the topology on Y, generated by p^*, then T^* is $\hat{T}_1|Y$-polar. Hence by Theorem 10.3.2, we find a sequence $\{z_n\}$ in Y with $z_1 = y$, $z_n = w_n - y$, $w_n = x_{\alpha_n}$, $\alpha_{n+1} > \alpha_n$, $n \geq 2$ such that $\{z_n\}$ is S.basic in (Y,T^*) and $\inf p^*(z_n) > 0$. Let $Y_1 = [z_n]$ and $Y_2 = [w_n : n \geq 2]$, where the closures are considered with respect to T^*. Since Y_2 is a maximal closed subspace of (Y_1,T^*), we find an f in $(Y_1,T^*)^*$ with $f(z_1) = 1$ and $f(w) = 0$ for w in Y_2; cf. [146], p. 156. Define $F:Y_1 \to Y_1$ by $F(z) = z - f(z)z_1$ and $H:Y_1 \to Y_1$ by $H(\sum_{n\geq 1} t_n z_n) = \sum_{n\geq 2} t_n z_n$. Then $HF(z_n) = z_n$, $n \geq 2$ and, by the usual extension theorem, $[z_n:n \geq 2] \simeq Y_2$ under F with $F(z_n) = w_n$, $n \geq 2$. In particular, $\{w_n:n \geq 2\}$ is S.basic in (X,T_2) and denote by $\{f_n:n \geq 2\}$ the s.a.c.f. corresponding to $\{w_n:n \geq 2\}$. Since $w_n \in M$, $n \geq 2$, it has a T_1-adherent point $w_o \in Y_2$ by the compatibility of T_1 and T_2. Therefore, $w_o = \sum_{n\geq 2} f_n(w_o)w_n$ and as each f_n is T_1-continuous, the T_1-adherence character of w_o yields $w_o = 0$. Consequently v contains $\{x_\alpha\}$ frequently, a contradiction.

Thus $y \in X$ and $p^*(x_\alpha - y) \not\to 0$. Observe that $x_\alpha - y \to 0$ in T_1 and $x_\alpha - y \not\to 0$ in T_2. Hence there exists an S.basic sequence $\{z_n\}$ in (X,T_2) with $z_1 \neq 0$, $z_n = w_n - y$, $w_n = x_{\alpha_n}$, $\alpha_n > \alpha_{n+1}$, $n \geq 2$ and $\inf p(z_n) > 0$. Starting with w_o as the T_1-adherent point of $\{w_n:n \geq 2\} \subset M$, we find that $w_o - y$ lies in T_2-closure of $\{sp\ z_n:n \geq 2\}$. Hence $w_o - y = 0$ and this gives $x_\alpha \to y$ in (M,T_1).

(iii) ⇒ (ii) If $\{x_n\}$ is any sequence in M, then there exist a subnet $\{y_\alpha\}$ of $\{x_n\}$ and an x_o in M with $y_\alpha \to x_o$ in T_1. If $y_\alpha \to x_o$ in T_2, then there is nothing to prove by the metrizability of T_2. On the other hand, if $y_\alpha - x_o \not\to 0$ in T_2, then on the basis of getting an S.basic sequence $\{z_n\}$, $z_n = y_{\alpha_n} - x_o$, $n \geq 2$ and arguing as in (iii) ⇒ (i) above, we find that x_o is the only T_1-adherent point of a subsequence $\{y_{\alpha_n}\}$ of $\{x_n\}$; that is, $y_{\alpha_n} \to x_o$ in T_1. □

10.6 BLOCK EXTENSIONS

Toward the end of this chapter, we treat the following natural

Problem 10.6.1: Does there exist an S.b. $\{x_n\}$ in a TVS X with an S.b. such that $x_{n_k} = y_k$, where $\{y_k\}$ is an S.basic sequence in X?

The above problem was raised for Banach spaces in [182] and was subsequently negated for the space $L^p[0,1]$; $p = 1$, $2 < p < \infty$ in [183]; for similar results, see [220] as well.

On the other hand, a complete solution of this problem in Banach spaces was provided in [238] for a specific class of S.basic sequences and, in the process of extending these results to Fréchet spaces, Robinson [194], whom we follow hereafter in this section also showed that the same is not true for all nuclear Fréchet spaces. Indeed, he established a characterization for extending Schauder block sequences (S.bl.s.) in a Fréchet space to S.b. for the entire space. It was also shown by him that all S.bl.s. in ω can be extended to S.b. for ω.

Definition 10.6.2: Let $\{x_n\}$ be an S.b. for a TVS (X,T) and $\{y_n\}$ a bl.s. corresponding to $\{m_n\}$; cf. Definition 9.1.8. An S.*block extension* (S.bl.e.) of $\{y_n\}$ is an S.b. $\{z_n\}$ such that (i) $z_{m_n} = y_n$, $n \geq 1$ and (ii) $z_i \in X_n$ for $m_{n-1}+1 \leq i \leq m_n$, $n \geq 1$, where $X_n = [x_{m_{n-1}+1},\ldots,x_{m_n}]$ is called the n-th *block space* corresponding to $\{x_n\}$ and determined by $\{m_n\}$.

PROPOSITION 10.6.3: Let (X,T) be a barrelled space containing an S.b. $\{x_n\}$. Suppose $\{z_n\}$ is a sequence in X such that $X_n = [z_{m_{n-1}+1},\ldots,z_{m_n}]$, $n \geq 1$. Then $\{z_n\}$ is an S.b. for (X,T) if and only if, for each p in D_T, there exist q in D_T and $M(p) > 0$ such that

$$p\Big(\sum_{i=m_{n-1}+1}^{r} \alpha_i z_i\Big) \leq M(p)q\Big(\sum_{i=m_{n-1}+1}^{m_n} \alpha_i z_i\Big), \tag{10.6.4}$$

valid for all r, $m_{n-1} + 1 \leq r \leq m_n$ and all scalars $\alpha_{m_{n-1}+1},\ldots,\alpha_{m_n}$, $n \geq 1$.

Proof: Cf. Theorem 5.1.7 for the necessary part. Conversely, let (10.6.4) be satisfied. Then the elements $z_{m_{n-1}+1},\ldots,z_{m_n}$ are linearly independent (1.i) in X_n.

Pick up p in D_T, arbitrary integers j,k with $j \leq k$ and put $Z^j = \sum_{i=1}^{j} \alpha_i z_i$ and $Z^k = \sum_{i=1}^{k} \alpha_i z_i$ for any given choice of scalars $\alpha_1,\ldots,\alpha_k$. There exist n_1 and n_2 in $\mathbb{N}$, $n_1 \leq n_2$, such that $1+m_{n_1-1} \leq j \leq m_{n_1}$ and $1+m_{n_2-1} \leq k \leq m_{n_2}$. Hence

$$Z^j = \sum_{i=1}^{m_{n_1}} t_i x_i;\ Z^k = Z^j + \sum_{i=m_{n_1}+1}^{m_{n_2}} t_i x_i$$

for a suitable choice of scalars $t_1,\ldots,t_{m_{n_1}},\ldots,t_{m_{n_2}}$. Making use of (5.1.3) with respect to $\{x_n\}$ with $m = m_{n_1}$ and $n = m_{n_2}$ (Theorem 5.1.7) and then using

Theorem 5.1.2, we find that $\{z_n\}$ is an S.b. for $[z_n] = X$. □

Theorem 10.6.5: Let (X,T) be a Fréchet space with an S.b. $\{x_n\}$. Let $\{y_n\}$ be a bl.s. corresponding to $\{x_n\}$ and $\{m_n\}$. Then $\{y_n\}$ has an S.bl.e. $\{z_n\}$ if and only if for each $n \geq 1$, there exists an isomorphism $A_n : X_n \to X_n$, X_n being the n-th block space such that $A_n x_{m_n} = y_n$ and that to every k in $\mathbb{N}$ there exists $m \equiv m(k)$ in $\mathbb{N}$ and an $M \equiv M(k) > 0$ satisfying

$$p_k(A_n \circ \pi_r^n \circ A_n^{-1}(y)) \leq M p_m(y), \quad \forall y \in X_n \tag{10.6.6}$$

valid for all $r \leq m_n - m_{n-1}$ and $n \geq 1$, $\pi_r^n : X_n \to X_n$ with

$$\pi_r^n \Big(\sum_{i=m_{n-1}+1}^{m_n} t_i x_i \Big) = \sum_{i=m_{n-1}+1}^{m_{n-1}+r} t_i x_i ,$$

it being understood that T is generated by a nondecreasing sequence $\{p_k : k \geq 1\}$ of seminorms on X.

Proof: Let $\{z_n\}$ be an S.bl.e. of $\{y_n\}$. Then for each $n \geq 1$, $\{z_{m_{n-1}+1}, \ldots, z_{m_n}\}$ is l.i. and $X_n = [z_{m_{n-1}+1}, \ldots, z_{m_n}]$. Hence we can define isomorphisms $A_n : X_n \to X_n$, $n \geq 1$ by

$$A_n\Big(\sum_{i=m_{n-1}+1}^{m_n} t_i x_i \Big) = \sum_{i=m_{n-1}+1}^{m_n} t_i z_i .$$

By Proposition 10.6.3, to every $k \geq 1$ there exists $m \geq k$ and a constant $M \equiv M(k) > 0$ such that, for all r, $1 \leq r \leq m_n - m_{n-1}$,

$$p_k(A_n \circ \pi_r^n(x)) \leq M p_m(A_n(x)),$$

valid for all x in X_n, $n \geq 1$, with

$$x = \sum_{i=m_{n-1}+1}^{m_n} \alpha_i x_i .$$

If $A_n x = y$, then

$$p_k(A_n \circ \pi_r^n \circ A_n^{-1}(y)) \leq M p_m(y); \ \forall y \in X_n ,$$

and this yields (10.6.5); clearly $A_n x_{m_n} = z_{m_n} = y_n$.

Conversely, assume the existence of isomorphisms $A_n : X_n \to X_n$, $n \geq 1$, satisfying (10.6.6). Define $\{z_n\}$ in X by $z_i = A_n x_i$ for $m_{n-1}+1 \leq i \leq m_n$, $n \geq 1$. Then $z_{m_n} = y_n$ and $z_i \in X_n$ for $m_{n-1}+1 \leq i \leq m_n$, $n \geq 1$. With this choice of the sequence $\{A_n\}$, one easily verifies the truth of (10.6.5) with the help of (10.6.6). Hence $\{z_n\}$ is an S.b. for (X,T). □

Note: Since Proposition 10.6.3 is also true for any TVS possessing a set of second category, Theorem 10.6.5 is, therefore, valid for any F-space.

Finally, we show that Problem 10.6.1 is completely solved for the space ω in the form of

Proposition 10.6.7: Every bl.s. $\{y^n\}$ corresponding to the S.b. $\{e^n\}$ for the space $(\omega, \sigma(\omega, \phi))$ has an S.bl.e. and, *a fortiori*, each bl.s. corresponding to an arbitrary S.b. $\{x^n\}$ for ω has an S.bl.e.

Proof: Consider an arbitrary bl.s. $\{y^n\}$ corresponding to $\{e^n\}$ with

$$y^n = \sum_{j=m_{n-1}+1}^{m_n} c_j e^j.$$

For each i with $1 + m_{n-1} \leq i \leq m_n$, define

$$z^i = \sum_{j=m_{n-1}+1}^{m_n} \alpha_j^i e^j,$$

where α_j^i's are so chosen that the finite set $\{z^i : 1 + m_{n-1} \leq i \leq m_n\}$ forms a Hamel base for $\omega_n = [e^{m_{n-1}+1}, \ldots, e^{m_n}]$. For each $n \geq 1$, find a maximal index i_o, $m_{n-1} + 1 \leq i_o \leq m_n$, such that $c_{i_o} \neq 0$. Then put $z^{m_n} = y_n$, $n \geq 1$.

Now it remains to show that $\{z^n\}$ is an S.b. for $(\omega, \sigma(\omega, \phi))$. So, let $x \in \omega$; then

$$x = \sum_{n \geq 1} \sum_{i=m_{n-1}+1}^{m_n} x_i e^i.$$

By the construction of z^i for $1 + m_{n-1} \leq i \leq m_n$, $n \geq 1$, we can determine unique scalars β_i, $1 + m_{n-1} \leq i \leq m_n$, $n \geq 1$, so that

$$\sum_{i=m_{n-1}+1}^{m_n} x_i e^i = \sum_{i=m_{n-1}+1}^{m_n} \beta_i z^i .$$

Hence

$$x = \sum_{n \geq 1} \beta_n z^n ,$$

and this completes the proof of the first part. For the other part, let us observe that every S.b. in ω is equivalent to $\{e^n\}$; for instance, see [13], p. 140; cf. also [80] where this result is implicit in Section 5. □

Part Three

11 Biorthogonal systems and cones

11.1 INTRODUCTION

In Chapters 2 through 10, we were essentially occupied with the basics and numerous initial results, occasionally advanced, relating to several aspects in the theory of Schauder bases (S.b.) in topological vector spaces (TVS) over the field $\mathbb{K}$ of real numbers ($\mathbb{R}$) or complex numbers ($\mathbb{C}$). At different places, we discovered a couple of applications of this theory in the structural study of TVS. There are several other applications of a similar nature; however, lack of space prevents us from reproducing these here and we postpone this to [133].

There is another interesting and useful application of the presence of an S.b. in a real TVS. Indeed, each S.b. in a TVS X over $\mathbb{R}$ gives rise to a natural partial ordering compatible with the linear structure. This aspect of the geometrical study of S.b. was possibly first suggested by Fullerton [62] at the ICM, Amsterdam, 1954. Schaefer [203] seems to be one of the earliest not only to develop a general theory of partially ordered l.c. TVS but also to apply it to the setting of the S.b. theory in l.c. TVS.

The purpose of this chapter is to initiate a discussion on the preliminaries of the order structure generated in a real TVS having a total biorthogonal sequence (b.s.) and to study the impact of this b.s. being an S.b. for its cone. In particular, when the b.s. in question is chosen to be an S.b., we should accordingly expect a rich order structure and this is dealt with in subsequent chapters.

Throughout this and the rest of the chapters, $X \equiv (X,T)$ will stand for an arbitrary TVS or l.c. TVS over the field $\mathbb{R}$ of reals (in the case of a TVS (X,T), we necessarily assume that $\{0\} \subsetneq X^*$). Also, we write $\{x_\alpha;f_\alpha\}$ or $\{x_\alpha;f_\alpha:\alpha \in \Lambda\}$ (resp. $\{x_n,f_n\}$) for an arbitrary biorthogonal system = b.sy. (resp. biorthogonal sequence = b.s.). For further notation and terminology relating to biorthogonal systems, the reader is referred to Chapter 7.

11.2 CONES AND WEDGES

Basically, we confine our attention to finding a criterion for the regularity of a cone generated by a total b.s. or the same as an M-generalized base. We also discuss an absolute b.s. At the outset, let us introduce

Definition 11.2.1: Corresponding to a b.s. $\{x_n;f_n\}$ for $\langle X,X^*\rangle$, $X \equiv (X,T)$ being a real TVS,

$$K_c = \{x \in X: x = \sum_{n\geq 1} f_n(x)x_n \text{ with } f_n(x) \geq 0, n \geq 1\},$$

$$K_o = \{x \in X, x \in [\sum_{i=1}^{n} \alpha_i x_i : \alpha_n \geq 0, n \geq 1]^T\},$$

and

$$K = \{x \in X: f_n(x) \geq 0, n \geq 1\}.$$

The properties of K_c, K_o and K and their interrelationship is summarized in the following straightforward

Proposition 11.2.2: The sets K_c, K_o and K are all wedges with $K_c \subset K_o \subset K$ and K_o, K being closed in (X,T). If $\{x_n;f_n\}$ is an M-generalized base, then each of the sets K_c, K_o and K is a cone, and if K is a cone, then $\{x_n;f_n\}$ is an M-generalized base. Finally, if $\{x_n;f_n\}$ is an S.b., then $K_c = K_o = K$.

Note: The dual wedges corresponding to K_c, K_o and K will henceforth be denoted by K_c^*, K_o^* and K^*. If $\{x_\alpha;f_\alpha\}$ is a b.sy., it is also possible to introduce wedges $K_o^\wedge$ and $K^\wedge$ by $K_o^\wedge = \{x \in X: x \in [\sum_{\alpha\in F} a_\alpha x_\alpha : a_\alpha \geq 0, F \in \Phi_\Lambda]^T\}$ and $K^\wedge = \{x \in X: f_\alpha(x) \geq 0, \forall\, \alpha \in \Lambda\}$.

The following two examples suggest that the hypothesis of 'S.b.' in Proposition 11.2.2 cannot be dropped; also they exhibit the situation when $K_c = K_o$ and $K_o = K$ respectively.

Example 11.2.3: Consider the usual b.s. $\{e^n;e^n\}$ for $\langle \ell^\infty, (\ell^\infty)^*\rangle$, which is obviously not a base for $(\ell^\infty, \|\cdot\|_\infty)$. Here $K_c = K_o \subsetneq K$.

Example 11.2.4: Recall the biorthogonal sequence $\{e^n;e^n\}$ which is not a Schauder base for $(k,\tau(k,\ell^1))$; cf. Example 6.2.1. Since $\bar{\phi} = k$, $K_o = K$ and contains e. But $e \notin K_c$.

Note: Where we wish to emphasize a particular b.s. that yields the wedges K_c, K_c^* etc., we will write the latter as $K_c\{x_n;f_n\}$, $K_c^*\{x_n;f_n\}$. Our next important note is that, unless specified otherwise, the natural ordering in a real TVS (X,T) as well as in X*, equipped with a b.s. $\{x_n;f_n\}$, will henceforth be understood as having been induced by K and K* respectively.

Let us now begin with (cf. [151])

Lemma 11.2.5: Let a TVS (X,T) contain a b.s. $\{x_n;f_n\}$. Then, for each σ in Φ, the collection of all finite subsets of $\mathbb{N}$,

$$\sum_{i\in\sigma} f_i(x)x_i \in [0,x]; \quad \sum_{i\in\sigma} f(x_i)f_i \in [0,f],$$

where $x \in K$ and $f \in K^*$.

Proof: Note that $0 \leq f_j(\sum_{i\in\sigma} f_i(x)x_i) \leq f_j(x)$, $j \geq 1$ and this proves the first part. The second part follows from the first. □

Next, after [167] we have

Lemma 11.2.6: Let $\{x_n;f_n\}$ be an M-generalized base for a TVS (X,T) and $Y = \{x \in X: x = \sum_{n\geq 1} f_n(x)x_n\}$. If B is a nonempty subset of Y, then the sup B (resp. inf B) exists in Y if and only if $\sum_{n\geq 1} (\sup_{x\in B} f_n(x))x_n$ (resp. $\sum_{n\geq 1} (\inf f_n(x))x_n$) converges in Y and this value is the sup B (resp. inf B).

Proof: We prove the result for sup B and similarly the other part follows. Let us first assume that sup B = z ∈ Y. Then $f_n(z) \geq \sup\{f_n(x):x \in B\}$ for $n \geq 1$. If, for some n_o in $\mathbb{N}$, $f_{n_o}(z) > \sup\{f_{n_o}(x):x \in B\}$, then choosing z_o in Y with

$$z_o = \sum_{\substack{n\geq 1\\ n\neq n_o}} f_n(z)x_n + \{\sup_{x\in B} f_{n_o}(x)\}x_{n_o},$$

we find $z \leq z_o$, a contradiction, and this proves the necessity part.

Conversely, let z denote the sum of the series in question. Then $x \leq z$ for each x in B. If z' is an upper bound of B, we easily find that $z \leq z'$ and so z = sup B. □

With this background, it is now possible to characterize the regularity of a cone in terms of compactness of positive intervals and other equivalent conditions. One of the main results in this direction is due to McArthur [167] and runs as follows:

<u>Theorem 11.2.7</u>: Let an l.c. TVS (X,T) be ordered by the cone K associated with an M-generalized base $\{x_n;f_n\}$. Then the following statements are equivalent:

(i) For each x in K and α in ℓ^∞, $\alpha_n \in \mathbb{R}$, with $\alpha_n \geq 0$ for $n \geq 1$, the series $\sum_{n\geq 1} \alpha_n f_n(x) x_n$ converges in (X,T).

(ii) For x in K, $[0,x]$ is compact in (X,T).

(iii) For x in K, $[0,x]$ is compact in $(X,\sigma(X,X^*))$.

(iv) For x in K, $[0,x]$ is $\sigma(X,X^*)$-sequentially complete and bounded in X.

(v) K is regular.

(vi) K is $\sigma(X,X^*)$-regular.

(vii) K is $\sigma(X,X^*)$-ω-regular.

(viii) K is ω-regular in (X,T).

Moreover, any of the above equivalent conditions implies that (a) sup B exists in K for any order bounded subset B of K; and (b) each order bounded $\sigma(X,X^*)$-Cauchy net in X converges in (X,T).

<u>Proof</u>: We proceed to prove the result in two stages: I and II as exhibited below.

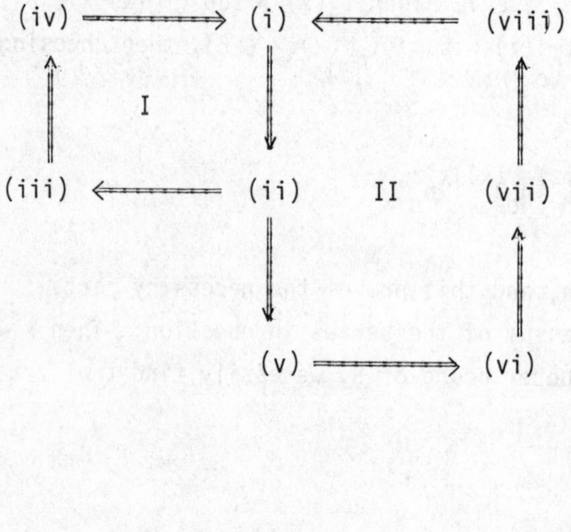

<u>Stage I</u>: (i) ⇒ (ii) Let $x \in K$ and B be the closed sphere in real ℓ^∞. If

$$S = \{ \sum_{i\in\sigma} \alpha_i f_i(x)x_i, \ \sigma \in \Phi, \ \alpha \in B\},$$

then $[0,x] \subset \bar{S}$ and so $[0,x]$ is precompact by Proposition 1.4.4. If $\{y_\delta : \delta \in \Lambda\}$ is a Cauchy net in $[0,x]$, then defining $\alpha_n = \lim f_n(y_\delta)$; β_n, $\beta_n^\delta \in [0,1]$ with $\alpha_n = \beta_n f_n(x)$, $f_n(y_\delta) = \beta_n^\delta f_n(x)$ for $n \geqslant 1$, $\delta \in \Lambda$, we find that $\sum_{n\geqslant 1} \alpha_n x_n$ converges, say, to y. Clearly $0 \leqslant y \leqslant x$.

By using Proposition 1.4.3, we have for any p in D_T and σ in Φ,

$$p(\sum_{i\in\sigma} (\beta_i - \beta_i^\delta) f_i(x)x_i) \leqslant 2 \sup_{i\in\sigma} |\beta_i - \beta_i^\delta| \sup_{\sigma_1 \subset \sigma} p(\sum_{i\in\sigma_1} f_i(x)x_i).$$

Now to each p in D_T and $\varepsilon > 0$, we find N in $\mathbb{N}$ such that

$$p(\sum_{i\in\sigma} f_i(x)x_i) \leqslant \frac{\varepsilon}{8}, \quad \forall \ \sigma \in \Phi \ni \sigma \cap [1,N] = \emptyset.$$

Therefore

$$p(\sum_{i\geqslant n} (\beta_i f_i(x) - f_i(y_\delta))x_i) \leqslant \frac{\varepsilon}{2}; \ \forall \ n > N, \quad \delta \in \Lambda.$$

Also

$$p(\sum_{i=1}^{N} (\beta_i f_i(x) - f_i(y_\delta))x_i) \leqslant \frac{\varepsilon}{2}, \quad \forall \ \delta \geqslant \delta_0 \equiv \delta_0(\varepsilon,N,p)$$

and so $p(y - y_\delta) \leqslant \varepsilon$, for $\delta \geqslant \delta_0$. Hence $[0,x]$ is complete, thus showing that $[0,x]$ is compact.

(ii) ⇒ (iii) ⇒ (iv) This is obvious.

(iv) ⇒ (i) Let α and x be as in (i) and $\lambda = \|\alpha\|_\infty$.

Since $0 \leqslant \sum_{i\in\sigma} \alpha_i f_i(x)x_i \leqslant \lambda x$, $\sigma \in \Phi$, the series $\sum_{n\geqslant 1} \alpha_n f_n(x)x_n$ is weakly unconditionally convergent and hence convergent in $[0,\lambda x]$ by Theorem 1.4.6 and the hypothesis.

<u>Stage II</u>: (ii) ⇒ (v) Here use Theorem 1.5.5(ii).

(v) ⇒ (vi) ⇒ (vii) Straightforward.

(vii) ⇒ (viii) Use Theorem 1.5.5(i).

(viii) ⇒ (i) For x and α as in (i), we find

$$0 \leq \sum_{i=1}^{n} \alpha_i f_i(x) x_i \leq \sum_{i=1}^{n+1} \alpha_i f_i(x) x_i \leq \|\alpha\|_\infty x, \ \forall\, n \geq 1,$$

and now use the ω-regularity of K.

(a) Let z be an upper bound of B in K. Choose α_n with $0 \leq \alpha_n \leq 1$ for $n \geq 1$ so that $\sup \{f_n(x) : x \in B\} = \alpha_n f_n(z)$. Then $\sum_{n \geq 1} \sup \{f_n(x) : x \in B\} x_n$ converges in K and apply Lemma 11.2.6.

(b) Let $\{x_\delta\}$ be a $\sigma(X,X^*)$-Cauchy net such that $y_1 \leq x_\delta \leq y_2$ for each δ. Since $[y_1,y_2]$ is compact by (ii), $\{x_\delta\}$ has a unique cluster point and this proves (b). □

Exercise 11.2.8: (i) In an arbitrary OTVS $(X, \leq, T)$ ordered by a wedge K, show that $[x_1,x_2]$ is T-compact if $[0,x]$ is T-compact for each x in K. (ii) Let (X,T) be a semireflexive space ordered by a cone associated with an M-generalized base; prove that each T-bounded order interval is T-compact.

Exercise 11.2.9: Let $\{x_n;f_n\}$ be an M-generalized base for a P-space (X,T) and let the associated cone K be generating. Show that $\sum_{n \geq 1} \alpha_n f(x_n) f_n$ converges in $(X^*, \beta(X^*,X))$ for each f in K^* and $\{\alpha_n\}$ in ℓ^∞ with $\alpha_n \geq 0$, for each $n \geq 1$.

Cones associated with absolute biorthogonal sequences

In this subsection, we consider order properties of bounded and equicontinuous subsets of X^*. To be precise in our discussion, let us first of all recall

Definition 11.2.10: A b.s. $\{x_n;f_n\}$ in an l.c. TVS (X,T) is said to be *absolute* if, for each x in X and p in D_T, $\sum_{n \geq 1} |f_n(x)| p(x_n) < \infty$.

We essentially follow [151] for the results of this subsection and begin with

Proposition 11.2.11: If the wedge K associated with a complete b.s. $\{x_n;f_n\}$ in an S-space (X,T) is generating, then sup A exists for each order bounded subset A of X^*.

Proof: Without loss of generality we may assume that $A \subset K^*$ and so $A \subset [0,g]$ for some g in K^*. For x in X, $x = y-z$ with y,z in K. Put $\alpha_n = \sup \{f(x_n): f \in A\}$. It is easily seen that, for $n \geq 1$,

$$\sum_{i=1}^{n} \alpha_i |f_i(x)| \leq \sum_{i=1}^{n} \alpha_i f_i(y) + \sum_{i=1}^{n} \alpha_i f_i(z) \leq g(y) + g(z) < \infty,$$

and so there is h in X^* with $h = \sigma(X^*,X)\text{-}\lim_n \sum_{i=1}^{n} \alpha_i f_i$. Clearly, $h(x_n) = \alpha_n$, $n \geq 1$ and so $h \leq g$.

Let $f \in [0,g]$ with $u \leq f$ for all u in A. Hence

$$\sum_{i=1}^{n} h(x_i) f_i \leq \sum_{i=1}^{n} f(x_i) f_i, \ \forall\, n \geq 1$$

Since $0 \leq f(x_n) \leq g(x_n)$ for $n \geq 1$ and so using the preceding arguments, we find an f_1 in X^* such that

$$f_1 = \sigma(X^*,X) - \lim_{n\to\infty} \sum_{i=1}^{n} f(x_i) f_i = f.$$

This shows that $h \leq f$. □

Proposition 11.2.12: Let the wedge K associated with $\{x_n;f_n\}$ in an l.c. TVS (X,T) be generating. If each equicontinuous subset A of X^* has an upper bound g in X^*, then $\{x_n;f_n\}$ is absolute.

Proof: For p in D_T, let $A = u^0$, where $u = \{x \in X: p(x) \leq 1\}$. By the hypothesis,

$$0 \leq p(x) = \sup \{|f(x)|: f \in A\} = \sup \{f(x): f \in A\} \leq g(x), \ \forall\, x \in K.$$

If x in X is arbitrary, then $x = y-z$ with $y,z \in K$ and it is easily seen that

$$\sum_{n\geq 1} f_n(y) p(x_n) \leq g(y); \quad \sum_{n\geq 1} f_n(z) p(x_n) \leq g(z).$$

This proves the absolute character of $\{x_n;f_n\}$. □

Proposition 11.2.13: If $\{x_n;f_n\}$ is a complete absolute b.s. in an S-space (X,T), then each equicontinuous subset M of X^* has an upper bound in X^*.

Proof: There exists p in D_T such that $|f(x)| \leq p(x)$ for all x in X and f in M. Also, by the ω-completeness of $(X^*,\sigma(X^*,X))$, $\sum_{n\geq 1} p(x_n)f_n(x) = g(x)$ for every x in X. Choose q in D_T so that $|f(x) - g(x)| \leq q(x)$ for all x in X, and f in M. As $\sum_{n\geq 1} q(x_n)|f_n(x)| < \infty$, $x \in X$, and $f(x_n) \leq g(x_n)$ for $n \geq 1$ and each f in M, we find that, for given f in M, there exists h_f in X* such that

$$\sum_{n\geq 1} (g(x_n) - f(x_n))f_n(x) = h_f(x), \quad \forall x \in X.$$

It is clear that $h_f = g-f$ and so $h_f \geq 0$, giving $f \leq g$ for each f in M. □

Now we arrive at the main result of this subsection, namely:

Theorem 11.2.14: Let the cone K associated with a complete b.s. $\{x_n;f_n\}$ in an S-space (X,T) be generating. Then the following are equivalent:

(i) $\{x_n;f_n\}$ is absolute.

(ii) Every equicontinuous subset M of X* has an upper bound in X*.

(iii) sup M exists for every equicontinuous subset M of X*.

Proof: (i) ⇒ (ii) This follows from Proposition 11.2.13.

(ii) ⇒ (iii) We have $M \subset u^o$ for some u in $\mathcal{B}_X$.

By (ii), there exists g in X* with $-g \leq f \leq g$ for every f in u^o. Thus M is order bounded and apply Proposition 11.2.11.

(iii) ⇒ (i) Apply Proposition 11.2.12. □

Biorthogonal systems generating the same cone

We now take up a very natural question, namely, the relationship between two different biorthogonal systems generating the same cone, and follow [221] for the rest of this subsection.

To begin with, let us consider an arbitrary b.sy. $\{x_\alpha;f_\alpha\}_{\alpha\in\Lambda}$ for $\langle X,X^*\rangle$, $X \equiv (X,T)$ being an arbitrary Hausdorff TVS. Further, let us recall from Chapter 1 the notation $R(0,x) = \{\alpha x:\alpha \geq 0\}$, $x \neq 0$ for a ray; then we have

Proposition 11.2.15: Let $\{x_\alpha,f_\alpha\}$ be an M-generalized base system and let K be the associated cone. Then, for x in K, R(0,x) is an extreme ray if and only if there exists β in Λ and $\lambda_\beta > 0$ such that $x = \lambda_\beta x_\beta$.

Proof: Let $R(0,x)$ be an extreme ray, $x \in K$, $x \neq 0$. Clearly there exists β so that $f_\beta(x) \neq 0$. As $[0,x] \subset R(0,x)$, we find that $f_\beta(x)x_\beta \in R(0,x)$ and so $x = \lambda_\beta x_\beta$ for some $\lambda_\beta > 0$.

Conversely, observe that $[0,x_\alpha] \subset R(0,x_\alpha)$ for each α and the result follows from Proposition 1.5.1. □

Proposition 11.2.16: Let K be the wedge associated with $\{x_\alpha;f_\alpha\}$. Then K has extreme rays if and only if $\{x_\alpha;f_\alpha\}$ is an M-generalized base system.

Proof: In view of Proposition 11.2.15, we need to prove the "only if" part. Assume therefore an x in X, $x \neq 0$, with $f_\alpha(x) = 0$ for all α in Λ. Thus x, -x are in K and so, if $y \in K$, $y \neq 0$, $\frac{1}{2}x + \frac{1}{2}(-x) \in R(0,y)$. However, this leads to a contradiction. □

Theorem 11.2.17: Let $\{x_\alpha;f_\alpha\}_{\alpha\in\Lambda}$ and $\{y_\beta;g_\beta\}_{\beta\in\Delta}$ be two b.sy in a TVS (X,T) with $\{x_\alpha;f_\alpha\}$ being an M-base system. If K_1 and K_2 are the wedges associated with $\{x_\alpha;f_\alpha\}$ and $\{y_\beta;g_\beta\}$ respectively, then $K_1 = K_2$ if and only if there exist a 1-1 mapping s of Λ onto Δ and a collection $\{\lambda_\alpha: \lambda_\alpha > 0, \alpha \in \Lambda\}$ such that $y_{s(\alpha)} = \lambda_\alpha x_\alpha$ and $g_{s(\alpha)} = \lambda_\alpha^{-1} f_\alpha$, $\alpha \in \Lambda$.

Proof: Let $K_1 = K_2$ and so, by Proposition 11.2.15, $\{R(0,x_\alpha):\alpha \in \Lambda\}$ = $\{R(0,y_\beta):\beta \in \Delta\}$ Hence define $s:\Lambda \to \Delta$, $s(\alpha) = \beta$ so that $R(0,x_\alpha) = R(0,y_\beta)$. s is clearly 1-1. Hence for each α in Λ, there exists $\lambda_\alpha > 0$ such that $y_{s(\alpha)} = \lambda_\alpha x_\alpha$. Consider the b.sy. $\{\lambda_\alpha x_\alpha;\lambda_\alpha^{-1} f_\alpha\}$ and $\{z_\alpha;h_\alpha\}$, where $z_\alpha = y_{s(\alpha)} = \lambda_\alpha x_\alpha$; $h_\alpha = g_{s(\alpha)}$, $\alpha \in \Lambda$. By Exercise 7.2.8, $g_{s(\alpha)} = \lambda_\alpha^{-1} f_\alpha$.

The converse is straightforward. □

In the rest of this subsection, we consider the relationship of b.sy. which gives rise to order isomorphic orderings. To begin with, we have

Theorem 11.2.18: Let there be two TVS (X_1,T_1) and (X_2,T_2) containing M-base systems $\{x_\alpha;f_\alpha\}_{\alpha\in\Lambda}$ and $\{y_\alpha;g_\alpha\}_{\alpha\in\Lambda}$ respectively. Let K_1 and K_2 be the respective cones in (X_1,T_1) and (X_2,T_2). Then we have (a) and (b), where

(a) If (X_1,T_1) and (X_2,T_2) are topologically order isomorphic, then there exists a permutation τ of Λ and a set $\{\lambda_\alpha: \lambda_\alpha > 0, \alpha \in \Lambda\}$ such that $\{x_\alpha;f_\alpha\}$ and $\{\lambda_\alpha y_{\tau(\alpha)}; \lambda_\alpha^{-1} g_{\tau(\alpha)}\}$ are equivalent systems.

(b) If $\{x_\alpha;f_\alpha\}$ and $\{y_\alpha;g_\alpha\}$ are equivalent, then X_1 and X_2 are order isomorphic.

Proof: (a) Let R denote the given isomorphism from X_1 onto X_2. If K_3 is the wedge associated with $\{R(x_\alpha);f_\alpha \circ R^{-1}\}$, then $K_2 = K_3$. Using Theorem 11.2.17, we find a permutation τ of Λ and a set $\{\lambda_\alpha : \lambda_\alpha > 0, \alpha \in \Lambda\}$ such that $R(x_\alpha) = \lambda_\alpha y_{\tau(\alpha)}$ and $f_\alpha \circ R^{-1} = \lambda_\alpha^{-1} g_{\tau(\alpha)}$ for α in Λ and this proves (a).

(b) By Proposition 7.3.3, there exists an algebraic isomorphism R from X_1 onto X_2 such that $R(x_\alpha) = y_\alpha$ for all α in Λ; also, $\mu_\Lambda = \nu_\Lambda$ by Proposition 7.3.2. Hence $R[K_1] = K_2$. □

Restricting X_1 and X_2, we derive

Theorem 11.2.19: Let the hypothesis of Theorem 11.2.18 be satisfied, where (X_1,T_1) and (X_2,T_2) are F-spaces. Then (X_1,T_1) and (X_2,T_2) are topologically order isomorphic if and only if there exist a permutation τ of Λ and a set $\{\lambda_\alpha:\lambda_\alpha > 0,\alpha \in \Lambda\}$ such that $\{x_\alpha;f_\alpha\}$ and $\{\lambda_\alpha y_{\tau(\alpha)};\lambda_\alpha^{-1} g_{\tau(\alpha)}\}$ are equivalent.

Proof: By virtue of Theorems 11.2.18(a), we now prove the "if" part only. However, invoking both the notation and proof of Theorem 7.3.6, we can construct a topological isomorphism R from (X_1,T_1) onto (X_2,T_2) such that, for x in X_1 and y in X_2,

$$R(x) = y \Leftrightarrow f_\alpha(x) = \lambda_\alpha^{-1} g_{\tau(\alpha)}(y),\ \forall\, \alpha \in \Lambda.$$

Thus $R[K_1] = K_2$ and this proves the result. □

11.3. BASES IN ASSOCIATED CONES

Examples of spaces are known to have b.s. which turn out to be S.b. only for their associated cones. It therefore becomes natural to unfold the properties of such a cone that enjoys the privilege of its members being uniquely represented in terms of the given b.s. Throughout this section, (X,T) denotes an arbitrary l.c. TVS having a b.s. $\{x_n;f_n\}$ and we follow [78] for the results of this section.

To begin with, let us recall a b.s. in a Banach space such that the former is an S.b. for its associated cone and not for the whole space.

Example 11.3.1: Consider $(c_0, \|\cdot\|_\infty)$ and define a b.s. $\{x^n;f^n\}$ for $\langle c_0,\ell^1\rangle$

as follows:

$$x^n = e^n + e^{n+1}; \quad f^n = \sum_{i=1}^{n} (-1)^{n-i} e^i, \quad n \geq 1.$$

Since $e^1 \in c_o$, $\{x^n;f^n\}$ is not an S.b. for $(c_o, \|\cdot\|_\infty)$. Here

$$K = \{\alpha \in c_o: \alpha_1 \geq 0 \text{ and } \alpha_n \geq \sum_{i=1}^{n-1} (-1)^{n-1-i}\alpha_i, \ n \geq 2\}.$$

Note that $e^1 \notin K$ and $x^n \in K$ for $n \geq 1$. For α in c_o,

$$\|\alpha - \sum_{i=1}^{n} \langle\alpha,f^i\rangle x^i\|_\infty = \|\{0,\ldots,0, \sum_{j=1}^{n+1} (-1)^{n+1-j}\alpha_j, \alpha_{n+2},\ldots\}\|_\infty.$$

In particular, for α in K,

$$\|\alpha - \sum_{i=1}^{n} \langle\alpha,f^i\rangle x^i\|_\infty = \sup_{i \geq n+2} |\alpha_i| \to 0, \text{ as } n \to \infty .$$

Hence $\{x^n;f^n\}$ is an S.b. for K.

We now pass on to an unconditional basis characterization of $\{x_n;f_n\}$ for K.

<u>Theorem 11.3.2</u>: Let $\{x_n;f_n\}$ be an S.b. for K. Consider the following statements:

(i) K is normal.

(ii) $\{x_n;f_n\}$ is an u-S.b. for K.

(iii) For each x in K, the series $\sum_{n \geq 1} f_n(x)x_n$ is weakly unconditionally Cauchy.

(iv) $[0,x]$ is bounded for each x in K.

(v) $[0,x]$ is topologically homeomorphic to a countable Hilbert cube for each x in K.

Then (i) ⇒ (ii) ⇒ (iii) ⇒ (iv) and (v) ⇒ (iv). If (X,T) is also a Fréchet space, then (i) through (v) are equivalent.

<u>Proof</u>: (i) ⇒ (ii) Here we may choose members of D_T to be monotone. Fix p in D_T, x in K and $\varepsilon > 0$. We can find N_o in $\mathbb{N}$ so that for any σ in Φ with $[1,N_o] \subset \sigma$, $p(\sum_{i \geq N_o} f_i(x)x_i) < \varepsilon$ and $0 \leq x - \sum_{i\in\sigma} f_i(x)x_i \leq \sum_{i \geq N} f_i(x)x_i$. This

proves (ii).

(ii) ⇒ (iii) Obvious.

(iii) ⇒ (iv) For f in X* and y in [0,x] with x in K, $|f(y)| \leq \sum_{n\geq 1} f_n(x) |f(x_n)|$. Hence (iv) follows.

(v) ⇒ (iv) Trivial, since the Hilbert cube is compact.

If (X,T) is a Fréchet space, then on account of Theorem 1.5.4, we need show only

(i) ⇒ (v) Let $x \in K$. If P is the Hilbert cube endowed with the product topology, we may then regard P as a subset of ω, having the topology induced by $\sigma(\omega,\phi)$, where, for α in P, $0 \leq \alpha_n \leq 1$, $n \geq 1$. Define $R:[0,x] \to P$ by $R(y) = \{f_n(y)/f_n(x)\}$, where 0/0 is regarded as 0. R is clearly 1-1. On the other hand, if $\alpha \in P$, we write $\beta_n = \alpha_n f_n(x)$, $n \geq 1$ and $z_n = \sum_{i=1}^{n} \beta_i x_i$. By the monotone character of members of D_T, we easily find that $\{z_n\}$ is Cauchy in (X,T). Hence $z_n \to z$ in (X,T), where $z = \sum_{n\geq 1} \beta_n x_n$. Clearly $0 \leq z \leq x$ and $R(z) = \alpha$. Thus R is onto. R is clearly continuous (since each f_n is continuous).

For proving the continuity of R^{-1}, let $\alpha^n, \alpha \in P$ with $\alpha^n \to \alpha$. Suppose $y_n = R^{-1}(\alpha^n)$ and $y = R^{-1}(\alpha)$. Then

$$y_n = \sum_{i\geq 1} \alpha_i^n f_i(x)x_i; \quad y = \sum_{i\geq 1} \alpha_i f_i(x)x_i.$$

For a given monotone seminorm p and ε > 0, we find an n_o in ℕ so that

$$p(\sum_{i\geq n} \alpha_i^q f_i(x)x_i),\ p(\sum_{i\geq n} \alpha_i f_i(x)x_i) \leq p(\sum_{i\geq n} f_i(x)x_i) \leq \frac{\varepsilon}{3},$$

$$\forall\, n \geq n_o,\ q \geq 1.$$

Hence

$$p(y_q - y) \leq \sum_{i=1}^{n_o - 1} |\alpha_i^q - \alpha_i| f_i(x)p(x_i) + \frac{2\varepsilon}{3},\ \forall\, q \geq 1,$$

and this yields the continuity of R^{-1}. □

<u>Note</u>: The above result [78] extends a similar result proved for Banach (resp. Fréchet) spaces in [173] (resp. [169]; here the result is stated without proof).

Remark: Let $\{x_n;f_n\}$ be an S.b. for an l.c. TVS (X,T). If $\{x_n;f_n\}$ is a u-S.b. for its cone, is it possible to conclude the unconditional character of $\{x_n;f_n\}$ for (X,T)? In general, the following two examples negate this question.

Example 11.3.3: Recall the conditional base $\{x^n;f^n\}$ of $(c_o, \|\cdot\|_\infty)$ from Example 8.2.19, where $x^n = e^{(n)}$, $f^n = e^n - e^{n+1}$, $n \geq 1$. Here $K = \{\alpha \in c_o : \alpha_n - \alpha_{n+1} \geq 0 \text{ for } n \geq 1\}$. Hence, for $\{\varepsilon_n\}$ in $\bar{e}$ (i.e. $\varepsilon_n = \pm 1$) and α in K,

$$\|\sum_{i=m}^{n} \varepsilon_i(\alpha_i-\alpha_{i+1})x^i\|_\infty = \sup_{m\leq j\leq n} |\sum_{i=j}^{n} \varepsilon_i(\alpha_i-\alpha_{i+1})| \leq \alpha_m - \alpha_{m+1}.$$

Consequently $\{x^n;f^n\}$ is a u-S.b. for K.

Example 11.3.4: Consider the space C[0,1] and denote by $\{x_n;f_n\}$ any S.b. for this space; for instance, one may consider the Schauder system of Example 5.2.11. Since all bases of C[0,1] are conditional ([218, p. 434), $\{x_n;f_n\}$ is a conditional base for C[0,1]. Here $K = K_c$ and the cone K is normal. Hence $\{x_n;f_n\}$ is a u-S.b. for K.

Remark: Let us note that (i) and (iv) are always equivalent in any F-space possessing a cone K. Concerning the relationship of (i), or, equivalently, (iv), in any Fréchet space with (ii) or (iii), one has to have the basis restriction of $\{x_n;f_n\}$ in K as shown in the next example. Also, (v) implies (iv) in general. Therefore, it is natural to question whether the basis restriction of $\{x_n;f_n\}$ for K is essential to infer (v) from (iv). The following example once again suggests that this is indeed the case.

Example 11.3.5: Recall the b.s. $\{e^n;e^n\}$ for ℓ^∞ ; cf. Example 11.2.3. Here $K = \{\alpha \in \ell^\infty: \alpha_n \geq 0, n \geq 1\}$ and so (i) and (iv) are satisfied but not (ii), (iii) and (v). Indeed, $\sum_{n\geq 1} \langle e,e^n\rangle e^n \not\to e \in K$ and

$$[0,e] = [0,1]^{\mathbb{N}} = \{\alpha \in \omega: 0 \leq \alpha_n \leq 1, n \geq 1\},$$

giving R as the identity map of $[0,1]^{\mathbb{N}}$. Since $e^{(n)} \to e$ in $\sigma(\omega,\phi)$ but $e^{(n)} \not\to e$ in $\|\cdot\|_\infty$, R^{-1} is not continuous.

Next, we have

Proposition 11.3.6: Let $\{x_n;f_n\}$ be an S.b. for the associated cone K. If K is sequentially complete and normal, then K is minihedral.

Proof: If $x,y \in K$, then $0 \leqslant \sup(f_n(x),f_n(y)) \leqslant f_n(x) + f_n(y)$, $n \geqslant 1$, and so $\{\sum_{i=1}^{n} \sup(f_i(x),f_i(y))x_i\}$ is Cauchy by using the normality of K. Let $z = \sum_{n\geqslant 1} \sup(f_n(x),f_n(y))x_n$, then $z = \sup\{x,y\}$. □

Remark: The converse of Proposition 11.3.6 is not necessarily true, for instance consider

Example 11.3.7: This is Example 8.2.22. Here $K = \{\alpha \in X: \sum_{i=1}^{n} \alpha_i \geqslant 0, n \geqslant 1\}$. Invoking the notation from this example, we find that for x in X, $|x| = x^* \in X$, thus giving the generating character of K. Also it is minihedral. Finally, choosing y^n, z^n in $X(n \geqslant 1)$ with

$$y^n = \frac{1}{n}\{(e^1 + e^3 + \dots + e^{2n-1}) - (e^2 + e^4 + \dots + e^{2n})\} = \frac{1}{n}\sum_{i=1}^{n} x^{2i-1};$$

$$z^n = \frac{1}{n}\ (e^1 - e^{2n}) = \frac{1}{n}\sum_{i=1}^{2n} x^i,$$

we find that $\|y^n\|_1 = 2$, $\|z^n\|_1 = 2/n$ and hence K is not normal.

Example 11.3.8: Consider the cone $K = \{\alpha \in \phi: \alpha_n \geqslant 0, n \geqslant 1\}$ associated with the S.b. $\{e^n;e^n\}$ for $(\phi, \|\cdot\|_\infty)$. Here K is generating, minihedral and normal. However, K is not sequentially complete (consider the nonconvergent Cauchy sequence $\{\sum_{i=1}^{n} \frac{1}{i} e^i\}$ in K).

Remark: Let us recall the part of Theorem 11.3.2 which asserts that, if the cone K is normal and $\{x_n;f_n\}$ is an S.b. for K, then $\{x_n;f_n\}$ is a u-S.b. for K. The next result, of which the Banach space analogue is to be found in [173], can be taken as a variant of the foregoing statement.

Theorem 11.3.9: Let the cone K associated with a b.s. $\{x_n;f_n\}$ in an l.c. TVS (X,T) be normal. If K is $\sigma(X,X^*)$-ω-complete, then $\{x_n;f_n\}$ is a u-S.b. for K and, if $\{\sum_{i=1}^{n} \alpha_i x_i\}$ is bounded in K for α in ω with $\alpha_n \geqslant 0$, then $\sum_{n\geqslant 1} \alpha_n x_n$ converges in K.

Proof: We first prove the second part and this will yield the first part.

By the hypothesis of normality on K, we may choose members of D_T to be monotone and hence $\sum_{n\geq 1} \alpha_n x_n$ is unordered bounded. Therefore, $\sum_{n\geq 1} |f(\alpha_n x_n)| < \infty$ for each f in X* and this gives the weak subseries convergence of $\sum_{n\geq 1} \alpha_n x_n$: consequently, it converges in K by Theorem 1.4.6.

For the first part, observe that

$$p\left(\sum_{i=1}^{n} f_i(x)x_i\right) \leq p(x);\ \forall\, x \in K,\ p \in D_T,\ n \geq 1.$$

Now apply the above arguments. □

Remark: The converse of Theorem 11.3.9 is not true in general; indeed consider (cf. [173])

Example 11.3.10: This is the same as Example 11.3.3 and it follows with the help of Theorem 11.3.2 that K is normal. As mentioned earlier, $\{x_n;f_n\}$ is a u-S.b. for K. Also, if $\alpha_n \geq 0$ for $n \geq 1$ and $\sup \|\sum_{i=1}^{n} \alpha_i x^i\|_\infty < \infty$, we easily find that

$$\sup_n \sup_{1\leq j\leq n} \left|\sum_{i=j}^{n} \alpha_i\right| < \infty,$$

and so $\alpha \in \ell^1$. Put $\beta_j = \sum_{i\geq j} \alpha_i$, then $\beta \in K \subset c_o$ and

$$\left\|\sum_{i=1}^{n} \alpha_i x^i - \sum_{i=1}^{n} \beta_i e^i\right\|_\infty = \sup_{1\leq i\leq n} \left|\sum_{j=1}^{n} \alpha_j - \beta_i\right| \to 0,$$

as $n \to \infty$. Hence $\sum_{n\geq 1} \alpha_n x^n$ converges to β in K. Finally note that K is not ω-complete; in fact, the sequence $\{x^n\} \subset K$ is weakly Cauchy without being convergent.

The next result due to Saxon [202] is also in the direction of Theorem 11.3.9.

Theorem 11.3.11: Let the cone K associated with a b.s. $\{x_n;f_n\}$ in an l.c. TVS (X,T) be normal. If K-K is barrelled, then $\{x_n;f_n\}$ is a u-S.b. for $[x_n]$.

Proof: For each σ in Φ, recall the operator $S_\sigma: Y \to Y$ mentioned after Lemma 8.2.5, where $Y = K-K$; in fact, $S_\sigma(x) = \sum_{i\in\sigma} f_i(x)x_i$ and suppose that $\bar{S}_\sigma$ denotes the extension of S_σ to $\bar{Y}$.

$$\bar{S}_\sigma(x) = \sum_{i\in\sigma} f_i(x)x_i, \ x \in \bar{Y}.$$

By the monotone character of p in D_T, the family $\{S_\sigma:\sigma \in \Phi\}$ is bounded on Y and so is $\{\bar{S}_\sigma:\sigma\in \Phi\}$ on $\bar{Y}$. Hence $\{\bar{S}_\sigma:\sigma\in \Phi\}$ is equicontinuous on $\bar{Y}$. Also note that $[x_n] \subset \bar{Y}$. Now proceed as in Theorem 7.4.4. □

After McArthur [167], let us characterize the regularity of a cone in the form of

Theorem 11.3.12: Let K be the cone associated with an M-generalized base $\{x_n;f_n\}$ in a Fréchet space (X,T). Then K is regular if and only if $\{x_n;f_n\}$ is a base for K and K is normal.

Proof: Let K be regular. Apply Theorem 11.2.7. (v) ⇒ (i), to conclude the basis character of $\{x_n;f_n\}$ for K; now make use of Theorem 1.5.5(i) to get the normality of K.

For the converse, use Theorems 11.3.2, 1.4.5 and 11.2.7(i) ⇒ (v). □

Remark: The basis character of $\{x_n;f_n\}$ for K in the above theorem cannot be dropped to infer the regularity of K.

Example 11.3.13: This is the same cone K considered in Example 11.3.5, for which $\{e^n;e^n\}$ is not a base. Now $e^{(n)} \in K$, $0 \prec e^{(n)} \prec e^{(n+1)} \prec e$. But $\{e^{(n)}\}$ does not converge in ℓ^∞ and so in K.

Finally, we characterize extreme subsets of K for which $\{x_n;f_n\}$ is a base; indeed, we have (cf. [221])

Theorem 11.3.14: Consider the cone K associated with a b.s. $\{x_n;f_n\}$ in a TVS (X,T) and let $\{x_n;f_n\}$ be the S.b. for K. A subset E of K is a closed extreme subset if and only if there exists a subset J of $\mathbb{N}$ such that $E = \{y \in K: f_n(y) = 0, \forall n \in J\}$.

Proof: To prove the necessity, let $J = \{n \in \mathbb{N}; f_n(y) = 0, \forall y \in E\}$ and

$B = \{y \in K;\ f_n(y) = 0\ \forall\ n \in J\}$. We need to show that $E = B$. If $y \in B$, then $y = \sum_{n \geq 1} f_n(y)x_n$ and $f_n(y) = 0$, $n \in J$, yield $y \in M \equiv \overline{\text{con}}\ \{\cup R(0,x_n):n \in \mathbb{N}\smallsetminus J\}$, the closure of the convex hull of $\{\cup R(0,x_n):n \in \mathbb{N}\smallsetminus J\}$. By the hypothesis, for n in $\mathbb{N}\smallsetminus J$, there exists z in E so that $f_n(z) > 0$. Using Proposition 1.5.1 twice and the fact that $f_n(z)x_n \in [0,z]$, we find that $R(0,x_n) \subset E$ and so $M \subset E$. Hence $B \subset E$. Clearly $E \subset B$; therefore $E = B$.

Conversely, let $E = \{y \in K:f_n(y) = 0,\ \forall\ n \in J\}$ for some $J \subset \mathbb{N}$. Observe that

$$E = K \cap \{\bigcap_{n \in J} f_n^{-1}(0)\},$$

and the result follows again by Proposition 1.5.1. □

12 Cones associated with bases

12.1 INTRODUCTION

In the last chapter, we discussed a number of properties of a cone associated with an M-generalized base $\{x_n;f_n\}$ for an l.c. TVS (X,T). Further restrictions on $\{x_n;f_n\}$ are likely to yield more order properties of the space X, and this is what we now investigate. Indeed, we take $\{x_n;f_n\}$ to be an S.b. for an l.c. TVS (X,T) and obtain a variety of results relating to the order structure of the space in question. In particular, we study the impact of the behaviour of an S.b. on the topological lattice structure of the space and vice-versa. More specifically, after establishing certain preliminary results, we confine our attention to the characterization of normal and regular cones.

Unless mentioned to the contrary, we assume throughout this section that $\{x_n;f_n\}$ is an S.b. for an arbitrary l.c. TVS (X,T).

12.2 INTERIOR POINTS AND LATTICE STRUCTURE

In this small section we exploit the presence of an S.b. to find that the corresponding cone has no interior points, characterize some special points associated with the cone and obtain conditions to ensure the vector lattice structure of the space in question. To begin with, let us observe the simple

Proposition 12.2.1: The cone $K \equiv K\{x_n;f_n\}$ is T-closed and $\overline{K-K} = X$; also (X,T) is an Archimedean regularly ordered space.

Proof: Straightforward. □

The next result [163] which extends a similar proposition in Banach spaces (cf. [218], p. 321) says that K is nowhere dense in (X,T).

Proposition 12.2.2: The cone $K \equiv K\{x_n;f_n\}$ contains no interior point.

Proof: Given x in K and u-x in $\mathcal{B}_X$, there exist v in $\mathcal{B}_X$ and m_0 in $\mathbb{N}$ so that $x + v \subset u$ and

$$x - \sum_{n=1}^{m} f_n(x)x_n \in \frac{1}{4}v, \ \forall m \geqslant m_o$$

$$\Rightarrow f_{m_o+1}(x)x_{m_o+1} \in \frac{1}{4}v + \frac{1}{4}v \subset \frac{1}{2}v.$$

Thus $x - 2f_{m_o+1}(x)x_{m_o+1} \in u$. But $x - 2f_{m_o+1}(x)x_{m_o+1} \notin K$, giving $u \not\subset K$. □

Next, observe that the dual wedge K^* of K is clearly a cone in X^*, associated with the $\sigma(X^*,X)$-Schauder base $\{f_n;\Psi(x_n)\}$ for X^*. After [163] let us have again a simple observation in the form of

Proposition 12.2.3: We have

$$K^* = \tilde{K} \equiv \{f \in X^*: f(x_n) \geqslant 0, \ \forall \ n \geqslant 1\}.$$

Proof: A consequence of direct computation. □

Concerning the special points of K, we once again recall from [163] the following

Proposition 12.2.4: For x in K, $x \neq 0$, the following are equivalent:

(i) x is a weak order unit.

(ii) $f_n(x) > 0$ for every $n \geqslant 1$.

(iii) x is a quasi-interior point of K.

Proof: (i) ⇒ (ii) For each n in $\mathbb{N}$, we find y_n in K, $y_n \neq 0$ such that $y_n \prec x, x_n$. Clearly $f_n(y_n) > 0$ for $n \geqslant 1$ and this proves (ii).

(ii) ⇒ (iii) Fix $n \geqslant 1$ and choose $\alpha_n > 0$ with $\alpha_n f_n(x) \geqslant 1$. If $y_n = \alpha_n x$, then $x_n \prec y_n$ and $\alpha_n^{-1} y_n \in [0,x]$. Hence $x_n \in sp\ \{[0,x]\}$, giving $sp\ \{x_n\} \subset sp\ \{[0,x]\}$. Thus $\overline{sp}\{[0,x]\} = X$.

(iii) ⇒ (i) It is easily seen that $f_n(x) > 0$ for each $n \geqslant 1$, otherwise, for some m, $f_m(y) = 0$ for all y in $[0,x]$ and, as $\overline{sp}\ \{[0,x]\} = X$, we find that $f_m(x_m) = 0$. Finally, for each y in K with $y \neq 0$, let $z_n = (\inf(f_n(x), f_n(y)))x_n$. Then $z_{n_o} \prec x,y$ for some n_o with $f_{n_o}(y) \neq 0$. Note that $z_{n_o} \in K$, $z_{n_o} \neq 0$. □

Concerning the order structure of X, we now prove the useful

Proposition 12.2.5: Let (X,T) be ω-complete. If the S.b. $\{x_n;f_n\}$ for (X,T)

is unconditional for K, then, for each x in X with $z \leqslant x \leqslant y$, where $y \in K$ and $z \in (-K)$, the elements x^+ and x^- exist in X, satisfying $x^+ \leqslant y$, $x^- \leqslant -z$. In particular, if $\{x_n; f_n\}$ is an u-S.b. for (X,T), then X is a vector lattice (VL).

<u>Proof</u>: With x,y and z as mentioned in the first statement, we find that $0 \leqslant f_n(x) \vee 0 \leqslant f_n(y)$, for all $n \geqslant 1$. Choose α in ω so that $0 \leqslant \alpha_n \leqslant 1$ and $f_n(x) \vee 0 = \alpha_n f_n(y)$, $n \geqslant 1$. Using Theorem 1.4.5 we easily verify that

$$x^+ = \sum_{n \geqslant 1} \alpha_n f_n(y) x_n = \sum_{n \geqslant 1} (f_n(x) \vee 0) x_n.$$

Observe that $x^+ \leqslant y$ and, as $-x \leqslant -z$, we get $x^- \leqslant -z$.

For the second part, let $x,y \in X$. Put $J = \{n \in \mathbb{N} : f_n(y) \leqslant f_n(x)\}$. By virtue of Theorem 1.4.5, we can find z in X with

$$z = \sum_{n \in J} f_n(x) x_n + \sum_{n \in \mathbb{N} \sim J} f_n(y) x_n.$$

Clearly, $z = x \vee y$. □

<u>Remark</u>: The condition that $z \leqslant x \leqslant y$ is essential to infer the existence of x^+ and x^- and so is the case with the unconditional character of $\{x_n; f_n\}$ for the VL conclusion of X. This is justified by the following

<u>Example 12.2.6</u>: We recall once again Example 11.3.3 (cf. also Example 8.2.19). Let $\alpha_n = (-1)^{n+1}/n$, $n \geqslant 1$. If $\beta = \alpha \vee 0 \in c_o$, then β, $\beta - \alpha \in K$ and so

$$\beta_1 \geqslant \beta_2 \geqslant \ldots \geqslant \beta_n \geqslant \beta_{n+1} \geqslant \ldots, \qquad (*)$$

and

$$\beta_1 - \alpha_1 \geqslant \ldots \geqslant \beta_n - \alpha_n \geqslant \beta_{n+1} - \alpha_{n+1} \geqslant \ldots . \qquad (**)$$

From (*) and (**),

$$\beta_1 \geqslant \beta_{2n} + \sum_{i=1}^{2n} \frac{1}{i}, \ \forall\, n \geqslant 1.$$

However, this is not possible and so β does not exist. Similarly, we cannot find β and γ in K such that $-\gamma \leqslant \alpha \leqslant \beta$; cf. (*) and (**).

12.3 NORMAL CONES

The importance of normal cones is well known not only to vector space pathologists but also to other experts working on extremal problems in differential and integral equations; cf. [148]. The purpose of this section is, therefore, to find minimum possible conditions on the S.b. $\{x_n;f_n\}$ and the underlying l.c. TVS (X,T), which may ensure the normality of the resulting cone $K \equiv \{x_n;f_n\}$. Besides, we also investigate the necessary and sufficient conditions for $\{x_n;f_n\}$ to be a u-S.b. for (X,T) in terms of the normality and other properties of K.

At the outset, we have (cf. [163])

Proposition 12.3.1: If (X,T) is an S-space and $\{x_n;f_n\}$ is a u-S.b. for (X,T), then K is $\sigma(X,X^*)$-normal.

Proof: Using Proposition 12.2.5, X* is a VL. Thus K* is generating. Now apply Proposition 1.5.7. □

Proposition 12.3.2: If $\{x_n;f_n\}$ is a u-S.b. and (X,T) is barrelled, then K is T-normal.

First proof: In view of Theorem 8.2.6 and Proposition 8.2.7, it suffices to prove that each $p_{V_{\bar{e}}}$ is monotone. Let now $0 \leqslant x \leqslant y$. Choose β_n with $0 \leqslant \beta_n \leqslant 1$, $n \geqslant 1$ such that $f_n(x) = \beta_n f_n(y)$. Then

$$p_{V_{\bar{e}}}(x) = \sup\{p_V(\sum_{i=1}^{n} \alpha_i\beta_i f_i(y)x_i) : n \geqslant 1, \alpha \in \bar{e}\}$$

$$\leqslant \sup\{p_V(\sum_{i=1}^{n} \gamma_i f_i(y)x_i) \quad : n \geqslant 1, \gamma \in \bar{e}\} = p_{V_{\bar{e}}}(y),$$

and we are done. □

Second proof: We follow [167] for this alternative proof. It is easily seen that T is also generated by $D^* = \{p^*: p \in D_T\}$, where for x in X,

$$p^*(x) = \sup\{p(S_\sigma(x)) : \sigma \in \Phi\}.$$

Clearly, $p^*(S_\sigma(x)) \leq p^*(S_\rho(x))$ if $\sigma \subset \rho$. Let $x,y \in K$ with $x \leq y$. Fix m in $\mathbb{N}$ and p^* in D^*; and let $z_m = S_m(x)$. A simple application of the Hahn-Banach theorem yields an f in X^* with $|f(z)| \leq p^*(z)$ for z in X and $p^*(z_m) = f(z_m)$. If $\sigma(m) = \{i \in \mathbb{N} : 1 \leq i \leq m, f(x_i) \geq 0\}$, and $\sigma(m) \neq \emptyset$, we get $p^*(S_m(x)) \leq p^*(S_\sigma(y))$ for all $\sigma \supset \sigma(m)$. Hence, for $n \geq m$

$$\begin{aligned} & p^*(S_m(x)) \leq p^*(S_n(y)) \\ & \Rightarrow p^*(S_m(x)) \leq p^*(y). \end{aligned} \tag{$*$}$$

If $\sigma(m) = \emptyset$, $(*)$ clearly holds. Thus $p^*(x) \leq p^*(y)$. □

Next, we have a characterization of unconditional bases in terms of the properties of the associated cone (cf. [167]).

Theorem 12.3.3: Let (X,T) be ω-complete and barrelled. Then $\{x_n;f_n\}$ is a u-S.b. if and only if $K \equiv K\{x_n;f_n\}$ is generating and normal.

Proof: For necessity, use Propositions 12.2.5 and 12.3.2.

For sufficiency, note that $X = K-K$ and $\{x_n;f_n\}$ is a u-S.b. for K by Theorem 11.3.2. □

Note: Theorem 12.3.3 includes a corresponding result of Ceĭtlin [22] who proved the same for ω-complete bornological spaces. Note that the sufficiency part is true in any arbitrary l.c. TVS.

Remark: It is essential to have the cone K generating and normal to infer the u-S.b. character of $\{x_n;f_n\}$, and vice versa, in the above theorem, as exhibited in the following examples.

Example 12.3.4: Recall Example 12.2.6 (cf. also Examples 11.3.3 and 8.2.19). Using Theorem 11.3.2, we conclude the normality of $K \equiv K\{x^n;f^n\}$. If K were generating, then there would exist α, β in K such that

$$\alpha_n - \beta_n = \frac{(-1)^{n+1}}{n}, \ \forall\, n \geq 1.$$

But

$$\alpha_1 = \beta_1 + 1 > \beta_2 + 1 = \alpha_2 + 1 + \frac{1}{2} > \ldots > \alpha_{2n} + \sum_{i=1}^{2n} \frac{1}{i},$$

for $n > 1$. This leads to absurdity. Note that $\{x^n;f^n\}$ is not a u-S.b. for (X,T).

Example 12.3.5: Here we confine ourselves to the conditional base $\{x^n;f^n\}$ for $(c_o, \|\cdot\|_\infty)$ discussed in Example 8.2.20. Observe that

$$K \equiv K\{x^n,f^n\} = \{\alpha \in c_o : \alpha_n + \alpha_{n+1} > 0, \forall n > 1\}.$$

Clearly K is generating. But K is not normal, for choose α and β in c_o with

$$\alpha_1 = 4; \quad \alpha_2 = 2; \quad \alpha_n = \frac{1}{n-1}, \quad n > 3;$$

$$\beta_1 = 5; \quad \beta_2 = \frac{1}{2}; \quad \beta_n = \frac{1}{n}, \quad n > 3.$$

Then $0 < \beta < \alpha$ and $4 = \|\alpha\|_\infty < \|\beta\|_\infty = 5$.

A variation of Theorem 12.3.3 is contained in (cf. [169])

Theorem 12.3.6: Let $\{x_n,f_n\}$ be a b.s. for a reflexive space (X,T). Then $\{x_n;f_n\}$ is a u-S.b. for (X,T) if and only if the wedge $K \equiv K\{x_n;f_n\}$ is generating and normal.

Proof: The necessity follows by Theorem 12.3.3.

Conversely, let K be generating and normal. By the normality of K, K is a cone and so $\{f_n\}$ is total on X. Since K is closed and normal, $[0,x]$ is $\sigma(X,X^*)$-compact for each x in K. Hence, using Theorems 11.2.7 and 11.3.2, $\{x_n;f_n\}$ is an u-S.b. for K and hence for (X,T), since $X = K-K$. □

Alternative proof for sufficiency: As before, we infer that K is a cone and $\{f_n\}$ is total on X. Since $\{f_i(S_n(x))\}$ is nondecreasing and bounded in $[0,f_i(x)]$ for each $i > 1$ and x in K, $\lim f(S_n(x))$ exists for each x in K and f in sp $\{f_i: i > 1\}$. The totality of $\{f_n\}$ and the reflexivity of (X,T) yields $X^* = [f_n]^\beta$, $\beta \equiv \beta(X^*,X)$. Hence, for $\varepsilon > 0$, x in K and g in X^*, there exists f in sp$\{f_i\}$ such that

$$|(f-g)(S_n(x))|, \ |(f-g)(x)| < \frac{\varepsilon}{3}, \ \forall n > 1;$$

also $|f(S_n(x)-S_m(x))| < \varepsilon/3$ for all $m,n > N$. Thus $\{S_n(x)\}$ is $\sigma(X,X^*)$-Cauchy, thereby making $\{x_n;f_n\}$ a weak Schauder base for K and hence for X. Use Proposition 6.1.5 and Theorem 11.3.2 to reach the desired conclusion. □

<u>Note</u>: The Banach space analogue of Theorem 12.3.6 is to be found in [84]. For the next exercise, which may be considered a slight variation of the preceding theorem, we refer to [63] and [141].

<u>Exercise 12.3.7</u>: Let an ω-complete l.c. TVS (X,T) contain an M-generalized base $\{x_n,f_n\}$. Prove that $\{x_n;f_n\}$ is a u-S.b. for (X,T) if and only if $K \equiv K\{x_n;f_n\}$ is generating and $[0,x]$ is compact in some compatible topology for each x in K.

Several order theoretic characterizations of an unconditional base are contained in (cf. [151])

<u>Theorem 12.3.8</u>: Let $\{x_n;f_n\}$ be an M-semibase for a barrelled space (X,T) and let K (resp. K*) be the associated wedge in X (resp. X*). Then the following statements are equivalent:

(i) $\{x_n;f_n\}$ is a u-S.b. for (X,T).

(ii) K is a T-normal b-cone.

(iii) K is T-normal and K* is $\beta(X^*,X)$-normal.

(iv) K* generates X* and $[0,f]$ is $\beta(X^*,X)$-bounded for each f in K*.

Each of the following conditions implies (i) through (iv):

(v) K generates X and $[0,x]$ is T-bounded for each x in K.

(vi) K generates X and K* generates X*.

(vii) K is generating and normal.

(viii) K is a normal strict b-cone.

Finally, if (X,T) is also ω-complete, then (i) through (viii) are equivalent.

<u>Proof</u>: We follow the following scheme of proof as shown below:

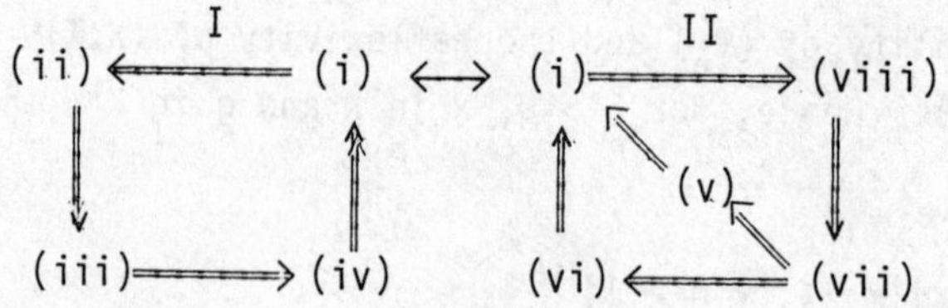

Step I: (i) ⇒ (ii) In view of Proposition 12.3.2, we need to show that K is a b-cone. So, let B be an arbitrary balanced bounded set in (X,T). If $S = \{S_\sigma(x) : x \in B \text{ and } \sigma \in \Phi\}$, then S is bounded (use barrelledness of (X,T)). For x in B and σ in Φ, let $\sigma^+ = \{i \in \sigma : f_i(x) \geq 0\}$ and $\sigma^- = \{i \in \sigma : f_i(x) < 0\}$ and suppose $y_\sigma = S_{\sigma^+}(x)$ and $z_\sigma = S_{\sigma^-}(x)$. Then $S_\sigma(x) = y_\sigma + z_\sigma \in S \cap K - S \cap K$ for every σ in Φ. Thus

$$B \subset \overline{S \cap K - S \cap K}$$

(ii) ⇒ (iii) To prove normality of K*, let $u \in \mathcal{B}_{X^*}$, where $X^* \equiv (X^*, \beta(X^*, X))$. Then, for some balanced, convex and bounded subset B of X,

$$W \equiv (\overline{B \cap K - B \cap K})^0 \subset u.$$

Set $V = (1/2)W$, then $V \in \mathcal{B}_{X^*}$. Let f ∈ V and consider g in X* with $0 \leq g \leq f$. For x in B ∩ K - B ∩ K, x = y-z where y,z ∈ B ∩ K. Since f(y) and $f(z) \leq 1/2$, $|g(x)| \leq 1$. Hence g ∈ W and consequently [0,f] ⊂ u, for every f in V. Thus K* is normal.

(iii) ⇒ (iv) See Theorem 1.5.16 for the first statement and the second one is an immediate consequence of normality.

(iv) ⇒ (i) By writing each f in X* as f = g - h with g,h being in K* and using Lemma 11.2.5, we easily find that $\{\sum_{i \in \sigma} f(x_i) f_i : \sigma \in \Phi\}$ is $\beta(X^*, X)$-bounded in X*. Therefore $\{x_n; f_n\}$ is unordered quasi-regular and as $[x_n] = X$, (i) follows from Proposition 8.2.9.

Step II: (v) ⇒ (i) The proof is similar to (iv) ⇒ (i).

(vi) ⇒ (i) Here the technique is the same as in (iv), (v) ⇒ (i). Indeed, it is easily seen that $\{S_\sigma(x) : \sigma \in \Phi\}$ is weakly bounded and hence bounded in (X,T). Now follow Proposition 8.2.9.

(vii) ⇒ (i) Observe that (vii) ⇒ (v), and now follow the proof of the sufficiency part of Theorem 12.3.3. Here we do not use the restriction on (X,T).

(viii) ⇒ (i) Indeed (viii) ⇒ (vii) and hence (i) follows.

Finally, assume that (X,T) is also ω-complete. Since (vii) ⇒ (vi) is true from Theorem 1.5.16, we have to show only

(i) ⇒ (viii) By Theorem 12.3.3, K is generating and normal. Hence we need to show that K is a strict b-cone. If B is any balanced bounded subset of (X,T), then, following (i) ⇒ (ii), the set $D = \bar{C}$ is also bounded, where $C = \{S_\sigma(x):x \in B, \sigma \in \Phi\}$. Introducing y_σ and z_σ as in (i) ⇒ (ii), and noting that $x^+ = \lim y_\sigma$ and $x^- = \lim z_\sigma$ with $x^+ \in D \cap K$ and $x^- \in -D \cap K$, we conclude that $B \subset D \cap K - D \cap K$, since $x = x^+ - x^-$ (cf. Proposition 12.2.5). □

Note: The above theorem includes similar results of Ceǐtlin [22], Gurevich [79], Levin and McArthur [151] and Schaefer [203]; p. 139.

Theorem 12.3.9: Let $\{x_n;f_n\}$ be an M-semibase for a reflexive space (X,T). Then statements (i) through (viii) of Theorem 12.3.8 are equivalent to

(ix) $\{f_n;\Psi(x_n)\}$ is a u-S.b. for $(X^*,\beta(X^*,X))$.

Proof: It suffices to prove that (i) ⇔ (ix).

(i) ⇒ (ix) By Theorem 8.3.17, $X^* = [f_n]^\beta$, $\beta = \beta(X^*,X)$ and now use Proposition 8.2.9.

(ix) ⇒ (i) Since $(X^*,\beta(X^*,X))$ is reflexive (cf. [93], p. 229), this follows as in (i) ⇒ (ix). □

Earlier we have seen (at least Theorem 11.3.2) that an important consequence of the normality of the cone K generated by an S.b. is the unconditional character of this base for K. In the direction of order structure of the space, we have the following application of normal cones contained in (cf. [163])

Proposition 12.3.10: Let (X,T) be ω-complete and $K \equiv K\{x_n;f_n\}$ be normal. Then X has the decomposition property.

Proof: It suffices to show that $[0,x+y] \subset [0,x] + [0,y]$ for all x,y in K. Suppose $u \in [0,x+y]$ and let

$$y_n = \sum_{i=1}^{n} \alpha_i f_i(u)x_i;\; z_n = \sum_{i=1}^{n} (1-\alpha_i)f_i(u)x_i,\; n \geq 1,$$

where

$$\alpha_i = \begin{cases} f_i(x)/f_i(x+y), & \text{if } f_i(x+y) \neq 0, \\ 0, & \text{if } f_i(x+y) = 0. \end{cases}$$

The monotone character of each p in D_T and the convergence of $\sum_{n \geq 1} f_n(x+y)x_n$ in (X,T) force each of the sequence $\{y_n\}$ and $\{z_n\}$ to the T-Cauchy in X. Hence there exist u_1 and u_2 in X with $y_n \to u_1$ and $z_n \to u_2$. Clearly, $u = u_1 + u_2$; $0 \leq u_1 \leq x$, $0 \leq u_2 \leq y$. □

12.4 REGULAR CONES

Regularity of a cone, which is a very useful property of order structure, has already figured in our earlier discussion; see, for instance, Theorems 11.2.7 and 11.3.12. It is clear that the regularity of a given cone $K \equiv K\{x_n;f_n\}$ is to some extent related to the unconditional character of $\{x_n;f_n\}$ and the latter has a close relationship with the normality of K. In the previous section we paid a good deal of attention on the normality of K, whereas in this section we study the regularity of K in terms of the unconditional character of the base, normality of K and the properties of the space. However, in general, a normal cone K associated with a b.s. $\{x_n;f_n\}$ is not necessarily regular, e.g. Examples 11.3.5 and 11.3.13 (here the cone is normal without being ω-regular).

To begin with, we have

<u>Proposition 12.4.1</u>: Let $\{x_n,f_n\}$ be a u-S.b. for (X,T). If (X,T) is ω-complete (resp. weakly ω-complete barrelled), then $K \equiv K\{x_n;f_n\}$ is regular (resp. fully ω-regular).

<u>Proof</u>: For the first (resp. second) part, use Theorem 1.4.5 and 11.2.7 (resp. Propositions 3.2 and 1.5.6). □

A characterization of a u-S.b. in terms of the regularity of its cone is contained in [169]; cf. also [84] for the Banach space analogue of this result.

<u>Theorem 12.4.2</u>: Suppose $\{x_n;f_n\}$ is an M-generalized base for a ω-complete l.c. TVS (X,T). Then $\{x_n;f_n\}$ is a u-S.b. for X if and only if $K \equiv K\{x_n;f_n\}$ is generating and regular.

<u>Proof</u>: The necessity part is immediate from Propositions 12.2.5 and 12.4.1.

For sufficiency, observe that [0,x] is σ(X,X*)-compact for each x in K;

cf. Theorem 11.2.7, (v) ⇒ (iii). Now proceed as in the proof of Proposition 12.3.6. □

Note: Observe that the regularity of K in Theorem 12.4.2 can be replaced by any other kind of regularity discussed in Theorem 11.2.7.

Remark: The ω-completeness in Proposition 12.4.1 and in the necessity part of Theorem 12.4.2 is indispensable; for consider

Example 12.4.3: This is the space $(k,\sigma(k,\ell^1))$ having the u-S.b. $\{e^n;e^n\}$; cf. Proposition 8.2.1. By Theorem 1.3.2 and the fact $k \subsetneq k^{\times\times}$, the space in question is not ω-complete. Here $K = \{\alpha \in k : \alpha_n \geq 0, n \geq 1\}$ is not ω-regular; indeed, if $x^n = \sum_{i=1}^{n} (1/i)e^i$, then $0 \leq x^n \leq x^{n+1} \leq e$ for $n \geq 1$, but $\{x^n\}$ does not converge in $(k,\sigma(k,\ell^1))$.

Exercise 12.4.4: Let the cone K associated with an M-generalized base for an l.c. TVS (X,T) be generating and regular. Without recourse to Theorem 11.2.7, show that $\{x_n;f_n\}$ is a u-S.b. for (X,T). [Hint: For x in K and σ, σ_o in Φ with $\sigma_o \subset \sigma$, $0 \leq S_{\sigma_o}(x) \leq S_\sigma(x) \leq x$.]

A characterization for a closed normal cone to be regular is reproduced below from [168]:

Theorem 12.4.5: In a Fréchet space (X,T), the following conditions are equivalent:

(i) Each closed normal cone K in X is fully regular.

(ii) Each closed normal cone K in X is regular.

(iii) X has no subspace topologically isomorphic to c_o.

(iv) Each closed subspace of X with an unconditional base is weakly ω-complete.

(v) Each unconditionally S. basic sequence in X is γ-complete.

Proof: We follow the scheme given below:

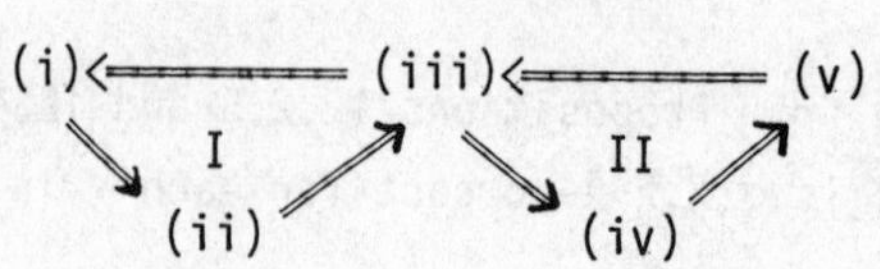

Stage I: (i) ⇒ (ii) Obvious.

(ii) ⇒ (iii) Suppose on the contrary that there exists a closed subspace $Y \subset X$ with $Y \simeq c_0$. Since $c \simeq c_0$ (cf. [5], p. 181), $Y \simeq c$, say, under $R:Y \to c$. The usual cone K of c is $\{\alpha \in c:\alpha_n \geqslant 0, n \geqslant 1\}$. Since $0 \leqslant e^{(n)} \leqslant e^{(n+1)} \leqslant e$, $n \geqslant 1$ and $\{e^{(n)}\}$ does not converge in c, K is not regular. If $K_1 = R[K]$, then K_1 is a cone in X such that K_1 is not regular, a contradiction.

(iii) ⇒ (i) If (i) were not true, there would exist a closed normal cone K in X such that K is not fully regular. Hence there exists a nonconvergent bounded sequence $\{x_n\}$ with $0 \leqslant x_1 \leqslant \ldots \leqslant x_n \leqslant \ldots$, where the ordering is induced by K. There exist u in $\mathcal{B}_X$ and increasing sequences $\{m_k\}$ and $\{n_k\}$, $m_k < n_k < m_{k+1}$ $(k \geqslant 1)$ such that $y_k \notin u$, $k \geqslant 1$, where $y_k = x_{n_k} - x_{m_k}$. If $\sigma \in \Phi$, say, $\sigma = \{p_1,\ldots,p_i\}$, then

$$\sum_{k\in\sigma} y_k \leqslant x_{n_{p_i}} \Rightarrow 0 \leqslant f(\sum_{k\in\sigma} y_k) \leqslant f(x_{n_{p_i}}), \ \forall f \in K^*$$

$$\Rightarrow 0 \leqslant f(\sum_{k\in\sigma} y_k) \leqslant M_f \equiv \sup_n f(x_n), \ \forall f \in K^*.$$

Hence, using Theorem 1.5.16, we conclude that $\{\sum_{k\in\sigma} y_k : \sigma \in \Phi\}$ is bounded in (X,T). However, this contradicts (iii); cf. Proposition 1.4.8.

Stage II: (iii) ⇒ (iv) Use Theorem 8.3.16.

(iv) ⇒ (v) If $\{x_n\}$ is an unconditionally S.basic sequence in X, then $Y = [x_n]$ is weakly ω-complete. Now apply Theorem 8.3.16.

(v) ⇒ (iii) On the other hand, let Y be a subspace of X with $c_0 \simeq Y$, say, under R. Observe that $\{R(e^n):n \geqslant 1\}$ is a u-S.b. for Y. But $\{e^n\}$ is not γ-complete and this will violate (v). □

The following two propositions (cf. [168]) may be considered as applications of Theorem 12.4.5.

Proposition 12.4.6: If X is a Banach space with X* being $\beta(X^*,X)$-separable, then each closed normal cone in $(X^*,\beta(X^*,X))$ is fully regular.

Proof: Since X* has no subspace isomorphic to c_0 (cf. [10], p. 155), the result follows by an application of Theorem 12.4.5. □

Proposition 12.4.7: If a Banach space X contains an unconditional base, then the following are equivalent:

(i) Each closed normal cone in X is fully regular.

(ii) X is $\sigma(X,X^*)$-ω-complete.

(iii) $X \simeq (Y^*,\beta(Y^*,Y))$ for some Banach space Y.

Proof: (i) ⇒ (ii) Apply Theorem 12.4.5, (i) ⇒ (iv).

(ii) ⇒ (i) See Proposition 1.5.6.

(ii) ⇒ (iii) By (ii) and Proposition 8.3.11, the base of X is γ-complete and so (iii) follows (cf. [162], p. 37).

(iii) ⇒ (i) Here Y* is $\beta(Y^*,Y)$-separable and now make use of Proposition 12.4.6. □

The next few results are concerned with the regularity of the associated dual cone and we follow [167] for the rest of this section.

Proposition 12.4.8: Let $\{x_n;f_n\}$ be an S.b. for a P-space (X,T). If $K \equiv K\{x_n;f_n\}$ is generating, then K* is regular.

Proof: Follows from Exercise 11.2.9 and Theorem 11.2.7. □

Proposition 12.4.9: Let $\{x_n;f_n\}$ be an S.b. for a Banach space (X,T) which is also a P-space. Then $K \equiv K\{x_n;f_n\}$ is generating if and only if $\{f_n;\Psi(x_n)\}$ is a u-S.b. for (K^*,β^*), $\beta^* \equiv \beta(X^*,X)|K^*$.

Proof: For necessity, use Exercise 11.2.9 and Theorem 1.4.5.

Conversely, observe first of all that K* is a cone. Since X* is $\beta(X^*,X)$-complete, $\sum_{n\geq 1} \alpha_n f(x_n) f_n$ converges in $(X^*,\beta(X^*,X))$ for each f in K* and α in ℓ^∞ with $\alpha_n \geq 0$ for $n \geq 1$. Now apply Theorems 11.2.7 and 1.5.8. □

Finally, we derive

Proposition 12.4.10: Let (X,T) be an ω-complete barrelled space containing a u-S.b. $\{x_n;f_n\}$. Then X* is $\beta(X^*,X)$-separable if and only if $\{f_n;\Psi(x_n)\}$ is a u-S.b. for $(X^*,\beta(X^*,X))$.

Proof: We need prove only the necessity part. Since (X,T) is also a P-space ([205], p. 143) and $K \equiv K\{x_n;f_n\}$ is generating and normal (Theorem 12.3.8),

$\{f_n;\Psi(x_n)\}$ is a u-S.b. for $(X^*,\beta(X^*,X))$ by Exercise 11.2.9, Theorems 1.4.5 and 1.5.16. □

Note: Proposition 12.4.10 extends a variation of a similar result of James [98], p. 523. It is also given in [232], p. 38.

13 Order structure

13.1. INTRODUCTION

With restrictions of somewhat mild nature on the space and its S.b., we discussed in the last chapter a number of analytical properties of the associated cone. In the course of this discussion, it was discovered that if (X,T) is ω-complete and the underlying S.b. is a u-S.b., the space X possesses a rich order structure, namely, X becomes a vector lattice (VL); cf. Proposition 12.2.5.

In this chapter, we intend to exploit this additional VL structure of the space. Indeed, specific attention will be paid on several notions related to the VL structure of an l.c. TVS containing an S.b.

As before, we assume throughout this chapter, unless the contrary is spelled out specifically, that (X,T) is an l.c. TVS over $\mathbb{R}$ containing an S.b. $\{x_n;f_n\}$ and the subset K of X is the cone associated with $\{x_n;f_n\}$; that is, $K \equiv K\{x_n,f_n\}$. We also assume that the order on X is induced by K. Further, we follow [22] and [91] for most of the results of this chapter.

13.2. LATTICE STRUCTURE

Necessary and sufficient conditions are obtained to obtain a VL and an l.c. TVL structure of an l.c. TVS containing an S.b. We begin with the following result which finds a different proof in [20].

Theorem 13.2.1: If (X,T) is ω-complete, then X is an order complete VL if and only if $\{x_n;f_n\}$ is a u-S.b.

Proof: (Necessity). For α in ℓ^∞, $0 \leqslant \alpha_n \leqslant 1$, $n \geqslant 1$ and x in K, let $A(\alpha;x) = \{\sum_{i=1}^{n} \alpha_i f_i(x)x_i : n \geqslant 1\}$. Then $A(\alpha;x) \subset K$ and is majorized by x. Put $z = \sup A(\alpha;x)$. If, for some i, $f_i(z) > \alpha_i f_i(x)$, then $y = \sum_{n\neq i} f_n(z)x_n + \alpha_i f_i(x)x_i$ is an upper bound of $A(\alpha;x)$ and so $y \geqslant z$. Hence $\alpha_i f_i(x) \geqslant f_i(z)$, a contradiction. Therefore, $f_n(z) = \alpha_n f_n(x)$, $n \geqslant 1$. The required conclusion follows from the fact that K is generating and $\sum_{n\geqslant 1} \alpha_n f_n(x)x_n \in K$.

Conversely, let $B \subset K$ be such that B is majorized by an element z in K. Choose α_n, $0 \leq \alpha_n \leq 1$, so that $\sup \{f_n(x) : x \in B\} = \alpha_n f_n(z)$. By Theorem 1.4.5, there exists y in X such that $y = \sum_{n \geq 1} \alpha_n f_n(z) x_n$. Clearly $y \in K$ and $y = \sup B$. Since X is already a VL (Proposition 12.2.5), we find that X is an order complete VL. □

Restricting (X,T) further, we have

Proposition 13.2.2: Let (X,T) be ω-complete and barrelled. If $\{x_n; f_n\}$ is a u-S.b., then (X,T) is an order complete l.c. TVL.

Proof: We need to show that (X,T) is an l.c. TVL. In view of Theorem 8.2.10, let us observe that T is also given by $\{P_p : p \in D_T\}$. Thus, if $|x| \leq |y|$ then $|f_n(x)| \leq |f_n(y)|$ for $n \geq 1$. We can find α_n, $|\alpha_n| \leq 1$, so that $f_n(x) = \alpha_n f_n(y)$, $n \geq 1$. It is clear that $P_p(x) \leq P_p(y)$ and so each P_p is solid and hence (X,T) is an l.c. TVL; cf. Section 1.5(d). □

Note: Proposition 13.2.2 includes, in particular, the results of Ceĭtlin ([20], p. 18) and Marti ([163], p. 31) who proved the same under more stringent conditions on (X,T). For an l.c. TVS (X,T) which is also a VL, let $p_*(x) = \sup \{p(y) : |y| \leq |x|\}$ for p in D_T, and call p_* a *lattice seminorm* corresponding to p. The next result is due to Carter ([19], p. 135) who proved it using Ceĭtlin's arguments in [20].

Exercise 13.2.3: Let (X,T) and $\{x_n; f_n\}$ be as in Proposition 13.2.2. Prove that T is given by lattice seminorms corresponding to members of D_T.

The following variation of Proposition 13.2.2 is due to Hofler [91].

Theorem 13.2.4: Let (X,T) be barrelled and $\{x_n; f_n\}$ a u-S.b. If K induces a vector lattice structure, i.e., X is a VL, then the lattice operations are continuous so that (X,T) is an l.c. TVL.

Proof: Invoking both the notation and the second proof of Proposition 12.3.2, we find that T is also generated by $D^* = \{p^* : p \in D_T\}$ with $p^*(S_\sigma(x)) \leq p^*(S_\rho(x))$ if $\sigma \subset \rho$ and $x \in X$. For x in X,

$$x^+ = \sum_{n \geq 1} \varepsilon_n f_n(x) x_n,$$

where $\varepsilon_n = 0$, if $f_n(x) < 0$, and $\varepsilon_n = 1$, if $f_n(x) \geq 0$. Hence for p^* in D^* and $n \geq 1$, we get (cf. Proposition 1.4.3)

$$p^*(\sum_{i=1}^{n} \varepsilon_i f_i(x)x_i) \leq 4 \sup_{\sigma\subset[1,n]} p^*(\sum_{i\in\sigma} f_i(x)x_i)$$

$$\leq 4p^*(\sum_{i=1}^{n} f_i(x)x_i)$$

$$\Rightarrow \quad p^*(x^+) \leq 4p^*(x).$$

Therefore, $x \to x^+$ is continuous. Now apply Proposition 12.3.2 and use Section 1.5(d). □

Remark: The conclusions in Proposition 13.2.2 and Theorem 13.2.4 are the same; however, the hypothesis in the former is stronger than in the latter. Consequently, Theorem 13.2.4 is an extension of Proposition 13.2.2. This is justified by the following

Example 13.2.5: Here we construct a subspace Y of ℓ^1 such that Y is barrelled but not ω-complete, possesses $\{e^n;e^n\}$ as its S.b. and is a VL. The construction of Y we follow hereafter is due to Saxon [202].

Fix an arbitrary element b in ℓ^1, $b_n \neq 0$ for infinitely many indices n with $|b_n| \leq 1$, $n \geq 1$ (for instance, one may take $b_n = 1/2^{2n}$) and let

$$Y \equiv Y_b = sp\,\{\{e^n : n \geq 1\} \cup \{\sum_{n\geq 1} b_n e^{m_n} : \{m_n\} \subset \mathbb{N}\}\}.$$

We proceed to show that Y is a barrelled subspace of ℓ^1, it being clearly true that Y is not ω-complete relative to the induced norm $\|\cdot\|_1$.

If Y were not a barrelled subspace of $(\ell^1, \|\cdot\|_1)$, there would exist a $\sigma(Y^*,Y)$-bounded subset M of Y^* such that M is not equicontinuous on $(Y,\|\cdot\|_1)$. If $A_n = \{|\langle e^n,f\rangle| : f \in M\}$, then $K_n \equiv \sup A_n < \infty$ for $n \geq 1$. Also observe that $A = \{|\langle e^n,f\rangle| : f \in M,\ n \geq 1\} = \cup\,\{A_n : n \geq 1\}$ is unbounded.

We now claim that there exist an increasing subsequence $\{m_n\}$ of $\mathbb{N}$, a strictly increasing function $\sigma:\mathbb{N} \to \mathbb{N}$ and a sequence $\{g_n\}$ from M such that

$$|g_n(\sum_{i=1}^{m_n} b_i e^{\sigma(i)})| \geq n; \qquad (*)$$

and

$$\left\| \sum_{i \geq m_n+1} b_i e^i \right\|_1 < \frac{1}{\|g_n\|}, \quad n \geq 1. \tag{**}$$

We prove (*) and (**) by induction. Choose $r(> 1)$ in $\mathbb{N}$ as the smallest number so that $|b_r| = \delta > 0$ and $\sum_{i=1}^{r} K_i \geq 1$. Since A is unbounded, we find g_1 in M and $p \geq 1$ such that $|\langle e^p, g_1\rangle| \geq 6\delta^{-1} \sum_{i=1}^{r} K_i$. It follows that $p \geq r+1$. Choose s in $\mathbb{N}$ so that $p \leq r+s$ and $\| \sum_{i \geq r+s+1} b_i e^i \|_1 \leq \|g_1\|^{-1}$. If $m_1 = r+s$, we readily obtain (**).

To verify (*), let us consider two cases.

Case 1. Assume the truth of

$$|g_1(\sum_{j=0}^{s} b_{j+r} e^{j+p})| \geq 3 \sum_{i=1}^{r} K_i. \tag{+}$$

Define $\sigma_1 : [1, m_1] \to \mathbb{N}$ by $\sigma_1(i) = i (1 \leq i \leq r-1)$ and $\sigma_1(j+r) = j+p$ $(0 \leq j \leq s)$. Then

$$|g_1(\sum_{j=1}^{m_1} b_j e^{\sigma_1(j)})| \geq |g_1(\sum_{j=0}^{s} b_{j+r} e^{j+p})| - \sum_{i=1}^{r-1} K_i$$

$$> 1.$$

Case 2. Suppose (+) is not true. Then define $\sigma_1 : [1, m_1] \to \mathbb{N}$ by $\sigma_1(i) = i$ $(1 \leq i \leq r)$ and $\sigma_1(j+r) = j+p$ $(1 \leq j \leq s)$. Hence

$$|g_1(\sum_{j=1}^{m_1} b_j e^{\sigma_1(j)})| \geq |g_1(\sum_{j=1}^{s} b_{j+r} e^{j+p})| - |g_1(\sum_{j=1}^{r} b_j e^j)|.$$

But

$$3 \sum_{i=1}^{r} K_i > |g_1(b_r e^p)| - |g_1(\sum_{j=1}^{s} b_{j+r} e^{j+p})|$$

$$\Rightarrow |g_1(\sum_{j=1}^{s} b_{j+r} e^{j+p})| > 3 \sum_{i=1}^{r} K_i,$$

and so

$$|g_1(\sum_{j=1}^{m_1} b_j e^{\sigma_1(j)})| > 2 \sum_{i=1}^{r} K_i > 1.$$

Hence (*) follows by choosing $\sigma|[1,m_1] = \sigma_1$.

Let us now assume the truth of (*) and (**) for $g_1,\ldots,g_n$ in M, increasing integers $m_1 < \ldots < m_n$ in $\mathbb{N}$ and a strictly increasing function $\sigma:\mathbb{N} \to \mathbb{N}$ with $\sigma|[1,m_i] = \sigma_i$ $(1 \leqslant i \leqslant n)$. Clearly

$$\sum_{i=1}^{\sigma_n(m_n)} K_i \geqslant n \geqslant 1.$$

Choose r in $\mathbb{N}$ as the smallest number such that $|b_{m_n+r}| > 0$. As before, we find g_{n+1} in M and $p \equiv p(n,r)$ in $\mathbb{N}$ such that

$$|\langle e^p, g_{n+1}\rangle| \geqslant 2(n+2)|b_{m_n+r}^{-1}| \sum_{i=1}^{\sigma_n(m_n)+r} K_i.$$

Then $p \geqslant \sigma_n(m_n) + r + 1$, otherwise this will contradict the preceding inequality. Choose s in $\mathbb{N}$ with $p \leqslant m_n + r + s$ and

$$\|\sum_{i\geqslant m_n+r+s+1} b_i e^i\|_1 \leqslant \frac{1}{\|g_{n+1}\|}.$$

If $m_{n+1} = m_n + r + s$, the preceding inequality yields (**). To prove (*), we consider once again two cases: Case 3 and Case 4.

<u>Case 3</u>. Suppose

$$|g_{n+1}(\sum_{j=0}^{s} b_{m_n+r+j} e^{p+j})| \geqslant (n+2) \sum_{i=1}^{\sigma_n(m_n)+r} K_i. \tag{++}$$

Define $\sigma_{n+1}:[1,m_{n+1}] \to \mathbb{N}$ such that $\sigma_{n+1}|[1,m_n] = \sigma_n$, $\sigma_{n+1}(m_n+j) = \sigma_n(m_n) + j$ $(1 \leqslant j \leqslant r-1)$ and $\sigma_{n+1}(m_n+r+j) = p+j(0 \leqslant j \leqslant s)$. Hence

$$|g_{n+1}(\sum_{j=1}^{m_{n+1}} b_j e^{\sigma_{n+1}(j)})| \geqslant (n+2) \sum_{i=1}^{\sigma_n(m_n)+r} K_i - \sum_{j=1}^{m_n+r-1} K_{\sigma_{n+1}(j)}$$

$$\geqslant (n+1) \sum_{i=1}^{\sigma_n(m_n)+r} K_i,$$

and this proves (*) by choosing $\sigma|[1,m_{n+1}] = \sigma_{n+1}$.

Case 4. Here we assume that (++) is not true. As in Case 2, define $\sigma_{n+1}:[1,m_{n+1}] \to \mathbb{N}$ such that $\sigma_{n+1}|[1,m_n] = \sigma_n$, $\sigma_{n+1}(m_n+j) = \sigma_n(m_n) + j$ $(1 \leqslant j \leqslant r)$ and $\sigma_{n+1}(m_n+r+j) = p+j$ $(1 \leqslant j \leqslant s)$. Note that

$$(n+2) \sum_{i=1}^{\sigma_n(m_n)+r} K_i > |b_{m_n+r}| \, |g_{n+1}(e^p)| - |g_{n+1}(\sum_{j=1}^{s} b_{m_n+r+j} e^{p+j})|$$

$$\Rightarrow |g_{n+1}(\sum_{j=1}^{s} b_{m_n+r+j} e^{p+j})| > (n+2) \sum_{i=1}^{\sigma_n(m_n)+r} K_i.$$

Therefore

$$|g_{n+1}(\sum_{j=1}^{m_{n+1}} b_j e^{\sigma_{n+1}(j)})| > (n+2) \sum_{i=1}^{\sigma_n(m_n)+r} K_i - \sum_{j=1}^{m_n+r} K_{\sigma_{n+1}(j)}$$

$$\geqslant n + 1,$$

and this again proves (*) as in Case 3.

Finally, we let x belong to Y with $x = \sum_{n \geqslant 1} b_n e^{\sigma(n)}$, where σ is the function occurring in (*). Then

$$|g_i(x)| \geqslant i - \|g_i\| \; \| \sum_{j \geqslant m_i+1} b_j e^{\sigma(j)} \|_1$$

$$\geqslant i-1,$$

showing thereby that M is not $\sigma(Y^*,Y)$-bounded, a contradiction.

Next, observe that $\{e^n;e^n\}$ is a subseries base for $(Y, \|\cdot\|_1)$ and so it is unconditional. Here $K \equiv K\{e^n;e^n\} = \{\alpha \in Y: \alpha_n \geqslant 0, n \geqslant 1\}$. Clearly $\{|\alpha_n|\} \in Y$ if $\{\alpha_n\} \in Y$. Hence, for α in Y, $\alpha^+ = (|\alpha| + \alpha)/2$; $\alpha^- = (|\alpha| - \alpha)2$ exist in K and so Y is a VL. This completes the construction of Y.

Exercise 13.2.6: Let (X,T) be barrelled and $\{x_n;f_n\}$ be a u-S.b. Suppose $K \equiv K\{x_n;f_n\}$ induces a VL structure on X. Show that $(X^*,\beta(X^*,X))$, $(X^*,\beta(X^*,X^{**}))$ and $(X^{**},\beta(X^{**},X^*))$ are l.c. TVL, where X^* and X^{**} are ordered by the dual cone K^* and the dual wedge K^{**} of K^* [Hint: Apply

Theorem 13.2.4 and Propositions 1.2.11, 1.5.9 and 1.5.10].

Weak lattice operations

In the next few pages, we show that the locally convex vector lattice structure on X or X* with respect to weak topologies can be characterized quite conveniently;cf. [91].

At the outset, let us first of all turn to a lemma (cf. [91]) which is a minor extension of a similar result proved earlier by Cook [24]. In what follows, the sequence spaces ϕ and ω are considered over the field $\mathbb{R}$.

Lemma 13.2.7: Let $\{x_n;f_n\}$ be an S.b. for an l.c. TVS (X,T). Then there eixsts a topological isomorphism R from (X,T) onto a subspace $(\lambda,\sigma(\omega,\phi)|\lambda)$ of $(\omega,\sigma(\omega,\phi))$ with $R(x_n) = e^n (n \geq 1)$ if and only if $\{f_n\}$ is a Hamel base for X*.

Proof: Since $\bar{\lambda} = \bar{\phi} = \omega$, $\lambda^* = \phi$, where $\lambda = R[X]$. Hence the necessity part is clear.

Conversely, let $\{f_n\}$ be a Hamel base for X*. Write $\lambda = \{\{f_n(x)\} : x \in X\}$ and suppose $R:X \to \lambda$, $R(x) = \{f_n(x)\}$. R is clearly an algebraic isomorphism from X onto λ, also R is T-$\sigma(\omega,\phi)|\lambda$ continuous (cf. Proposition 2.3.1). To prove the continuity of R^{-1}, let $\alpha^n \to 0$ in $(\lambda,\sigma(\omega,\phi)|\lambda)$. Choose an arbitrary f in X*. Then

$$f = \sum_{i=1}^{n} \beta_i f_i$$

and put $\beta = \{\beta_1,\ldots,\beta_n,0,0,\ldots\}$. There exists K_β such that for each $m \geq 1$

$$|\langle y^m,f\rangle| = |\sum_{i=1}^{n} \alpha_i^m \beta_i| \leq K_\beta, \quad \forall m \geq 1$$

where $R(y_m) = \alpha^m$. Therefore $\{R^{-1}(\alpha^m):m \geq 1\}$ is bounded in (X,T) and so R^{-1} is continuous by [146], p. 168 or [200], p. 23. □

Theorem 13.2.8: Let (X,T) be an S-space and $\{x_n,f_n\}$ a u-S.b. If $K \equiv K\{x_n,f_n\}$ defines a VL structure on X, then the following are equivalent:

(i) $(X,\sigma(X,X^*))$ is an l.c. TVL.

(ii) $\{f_n\}$ is a Hamel base for X^*.

(iii) There exists a topological isomorphism R from $(X.T)$ onto an S-subspace λ of $(\omega,\sigma(\omega,\phi))$ with $R(x_n) = e^n$; $n \geq 1$.

(iv) $\{S_n\}$ is $\sigma(X,X^*)$-$\sigma(X,X^*)$ equicontinuous.

Proof: (i) ⇒ (ii) On the other hand, let (ii) be not true. Then there exists f in X^* such that

$$f = \sigma(X^*,X) - \lim_{n\to\infty} \sum_{i=1}^{n} f(x_i)f_i,$$

where $f(x_{n_k}) \neq 0$ for some infinite subsequence $\{n_k\}$ of $\mathbb{N}$. Since $X^* = K^* - K^*$ (Proposition 12.2.5), we may assume that $f(x_n) \geq 0$ for $n \geq 1$. Let

$$g_k = \sum_{i=1}^{n_k} f(x_i)f_i,\ k \geq 1.$$

Then $\{g_k\}$ is linearly independent and is contained in $[0,f]$, that is, $[0,f]$ is not finite dimensional, a contradiction; cf. Proposition 1.5.9.

(ii) ⇒ (i) In view of Proposition 12.3.1, it suffices to prove the continuity at the origin of one of the lattice operations, say, $x \to x^+$. Let a net $\{y_\alpha\}$ converge to zero in $\sigma(X,X^*)$. By Lemma 11.2.6,

$$y_\alpha^+ = \sum_{n\geq 1} \varepsilon_n^\alpha(y_\alpha)x_n,$$

where $\varepsilon_n^\alpha = 0$, if $f_n(y_\alpha) < 0$ and $\varepsilon_n^\alpha = 1$, if $f_n(y_\alpha) \geq 0$. As $\{f_n\}$ is a Hamel base for X^*, for each f in X^*, $f(x_n)$ vanishes eventually in n and we easily conclude that $f(y_\alpha^+) \to 0$.

(ii) ⇔ (iii) Cf. Lemma 13.2.7.

(ii) ⇒ (iv) If $f \in X^*$, then $f = \sum_{i=1}^{n} f(x_i)f_i$

and so $|f(S_n(x))| = |f(x)|$ for each x in X and all $n \geq N$. This proves (iv).

(iv) ⇒ (ii) For any f in X^*, $\{f \circ S_n\}$ is finite dimensional; cf. [140], p. 161 (see also [132], p. 20). Hence there exist a linearly independent set $\{f \circ S_{n_i} : 1 \leq i \leq k\}$ and a set $\{\alpha_i^n : 1 \leq i \leq k\}$ of reals such that

$$f(S_n(x)) = \sum_{i=1}^{k} \alpha_i^n f(S_{n_i}(x));\ \forall\, n \geq 1,\ x \in X.$$

Hence $f(x_i) = 0$ for $i > n_k$ and so

$$f = \sum_{i=1}^{n_k} f(x_i)f_i. \qquad (*)$$

Therefore, each f in X* is representable as (*) for some finite n_k. But $\{f_n\}$ is clearly linearly independent. Consequently $\{f_n\}$ is a Hamel base for X*. □

Note: The proof of (ii) ⟺ (iv) is essentially taken from [24], while the proof of other implications in the above theorem are reproduced from [91].

Proposition 13.2.9: Let (X,T) be ω-complete and an S-space with $\{x_n;f_n\}$ being a u-S.b. If any of the lattice operations is $\sigma(X,X^*)$-continuous at the origin, then $(X,T) \simeq (\omega,\sigma(\omega,\phi))$.

Proof: By virtue of Propositions 12.2.5 and 12.3.1, we find the truth of Theorem 13.2.8(i) and hence that of (ii) and (iii) of this theorem. The required result now follows from Theorem 2.8(iii) provided that we show that $\lambda = \omega$. So, let $\alpha \in \omega$. Since $f(x_n) = 0$ eventually in n for each given f, the sequence $\{\sum_{i=1}^{n} \alpha_i x_i\}$ is $\sigma(X,X^*)$-Cauchy. Now $\sigma(X,X^*)$ is metrizable ([234], p. 150); therefore $T \approx \sigma(X,X^*)$. Thus for x in X, $x = \sum_{n\geq 1} \alpha_n x_n$ and $R(x) = \alpha$. □

The restriction on the space (X,T) in Proposition 13.2.9 cannot be dropped, for we have

Example 13.2.10: The space $(\ell^p,\sigma(\ell^p,\phi))$ having the u-S.b. $\{e^n;e^n\}$ is a VL with ordering induced by $K = \{\alpha \in \ell^p\colon \alpha_n \geq 0, n \geq 1\}$, where $1 \leq p < \infty$. Here the lattice operations are obviously continuous. If $\alpha^n = x^{(n)}$ with $x = \{1/n^2\}$, then $\{\alpha^n\}$ is $\sigma(\phi,\ell^p)$-Cauchy but does not converge in $(\phi,\sigma(\phi,\ell^p))$. Thus $(\ell^p,\sigma(\ell^p,\phi))$ is not an S-space. In a similar fashion, by considering the $\sigma(\ell^p,\phi)$-Cauchy sequence $\{e^{(n)}\}$, one can show that ℓ^p is not $\sigma(\ell^p,\phi)$-ω-complete. Finally, observe that $(\ell^p,\sigma(\ell^p,\phi)) \not\simeq (\omega,\sigma(\omega,\phi))$.

Exercise 13.2.11: Show that the space $(c_0,\sigma(c_0,\phi))$ serves the same purpose as the space considered in Example 13.2.10.

Exercise 13.2.12: Let (X,T) be ω-complete and an S-space possessing a u-S.b.

$\{x_n;f_n\}$. If K^* is the dual cone of $K \equiv K\{x_n;f_n\}$, show that $(X^*,\sigma(X^*,X))$ is an l.c. TVL if and only if $\{x_n\}$ is a Hamel base for X. [Hint: use Proposition 12.2.5 and 1.5.7 to conclude the $\sigma(X^*,X)$-normality of K^*, now proceed as in Theorem 13.2.8 (i) $\Leftrightarrow$ (ii).]

Exercise 13.2.13: Let (X,T) be an l.c. TVL ordered by $K \equiv K\{x_n;f_n\}$. If $\{x_n;f_n\}$ is γ-complete, prove that $\{x_n;f_n\}$ is b.m-S.b.

Boundedness property (BP)

At the end of this chapter, we characterize the BP of (X,T) in the form of the following result [22].

Theorem 13.2.14: Let (X,T) be an ω-complete S-space such that it is also an order complete l.c. TVL under its cone $K \equiv K\{x_n;f_n\}$. Then the following statements are equivalent:

(i) $\{x_n;f_n\}$ is γ-complete.

(ii) $(X,\sigma(X,X^*))$ is ω-complete.

(iii) X is a T-order complete l.c. TVL.

(iv) X has the BP.

Proof: At the outset, we may assume that T is generated by lattice seminorms.

(i) $\Rightarrow$ (ii) By Exercise 13.2.13 and Proposition 8.2.4, $\delta^\times = \mu$. Thus, using Propositions 2.3.1, 8.3.9 and Theorem 1.3.2, X is $\sigma(X,X^*)$-ω-complete.

(ii) $\Rightarrow$ (iii) Here we verify (a) and (b) of Definition 1.5.14. Since (a) is immediate, we proceed to establish (b). On the other hand, let there be an order unbounded increasing sequence $\{y_n\}$ in K such that $\{y_n\}$ is bounded in (X,T). Hence, for f in K^*, there exists $M \equiv M(f) > 0$ such that $0 \leqslant f(y_n) \leqslant f(y_{n+1}) \leqslant M$ for $n \geqslant 1$, giving $\{f(y_n)\}$ to be a convergent sequence. Therefore, $\{y_n\}$ is $\sigma(X,X^*)$-Cauchy (here $X^* = K^* - K^*$; cf. Theorem 1.5.16) and so, for some y in X, $f_i(y) = \sup \{f_i(y_n):n \geqslant 1\}$, $i \geqslant 1$. Consequently $\{y_n\}$ is order bounded and this proves (iii).

(iii) $\Rightarrow$ (iv) Since in l.c. TVS with a Schauder base clearly possesses the ω-property, (iv) follows by an application of Theorem 1.5.15(ii).

(iv) $\Rightarrow$ (i) On account of Theorems 13.2.1 and 1.4.5, $\{x_n;f_n\}$ is a b.m-S.b. If (i) were not true, we would get an α in ω (over $\mathbb{R}$) such that $\{\sum_{i=1}^{n} \alpha_i x_i\}$ is bounded but $\sum_{n \geq 1} \alpha_n x_n$ does not converge in (X,T). Without loss of generality, we may take all $\alpha_n > 0$ and let $B = \{\alpha_n x_n\}$. It is easily seen that B is not order bounded (use the b.m-S.b. nature of $\{x_n; f_n\}$ to arrive at a contradiction); and also that $\sum_{n \geq 1} \beta_n \alpha_n x_n$ converges in (X,T) for each β in c_o. In particular, if $\beta_n \downarrow 0$ and $\gamma_n = \sqrt{\beta_n}$, then, for some z in X,

$$z = \sum_{n \geq 1} \gamma_n \alpha_n x_n.$$

If $y_n = \gamma_n z$, then $y_n \downarrow 0$ and $|\beta_n \alpha_n x_n| \leq y_n$, $n \geq 1$. Hence $\beta_n \alpha_n x_n \xrightarrow{o} 0$ and this shows (by virtue of (iv)) that B is order bounded, a contradiction. □

Remark: Note that the S-character of the space is required only in Theorem 13.2.14 (i) ⇒ (ii). A variation of Theorem 13.2.14 (i) ⟺ (ii) is also given in [233], p. 479.

14 Order topology

14.1 INTRODUCTION

Having dealt at length with the impact of order structure on the behaviour of the corresponding S.b. and vice versa, we now turn in this chapter to the related study of the order topology T^{or} and exploit its composition with the help of the order structure induced by an S.b. present in an l.c. TVS (X,T) over $\mathbb{R}$. In particular, we pay specific attention to the natural problem of finding a comparison between T^{or} and T. A similar study is also made in obtaining the relationship between the order topology T^{*or} on the dual space and the related strong and other topologies. We follow [19] and [22] for most of the discussion of this chapter.

Without further reference, we will write (X,T) to mean an arbitrary real TVS or l.c. TVS having an S.b. $\{x_n;f_n\}$ and further assume that the ordering $\prec$ in X is induced by the cone $K \equiv K\{x_n;f_n\}$.

14.2 THE T^{or}-TOPOLOGY

In general, there is no relationship between the topologies T^{or} and T. Accordingly, conditions are investigated to find the inclusion relationship between T^{or} and T, which finally yield $T^{or} \approx T$. We begin with a useful though simple

Proposition 14.2.1: A decreasing net $\{y_\alpha : \alpha \in \Lambda\}$ of positive elements in (X,T) satisfies $y_\alpha \downarrow 0$ if and only if $f_n(y_\alpha) \to 0$ for each $n \succcurlyeq 1$. Also, if $y_\alpha \downarrow 0$, then there exists an increasing sequence $\{\alpha_n : n \succcurlyeq 1\} \subset \Lambda$ such that $y_{\alpha_n} \downarrow 0$.

Proof: Concerning the first part, the necessity is obvious.

For the converse, if $\inf y_\alpha \neq 0$, then there exists z in X such that $0 < z \prec y_\alpha$ for each α. We find an i_0 with $f_{i_0}(z) > 0$ and so $f_{i_0}(y_\alpha) \not\to 0$.

Coming to the second part, let us observe that $f_n(y_\alpha) \downarrow 0$ for $n \succcurlyeq 1$. Choose α_1 in Λ with $f_1(y_{\alpha_1}) < 1$; then choose $\alpha_2 \succcurlyeq \alpha_1$ so that $f_1(y_{\alpha_2})$,

$f_2(y_{\alpha_2}) < 1/2$. Hence by induction we get an increasing sequence $\{\alpha_n\}$ such that $f_i(y_{\alpha_n}) < 1/n$, $1 \leq i \leq n$; $n \geq 1$, and this proves the result. □

Exercise 14.2.2: Consider a net $\{x_\alpha : \alpha \in \Lambda\}$ in (X,T). If $x_\alpha \uparrow x$, show that $f_n(x_\alpha) \to f_n(x)$ for $n \geq 1$ and there exists an increasing net $\{\alpha_n\} \subset \Lambda$ so that $y_{\alpha_n} \uparrow x$. If $x_\alpha \xrightarrow{(o)} x$, prove also the existence of an increasing sequence $\{\alpha_n\} \subset \Lambda$ with $x_{\alpha_n} \xrightarrow{(o)} x$, and consequently establish that a set B in X is T^{or}-closed if and only if it contains the (0)-limits of its (0)-convergent sequences.

To establish comparability between the order and original topologies, we need an intermediary result in the form of a

Lemma 14.2.3: In an OVS Y, T^{or} is the finest topology on Y such that each (0)-convergent net is T^{or}-convergent.

Proof: Cf. Proposition 1.5.13 (vi), (vii). □

Remark: In the above lemma, if the ordering is given by an S.b., that is, if $Y = (X,T)$, then, by virtue of Exercise 14.2.2, nets can be replaced by sequences. This statement also applies to spaces having weak S.b.

Theorem 14.2.4: Let (X,T) be ω-complete and K be normal, then $T \subset T^{or}$. Conversely, if (X,T) is a Fréchet space and $T \subset T^{or}$, then K is normal.

Proof: We may assume the members of D_T to be monotone. Let now $\{y_n\}$ be a sequence in X with $y_n \downarrow 0$. Choose p in D_T and $\varepsilon > 0$ arbitrarily. There exists $N \equiv N(p,\varepsilon,y_1)$ such that (using $y_n \leq y_1$)

$$p\Big(\sum_{i \geq N+1} f_i(y_n)x_i\Big) \leq \frac{\varepsilon}{2}, \ \forall\, n \geq 1.$$

Let $\sigma = \{i : 1 \leq i \leq N,\ p(x_i) \neq 0\}$, $m = \#(\sigma)$ and $M = \max\,\{p(x_i) : i \in \sigma\}$. Using Proposition 14.2.1 we find n_o in $\mathbb{N}$ so that

$$|f_i(y_n)| \leq \frac{\varepsilon}{2mM}\ ;\ \forall\, n \geq n_o,\ i \in \sigma$$

$$\Rightarrow p(y_n) \leq \sum_{i\in\sigma} \frac{\varepsilon}{2mM}\, p(x_i) + \frac{\varepsilon}{2} \leq \varepsilon,\ \forall\, n \geq n_o.$$

Thus $y_n \to 0$ in (X,T).

In view of the remark made following Lemma 14.2.3, it is enough to show that each (0)-convergent sequence is T-convergent. Let therefore $y_n \xrightarrow{(o)} 0$. Then there exist sequences $\{u_n\}$ and $\{v_n\}$ with $u_n \uparrow 0$, $v_n \downarrow 0$ and $u_n \leq y_n \leq v_n$, $n \geq 1$. Since $y_n^+ \leq v_n$, $y_n^- \leq -u_n$, $p(y_n) \leq p(u_n) + p(v_n) \to 0$ for each p in D_T.

For the converse, it is easily seen that $x = \sup \{S_\sigma(x):\sigma \in \Phi\}$ for x in K, where Φ is considered a directed set under the usual set-theoretic inclusion. Thus $S_\sigma(x) \uparrow x$ and consequently, by Proposition 1.5.13, $S_\sigma(x) \xrightarrow{(o)} x$ and hence $S_\sigma(x) \to x$ in T^{or}, that is, $S_\sigma(x) \to x$ in T for x in K. In other words, $\{x_n;f_n\}$ is unconditional on K. Now apply Theorem 11.3.2. □

Corollary 14.2.5: Let (X,T) be an ω-complete barrelled l.c. TVS and $\{x_n;f_n\}$ an u-S.b., then $T \subset T^{or}$.

For proof, make use of Proposition 12.3.2.

Note: The above corollary in a slightly weaker form is due to Cešlin [22].

Concerning the inclusion $T^{or} \subset T$, we have the following two results:

Proposition 14.2.6: Let an OVS Y be equipped with a first countable topology S such that $y_n \xrightarrow{(*)} 0$ whenever $y_n \to 0$ in S. Then $T^{or} \subset S$. If $(Y,S) = (X,T)$, then the converse also holds good.

Proof: The first part is easily proved by showing that if $y_n \to y$ in S, then $y_n \to y$ in T^{or} with the help of Proposition 1.5.13(vii). The second part is an immediate consequence of the following proposition. □

Proposition 14.2.7: A sequence $\{y_n\}$ T^{or}-converges to y in (X,T) if and only if it (*)-converges to y.

Proof: In view of Proposition 1.5.13 (vii), it suffices to prove the necessity. So, let $y_n \to Y$ in T^{or} and suppose $y_n \xrightarrow{(*)} y$ is not true. Then there exists a subsequence $\{y_{n_i}\}$ of $\{y_n\}$ such that no subsequence of $\{y_{n_i}\}$ (0)-converges to y. We may therefore assume that $y_{n_i} \neq y$ for $i \geq 1$.

We consider now two cases:

Case 1: Suppose that there is no subsequence of $\{y_{n_i}\}$, (0)-converging to any member of X. Then the set $A = \{y_{n_i} : i \geq 1\}$ is T^{or}-closed (Exercise 14.2.2) and $x \in X \setminus A$. This gives contradiction to T^{or}-convergence of $\{y_n\}$ to x.

Case 2: Let a subsequence $\{y_{n_{i_j}}\}$ of $\{y_{n_i}\}$ (0)-converge to z in X. Clearly $z \neq y$. If $D = \{y_{n_{i_j}} : j \geq 1\} \cup \{z\}$, then $X \setminus D$ is an open set containing y, a situation as in case 1. □

Proposition 14.2.8: Let there be a Fréchet lattice (Y,S). Then $T^{or} \subset S$.

Proof: Arrange D_S as $\{p_1 \leq \ldots \leq p_n \leq \ldots\}$. Let now $y_n \to 0$ in S. Find a subsequence $\{y_{n_i}\}$ of $\{y_n\}$ so that $p_i(y_{n_i}) < i^{-3}$, $i \geq 1$. Fix m in $\mathbb{N}$, then

$$\sum_{i \geq 1} p_m(i|y_{n_i}|) \leq \sum_{i=1}^{m-1} p_m(i|y_{n_i}|) + \sum_{i \geq m} \frac{1}{i^2},$$

and, as m is arbitrary, we find y in X such that $y = \sum_{i \geq 1} i|y_{n_i}|$. Clearly $|y_{n_i}| \leq i^{-1} y$ for $i \geq 1$ and so $y_{n_i} \xrightarrow{(o)} 0$. In particular, repeating the above analysis with a subsequence of $\{y_n\}$, we conclude that $y_n \xrightarrow{(*)} 0$. Finally, we make use of Proposition 14.2.6. □

Now follows one of the main results of this section, namely,

Theorem 14.2.9: Let (X,T) be a Fréchet space and $\{x_n; f_n\}$ a u-S.b. Then $T \approx T^{or}$.

Proof: By Proposition 13.2.2, (X,T) is a Fréchet lattice. Now make use of Corollary 14.2.5 and Proposition 14.2.8. □

Note: Theorem 14.2.9 is due to Ceǐtlin [22] and answers a question raised earlier by Kantrovic, Vulih and Pinsker [137] in 1951.

Remark: The Fréchet character in Theorem 14.6.9 is indispensable; in other words, completeness and metrizability in Theorem 14.2.9 cannot be dropped, as is illustrated in the following two examples respectively.

Example 14.2.10: Consider the u-S.b. $\{e^n;e^n\}$ for the incomplete normed space $(\phi, \|\cdot\|_\infty)$; cf. Examples 2.2.14 and 11.3.8. If T_∞ denotes the topology generated by $\|\cdot\|_\infty$, we will show that $T_\infty \subsetneqq T_\infty^{or}$. Indeed, $T_\infty \subset T^{or}$ follows by Lemma 14.2.3 and the remark following it. For, let $\alpha^n \downarrow 0$. If ℓ is the length of α^1, then for any given $\varepsilon > 0$ we find an $m \equiv m(\varepsilon,\ell)$ in $\mathbb{N}$ so that $|\alpha_i^n| \leq \varepsilon$ for all $n \geq m$ and $1 \leq i \leq \ell$. Since $\alpha_i^n = 0$ for $i > \ell$ and $n \geq 1$, $\|\alpha^n\|_\infty \leq \varepsilon$ for all $n \geq m$. Thus $\alpha^n \to 0$ in T_∞. On the other hand, $e^n/n \to 0$ in T_∞ but $e^n/n \overset{(*)}{\not\longrightarrow} 0$, since no subsequence of $\{e^n/n\}$ is order bounded in ϕ. Therefore, by Proposition 14.2.6, $T^{or} \not\subset T_\infty$, giving $T_\infty \subsetneqq T^{or}$.

Example 14.2.11: Here we consider the ω-complete nonmetrizable space $(\ell^1,\sigma(\ell^1,\ell^\infty))$ having the unconditional base $\{e^n;e^n\}$. Observe that $\|\cdot\|_1$ generates the topology $\eta(\ell^1,\ell^\infty)$ and, by Theorem 14.2.9, $\eta(\ell^1,\ell^\infty) \approx T^{or}$. However $\sigma(\ell^1,\ell^\infty) \subsetneqq \eta(\ell^1,\ell^\infty)$ and so $\sigma(\ell^1,\ell^\infty) \subsetneqq T^{or}$.

Remark: Theorem 14.2.9 may also fail to hold good if we delete the S.b. character of $\{x_n;f_n\}$ from the hypothesis of this theorem.

Example 14.2.12: Consider the Banach lattice $(\ell^\infty, \|\cdot\|_\infty)$ ordered by the b.s. $\{e^n;e^n\}$. From Proposition 14.2.8, $T^{or} \subset T_\infty$. If $\alpha^n = e - e^{(n)}$, then $\alpha^n \not\to 0$ in T_∞ but $\alpha_n \downarrow 0$. Therefore, $\alpha_n \to 0$ in T^{or} and $\alpha^n \not\to 0$ in T_∞, giving $T_\infty \not\subset T^{or}$.

Now we have the main result of this section, namely,

Theorem 14.2.13: Let (X,T) be a Fréchet space. Then the following are equivalent:

(i) $T^{or} \approx T$.

(ii) $\{x_n;f_n\}$ is a u-S.b.

(iii) K is generating and normal.

(iv) K is generating and $T \subset T^{or}$.

(v) K is normal and $T^{or} \subset T$.

Proof: (i) ⇒ (ii) By Theorems 11.3.2 and 14.2.4, $\{x_n;f_n\}$ is a u-S.b. for K. To prove (ii), it suffices to establish x^+ and x^- for each x in X. Given x in X, let $y_n = x/n$, $n \geq 1$, then $y_n \overset{(*)}{\longrightarrow} 0$ by Proposition 14.2.7. Hence, for some subsequence $\{y_{n_i}\}$ of $\{y_n\}$, $y_{n_i} \overset{(o)}{\longrightarrow} 0$. Thus, for some u in K and v in -K, $v \leq y_{n_i} \leq u$, for $i \geq 1$. Now make use of Proposition 12.2.5.

(ii)$\Leftrightarrow$(iii) Cf. Theorem 12.3.8.

(iii) $\Rightarrow$ (iv) This follows from (ii) and Theorem 14.2.9.

(iv) $\Rightarrow$ (v) Make use of Theorems 14.2.4, 12.3.3 and 14.2.9.

(v) $\Rightarrow$ (i) This follows from Theorem 14.2.4. □

At the end of this subsection, we come to a very natural question raised by McArthur [170]: "When is T^{or} a linear topology?" In fact, this question originates in an important observation made earlier by Schaefer [205], pp. 230 and 253, that T^{or} is usually not linear on a VL. Potepun [187] gives some sufficient conditions in VL for T^{or} to be linear. Theorem 14.2.13 is also an answer in this direction and this result, along with the following theorem, yields a necessary and sufficient condition for T^{or} to be linear (cf. [19]).

<u>Theorem 14.2.14</u>: Let (X,T) be a Fréchet space with K normal. Then T^{or} is linear if and only if $\{x_n;f_n\}$ is a u-S.b.

For the proof we require

<u>Proposition 14.2.15</u>: Let (X,T) be Fréchet. If K is normal, then a sequence $\{y_n\}$ in X (0)-converges to y if and only if $y_n \xrightarrow{(r)} y$.

<u>Proof</u>: We may assume $D_T = \{p_1 \leq p_2 \leq \ldots \leq p_n \leq \ldots\}$, where each p_n is monotone.

Let $y_n \xrightarrow{(o)} y$. First consider a sequence $\{u_n\}$ in X with $u_n \downarrow 0$. Then $p_i(u_n) \to 0$ for each $i \geq 1$, cf. proof of the first part of Theorem 14.2.4. Hence $p_i(u_{n_i}) < 1/i^3$, $i \geq 1$. Following the proof of Proposition 14.2.8, we find u in K with

$$u = \sum_{i \geq 1} i u_{n_i}.$$

Then $u_{n_i} \leq u/i$ for $i \geq 1$ and so, to each $\varepsilon > 0$, we find an i_0 with $u_{n_i} \leq \varepsilon u$, for $i \geq i_0$. Therefore $u_n \leq \varepsilon u$ eventually in n, giving $u_n \xrightarrow{(r)} 0$. Secondly, suppose y = 0, so that $y_n \xrightarrow{(o)} 0$. Then there exist $\{u_n\}$ and $\{v_n\}$ in X with $u_n \downarrow 0$, $v_n \uparrow 0$ and $v_n \leq y_n \leq u_n (n \geq 1)$. By Proposition 12.2.5, $|y_n|$ exists and $|y_n| \leq u_n - v_n$ for $n \geq 1$. Since $u_n - v_n \downarrow 0$, $u_n - v_n \xrightarrow{(r)} 0$. It then easily follows that $y_n \xrightarrow{(r)} 0$. The case when $y \neq 0$ is now straightforward.

Conversely, if $y_n \xrightarrow{(r)} y$, then $|y_n - y| \leq \varepsilon u$ for $n \geq n_\varepsilon$, where $u \in K$. Let $z_n = y + u/n$, $w_n = y - u/n$. By Proposition 14.2.1, $z_n \downarrow y$ and $w_n \uparrow y$, yielding thereby $y_n \xrightarrow{(o)} y$. □

Remark: Observe that the converse of the preceding proposition is valid without restricting (X,T), $\{x_n;f_n\}$ and K.

Proof of Theorem 14.2.14: The sufficiency part is clear from Theorem 14.2.13.

For the converse, let $\{x_n;f_n\}$ be not a u-S.b. Hence K is not generating by the theorem referred to above and, in particular, there exists an x in X such that $|x|$ does not exist. By hypothesis and Proposition 14.2.7, $x/n \xrightarrow{(*)} 0$ and consequently there exists a sequence $\{n_k\}$ so that, using Proposition 14.2.15, $x/n_k \xrightarrow{(r)} 0$. Hence, for all sufficiently large k, $|x|/n_k$ exists and so does $|x|$, a contradiction. □

Further results on (0)- and T^{or}-convergence

We begin with a very special situation in terms of

Proposition 14.2.16: In the space $(\omega,\sigma(\omega,\phi))$ ordered by its cone $K \equiv K\{e^n;e^n\}$, $\alpha^n \to 0$ in $\sigma(\omega,\phi)$ if and only if $\alpha^n \xrightarrow{(o)} 0$.

Proof: If $\alpha^n \xrightarrow{(o)} 0$, then usual considerations from the classical analysis yield $\alpha_i^n \to 0$ for $i \geq 1$; that is, $\alpha^n \to 0$ in $\sigma(\omega,\phi)$.

Conversely, let $\alpha^n \to 0$ in $\sigma(\omega,\phi)$. Then $\alpha_i^n \to 0$ for each $i \geq 1$ and there exists $\{r_i\}$ in ω, $r_i \geq 0$, $i \geq 1$ such that $|\alpha_i^n| \leq r_i$ for $i, n \geq 1$. Given i and n, define $\beta_i^n = \sup\{|\alpha_i^j| : j \geq n\}$. Then $\{\beta^n\}$ is a decreasing sequence in ω with $\beta^n \downarrow 0$ and $|\alpha^n| \leq \beta^n$ for $n \geq 1$. Therefore, $\alpha^n \xrightarrow{(o)} 0$. □

More generally, we have

Theorem 14.2.17: Let (X,T) be ω-complete, $\{x_n;f_n\}$, a u-S.b. for K and $\{y_n\} \subset K$. Then $y_n \xrightarrow{(o)} 0$ if and only if $f_i(y_n) \to 0$ for each $i \geq 1$ and there exists y in K with $y_n \leq y$ for each $n \geq 1$.

Proof: Let $y_n \xrightarrow{(o)} 0$. Then there exists u in K such that $y_n \leq u$ for $n \geq 1$

and, by Proposition 14.2.1, $f_i(y_n) \to 0$ for each $i \geq 1$.

For the converse, observe that if $\alpha^n = \{f_i(y_n)\}$, then $\alpha^n \to 0$ in $(\omega,\sigma(\omega,\phi))$ where ω is regarded as having been ordered by the usual $\sigma(\omega,\phi)$-u-S.b. $\{e^n;e^n\}$. Let $\beta_i^n = \sup\{f_i(y_j):j \geq n\}$. Then $\beta^n \in \omega$ and $\beta^n \downarrow 0$. Write $\gamma^n = \beta^n \wedge \{f_i(y)\}$. If we fix $n \geq 1$ and choose $\mu_i^n(i \geq 1)$ with $0 \leq \mu_i^n \leq 1$ so that $\gamma_i^n = \mu_i^n f_i(y)$, then $\sum_{i \geq 1} \mu_i^n f_i(y) x_i$ converges in (X,T), say to z_n. Clearly $\gamma_i^n = f_i(z_n)$ for $i, n \geq 1$. Since $\alpha_i^n \leq f_i(y)$, β_i^n, $\alpha^n \leq \gamma^n$ for $n \geq 1$. Similarly, it is clear that $\gamma^n \downarrow 0(\gamma^{n+1} \leq \gamma^n$ and $\inf \gamma^n = 0)$. Hence $y_n \leq z_n$ and $z_n \downarrow 0$, so that $y_n \xrightarrow{(o)} 0$. □

We immediately derive (cf. [170])

<u>Corollary 14.2.18</u>: Let (X,T) be ω-complete and $\{x_n;f_n\}$ a u-S.b. Suppose $\{y_n\} \subset X$. Then $y_n \xrightarrow{(o)} 0$ if and only if $f_i(y_n) \to 0$ for each $i \geq 1$ and there exists y in K with $|y_n| \leq y$, $n \geq 1$.

A slight variation of Theorem 14.2.17 is contained in

<u>Proposition 14.2.19</u>: Let (X,T) be ω-complete and $\{x_n;f_n\}$ be a u-S.b. for K. For $\{y_n\} \subset X$, $y_n \xrightarrow{(o)} 0$ if and only if $f_i(y_n) \to 0$ for each $i \geq 1$ and there exist y in K, z in $-K$ and an N in $\mathbb{N}$ with $z \leq y_n \leq y$ for all $n \geq N$.

<u>Proof</u>: Necessity follows from the definition of (O)-convergence.

For sufficiency, let us observe that, on account of Proposition 12.2.5, y_n^+, y_n^- exist with $y_n^+ \leq y$ and $y_n^- \leq -z$ for all $n \geq N$. Since $f_i(y_n^+) = f_i(y_n) \vee 0$ and $f_i(y_n^-) = -f_i(y_n) \vee 0$, $f_i(y_n^+)$, $f_i(y_n^-) \to 0$ for $i \geq 1$. Hence, from Theorem 14.2.17, y_n^+, $y_n^- \xrightarrow{(o)} 0$. Consequently, we get sequences $\{u_n\}$ and $\{v_n\}$ in K satisfying $u_n \downarrow 0$, $v_n \downarrow 0$ and $y_n^+ \leq u_n$, $y_n^- \leq v_n$, $n \geq 1$. As $y_n = y_n^+ - y_n^-$, one gets $-v_n \leq y_n \leq u_n$ for $n \geq 1$ and this yields $y_n \xrightarrow{(o)} 0$. □

Concerning the second aspect of this subsection, namely, the characterization of T^{or}-closure of subsets, we need a simple observation in the form of

<u>Proposition 14.2.20</u>: Let (X,T) be a Fréchet space, then (X,T) possesses the property (R) relative to K.

<u>Proof</u>: Choose D_T as $\{p_1 \leq p_2 \leq \ldots \leq p_n \leq \ldots\}$ and consider any sequence $\{y_n\}$ in K. Let $\beta_n = p_n(y_n)$, $n \geq 1$ and define α in ω with $\alpha_n = 1$ if $\beta_n = 0$, and $\alpha_n = 1/n^2\beta_n$ if $\beta_n \neq 0$. Hence we find y in K such that $y = \sum_{n \geq 1} \alpha_n y_n$. Clearly

$f_i(\alpha_n y_n) < f_i(y)$ for $i, n \geq 1$. Therefore, $\{\alpha_n\}$ is the desired sequence. □

Lemma 14.2.21: In an OVS $(Y,<)$ having the property (R) and where r-convergence is additive, the r-convergence has the diagonal property.

Proof: Let $y_{m,n} \xrightarrow{(r)} y_m$ and $y_m \xrightarrow{(r)} y$ in Y. Denote by z_m the regulator of convergence for $\{y_{m,n}:n \geq 1\}$, for $m \geq 1$. The property (R) ensures the existence of a sequence $\{\alpha_m\}$ with $\alpha_m > 0$ and z in K such that $\alpha_m z_m < z$, $m \geq 1$. Hence we can find an increasing sequence $\{n_m\}$ from $\mathbb{N}$ so that

$$|y_{m,n_m} - y_m| < \frac{\alpha_m z_m}{m} < \frac{1}{m} z, \quad \forall\, m \geq 1.$$

Thus $y_{m,n_m} - y_m \xrightarrow{(r)} 0$ and this proves the result. □

Theorem 14.2.22: Let (X,T) be a Fréchet space and K a normal cone. For any subset A of X, the T^{or}-closure $\bar{A}$ of A is given by $\bar{A} = B \equiv \{x \in X: y_n \xrightarrow{(r)} x,$ for some sequence $\{y_n\}$ contained in A$\}$.

Proof: To begin with, let us observe that, on account of Proposition 12.2.5 and Theorem 11.3.2, $|x+y|$ exists whenever $|x|$ and $|y|$ exist for x,y in X. Hence the r-convergence is additive.

Clearly $A \subset B$. By Exercise 14.2.2 and Proposition 14.2.15, $B \subset \bar{A}$. It now remains to show that B is T^{or}-closed. Let therefore $y \in \bar{B}$, the T^{or}-closure of B. On account of Proposition 14.2.15, there exist y_m in B, $m \geq 1$ so that $y_m \xrightarrow{(r)} y$. We find sequences $\{y_{m,n}:n \geq 1\}$ in A for each $m \geq 1$ so that $y_{m,n} \xrightarrow{(r)} y_m$, $m \geq 1$. Therefore, using Lemma 14.2.21, we get a subsequence $\{n_m\}$ such that $y_{m,n_m} \xrightarrow{(r)} y$, that is, $y \in B$. □

As an application of the foregoing result, we obtain

Proposition 14.2.23: Let (X,T) and K be as in Theorem 14.2.22. Consider a net $\{x_\alpha:\alpha \in \Lambda\}$ in X. Then $x_\alpha \to 0$ in T^{or} if and only if every cofinal subnet $\{x_{\alpha_\beta}\}$ of $\{x_\alpha\}$ contains a sequence $\{x_{\alpha_{\beta_n}}\}$ with $x_{\alpha_{\beta_n}} \xrightarrow{(r)} 0$.

Proof: For necessity, observe that the set $\{x_{\alpha_\beta}\} \cup \{0\}$ is T^{or}-closed. Now apply Theorem 14.2.22.

For the converse, assume the existence of a net $\{x_\alpha\}$ satisfying the given condition but $x_\alpha \not\to 0$ in T^{or}. Hence we find a T^{or}-open set u containing zero such that for each β in Λ there exists α_β with $\alpha_\beta \geq \beta$ and $x_{\alpha_\beta} \notin u$. By the hypothesis, we find a sequence $\{x_{\alpha_{\beta_n}}\}$ with $x_{\alpha_{\beta_n}} \xrightarrow{(r)} 0$. By Theorem 14.2.22, $0 \in X \setminus u$, a contradiction. □

14.3 ORDER TOPOLOGY IN DUAL SPACES

We next come to the second major aspect of this chapter, namely, the relationship of T^{*or} with other dual topologies. As before, we write throughout (X,T) for an l.c. TVS containing an S.b. $\{x_n;f_n\}$; also we recall that K^* is the dual cone of $K \equiv K\{x_n;f_n\}$ and also the cone in X^* generated by the weak*-S.b. $\{f_n;\Psi(x_n)\}$, i.e. $K^* \equiv K^*\{f_n;\Psi(x_n)\}$; cf. Proposition 12.2.3. Further, it will be assumed that X^* is ordered by the dual cone K^*.

At the outset, we have

Proposition 14.3.1: If K is generating in X, then $\sigma(X^*,X) \subset T^{*or}$.

Proof: By the remark following Lemma 14.2.3, it is sufficient to prove $h_n \to 0$ in $\sigma(X^*,X)$ for any sequence $\{h_n\}$ in X^* with $h_n \xrightarrow{(o)} 0$.

First consider $h_n \downarrow 0$ in X^*. Fix x in K and $\varepsilon > 0$. Choose a positive integer N so that

$$\sum_{n>N} f_n(x)h_1(x_n) < \frac{\varepsilon}{2}.$$

Since $h_n(x_i) \to 0$ for each $i \geq 1$, we find an n_o such that

$$\sum_{i=1}^{N-1} f_i(x)h_n(x_i) < \frac{\varepsilon}{2}, \quad \forall\, n \geq n_o$$

$$\Rightarrow \quad h_n(x) < \varepsilon, \forall\, n \geq n_o.$$

Thus $h_n(x) \to 0$ for each x in K and hence for every x in $X = K-K$. Similarly, if $g_n \uparrow 0$, then $-g_n \downarrow 0$ and so $g_n(x) \to 0$ for each x in X. This discussion clearly yields the required result. □

Concerning the comparison of T^{*or} with $\beta(X^*,X)$, we have

Theorem 14.3.2: Let $(X, \|\cdot\|)$ be a Banach space and $\{x_n;f_n\}$ a u-S.b. for it. Then $T^{*or} \subset \beta(X^*,X)$.

Proof: Let $g_n \in K^*$, $n \geq 1$ and $\|g_n\| \to 0$. We find a subsequence $\{n_i\}$ so that $\|g_{n_i}\| \leq 1/i^3$, $i \geq 1$ and hence, for some g in K^*,

$$g = \sum_{i \geq 1} i g_{n_i} \quad \text{(convergence in } \beta(X^*,X)).$$

Thus $g_{n_i} \xrightarrow{(o)} 0$. Applying similar arguments to any subsequence of $\{g_n\}$, we conclude that $g_n \xrightarrow{(*)} 0$.

Consider now any arbitrary sequence $\{g_n\}$ in X^* with $\|g_n\| \to 0$. Observe that X is a Banach lattice and so we may assume $\|x\| \leq \|y\|$ whenever $|x| \leq |y|$. Also, for $x \leq z \leq y$, $|z| \leq |x| + |y|$; consequently, if $x \leq z \leq y$ with $\|x\|$, $\|y\| \leq 1$, then $\|z\| \leq \|x\| + \|y\| \leq 2$. Hence, from Proposition 1.5.7(ii), there exist $\{g_n^{(1)}\}$ and $\{g_n^{(2)}\}$ in K^* such that $g_n = g_n^{(1)} - g_n^{(2)}$ and $\|g_n^{(1)}\| + \|g_n^{(2)}\| \leq 2\,\|g_n\|$, $n \geq 1$. Thus $g_n^{(i)} \to 0$ for $i = 1,2$. It can be easily verified from the first paragraph, Proposition 1.6.2(v) that given any subsequence $\{n_k\}$, we can select a subsequence $\{n_{k_\ell}\}$ of $\{n_k\}$ such that $g_{n_{k_\ell}}^{(1)} - g_{n_{k_\ell}}^{(2)} \xrightarrow{(o)} 0$, giving thereby $g_n \xrightarrow{(*)} 0$. Now apply Proposition 14.2.6. □

Remarks: The nonavailability of Proposition 1.5.7(ii) for Fréchet spaces possibly prevents us from extending Theorem 14.3.2 even to Fréchet spaces. Besides, we are not aware of any infinite dimensional Banach space for which $\sigma(X^*,X) \approx T^{*or}$ nor do we know sufficient conditions on (X,T) and $\{x_n;f_n\}$ ensuring the equivalence of T^{*or} and $\sigma(X^*,X)$. However, for an l.c. TVS which is not a Banach space, we have the equivalence of T^{*or} and $\sigma(X^*,X)$; e.g., consider $(\phi,\eta(\phi,\omega))$ having the S.b. $\{e^n;e^n\}$ and apply Theorem 14.2.9 (here $\phi^* = \omega$ and $T^{*or} \approx \sigma(\omega,\phi)$). In the opposite direction, we have a result for Banach spaces for which $T^{*or} \approx \beta(X^*,X)$. To begin with, we have

Proposition 14.3.3: Let (X,T) be a Mazur space, $\{x_n;f_n\}$ a shrinking base and K^* a $\beta(X^*,X)$-normal cone. Then $\beta(X^*,X) \subset T^{*or}$.

Proof: This follows from Proposition 1.2.12 and Theorem 14.2.4. □

In particular, we have (cf. [19])

Proposition 14.3.4: Let (X,T) be a Banach space, $\{x_n;f_n\}$ a shrinking base and K a generating cone. Then $\beta(X^*,X) \subset T^{*or}$.

Proof: Since $(X^*,\beta(X^*,X))$ is a P-space ([205], p. 143), K^* is $\beta(X^*,X)$-normal by Theorem 1.5.8. Now apply Proposition 14.3.3. □

A sort of converse to Proposition 14.3.3 is contained in

Proposition 14.3.5: If K is normal in (X,T) and $\beta(X^*,X) \subset T^{*or}$, then $\{x_n;f_n\}$ is shrinking.

Proof: By Theorem 1.5.16, K^* is generating in X^*. Let $f \in K^*$. Then $\{\sum_{i=1}^{n} f(x_i)f_i\} \uparrow f$; that is,

$$\sum_{i=1}^{n} f(x_i)f_i \xrightarrow{(0)} f \Rightarrow \sum_{i=1}^{n} f(x_i)f_i \to f$$

in T^{*or} by Proposition 1.5.13 and hence in $\beta(X^*,X)$. □

Remark: The Banach space analogue of the foregoing proposition is to be found in [19] and, even in this case, the condition $\beta(X^*,X) \subset T^{*or}$ cannot be dropped. For, we have

Example 14.3.6: Consider $(\ell^1, \|\cdot\|_1)$ with its S.b. $\{e^n;e^n\}$. Here K is normal (and so is K^*). From Example 14.2.12, $\beta(\ell^\infty,\ell^1) \not\subset T^{*or}$, where $\beta(\ell^\infty,\ell^1) \approx T_\infty$. Further, by Example 8.3.1, $\{e^n;e^n\}$ is not shrinking.

Now we have the main result of this subsection (cf. [19]):

Theorem 14.3.7: Let (X,T) be a Banach space and $\{x_n;f_n\}$ a u-S.b. Then the following statements are equivalent:

(i) $\{x_n;f_n\}$ is shrinking.

(ii) $T^{*or} \approx \beta(X^*,X)$.

(iii) (X,T) is a non ℓ^1-space.

(iv) $(X^*,\beta(X^*,X))$ is separable.

(v) $(X^*,\sigma(X^*,X^{**}))$ is ω-complete.

Proof: To begin with let us observe that K is generating and normal in

(X,T); cf. Theorem 12.3.3.

(i) $\Rightarrow$ (ii) Use Theorem 14.3.2 and Proposition 14.3.4.

(ii) $\Rightarrow$ (i) Cf. Proposition 14.3.5.

(i) $\Leftrightarrow$ (iii) $\Leftrightarrow$ (iv) This is a consequence of Theorem 8.3.6:

(i) $\Rightarrow$ (v) By Propositions 8.3.21 and 12.4.10, $\{f_n, \Psi(x_n)\}$ is γ-complete and u-S.b. for $(X^*, \beta(X^*,X))$. Hence, from Proposition 13.2.2, $(X^*, \beta(X^*,X))$ is an order complete l.c. TVL under $K^* \equiv K\{f_n, \Psi(x_n)\}$. Therefore (v) follows from Theorem 13.2.14.

(v) $\Rightarrow$ (i) This follows from Exercise 8.3.8. □

15 Conic bases

15.1. INTRODUCTION

The advantage of studying conic bases (c.b.) lies in the fact that many geometrical properties of a cone may be identified through a smaller subset of the c.b. and this, in turn, has applications to the uniqueness theory of Choquet integral representations (cf. [184] and [185]). Besides, the existence of a c.b. for a cone guarantees the existence of strictly positive linear functionals.

Thus we are faced in this last chapter with two problems, namely, the existence of a c.b. and, secondly, the advantage derived by the presence of a c.b. for a cone.

Throughout this chapter we write, unless the contrary is specified, X or (X,T) to mean an arbitrary real TVS; also, we assume that $\{x_n;f_n\}$ is a b.s. for $\langle X,X'\rangle$ such that $\{x_n;f_n\}$ is a t.b. for its cone $K \equiv K\{x_n;f_n\}$.

We follow [202] for most of the results of this chapter.

15.2 COMPANION SEQUENCES

Each c.b. in K gives rise to a unique sequence in K, called a companion sequence, and every sequence of this nature in K yields a c.b. for K. Thus the study of c.b. is closely related with that of companion sequences. Let us, therefore, introduce

Definition 15.2.1: A sequence $\{y_n\}$ in X is called a *companion sequence* (c.s.) *of* $\{x_n\}$ if $y_n = \alpha_n x_n$, $n \geqslant 1$ for some α in ω with $\alpha_n > 0$, $n \geqslant 1$ and whenever $\sum_{n\geqslant 1} \beta_n y_n$ converges in K, then $\beta_n \geqslant 0$, $n \geqslant 1$ together with $\sum_{n\geqslant 1} \beta_n < \infty$.

Note: Observe that a c.s. $\{y_n\}$ is always a t.b. for K.

Proposition 15.2.2: Corresponding to a c.s. $\{y_n\}$ of $\{x_n\}$, let $B = \{y \in K: y = \sum_{n\geqslant 1} \beta_n y_n$ with $\beta_n \geqslant 0, \sum_{n\geqslant 1} \beta_n = 1\}$. Then B is c.b. for K.

Proof: If $x \in K$, then $x = \sum_{n \geq 1} \beta_n y_n$, where $\beta_n = f_n(x)/\alpha_n$ and $y_n = \alpha_n x_n$ with $\alpha_n > 0$, $n \geq 1$. The rest of the argument for showing B to be a c.b. is a routine exercise. □

Example 15.2.3: Consider the cone K in $(\ell^1, \|\cdot\|_1)$ associated with the S.b. $\{e^n; e^n\}$. It is known that the subset $B = \{\alpha \in K: \sum_{n \geq 1} \alpha_n = 1\}$ is a c.b. for K; cf. [99], p. 34. Since $\{e^n\}$ is trivially a c.s. of itself, B is the base of the type considered in Proposition 15.2.2.

Proposition 15.2.4: Let B be a c.b. for K. Then B contains a unique c.s. $\{y_n\}$ of $\{x_n\}$ and, if $\sum_{n \geq 1} \alpha_n y_n$ converges in B, then $\sum_{n \geq 1} \alpha_n \leq 1$.

Proof: The first statement follows by the fact that $x_n \in K$, $n \geq 1$ and the definition of B. For the last statement, let there be x in B with $x = \sum_{n \geq 1} \alpha_n y_n$ and $1 < \sum_{n \geq 1} \alpha_n < \infty$. There exists p in $\mathbb{N}$, so that $\sum_{i=1}^{p} \alpha_i > 1$. Clearly, $y = \sum_{i=1}^{p} \alpha_i y_i / \sum_{i=1}^{p} \alpha_i$ lies in B (use convexity of B). Note that $x \neq y$ and $x-y \in K \cap (B-B)$ and this violates (iii) of Theorem 1.5.18. □

The preceding discussion motivates to introduce

Definition 15.2.5: Let $\mathcal{F}$ and $\mathbb{G}$ denote respectively the collection of all c.s. of $\{x_n\}$ and the family of all c.b. for K. Define $F: \mathcal{F} \to \mathbb{G}$ and $G: \mathbb{G} \to \mathcal{F}$ by $F(\{y_n\}) = \{\sum_{n \geq 1} \alpha_n y_n \in K: \alpha_n \geq 0 \ \sum_{n \geq 1} \alpha_n = 1\}$ and $G(B) = \{z_n\}$, where $\{y_n\}$ (resp. $\{z_n\}$) is the unique c.s. of $\{x_n\}$ in a c.b. as given in Proposition 15.2.4.

Propositions 15.2.2 and 15.2.4 immediately yield

Proposition 15.2.6: $\mathcal{F} \neq \emptyset$ if and only if $\mathbb{G} \neq \emptyset$; that is, there exists a c.b. for K if and only if there exists a c.s. of $\{x_n\}$.

Proposition 15.2.7: If $\{\gamma_n x_n\}$ is regular for some γ in ω with $\gamma_n > 0$, $n \geq 1$, then K possesses a c.b.

Proof: Since $\{\gamma_n x_n\}$ is a t.b. for K, $f_n(x)/\gamma_n \to 0$ for each x in K. Hence $\sum_{n \geq 1} f_n(x)/2^n \gamma_n < \infty$ for each x in K and this shows that $\{2^n \gamma_n x_n\}$ is a c.s. of

$\{x_n\}$. Now apply Proposition 15.2.6. □

<u>Corollary 15.2.8</u>: If (X,T) is normable, then K always has a c.b.

Indeed, $\{x_n/\|x_n\|\}$ is regular in (X,T). On the other hand, if (X,T) is non-normable, K may never have a c.b., for consider

<u>Example 15.2.9</u>: Consider the cone $K \equiv K\{e^n;e^n\}$ in the space $(\omega,\sigma(\omega,\phi))$. Let $y^n = \alpha_n e^n$; $\alpha_n > 0$, $n \geq 1$. Note that $\sum_{n\geq 1} 2^n y^n$ converges in K but $\sum_{n\geq 1} 2^n = \infty$. Hence there cannot be any c.s. of $\{e^n\}$ and so K does not have a c.b.

At the end of this section, we pass on to an example which negates the converse of Proposition 15.2.7.

<u>Example 15.2.10</u>: Consider the cone K associated with the S.b. $\{e^n;e^n\}$ for $(\ell^1,\sigma(\ell^1,\phi))$; K, being the same as the cone associated with the S.b. $\{e^n;e^n\}$ for $(\ell^1,\|\cdot\|_1)$, has therefore a c.b. by Corollary 15.2.8. However, for no strictly positive sequence $\{\gamma_n\}$, $\{\gamma_n e^n\}$ is regular in $(\ell^1,\sigma(\ell^1,\phi))$.

15.3 <u>TYPES OF CONIC BASES AND CONES</u>

Let (X,T), $\{x_n;f_n\}$ and K be as mentioned in Section 15.1. Then, we have

<u>Definition 15.3.1</u>: A c.b. B for K is called (i) a *canonical conic base* (c.c.b.) if F(G(B)) = B (cf. Definition 15.2.5) and (ii) a *hyper conic base* (h.c.b) if K is closed and $B = K \cap f^{-1}(1)$ for some strictly positive f in X*.

<u>Definition 15.3.2</u>: The cone K is called (i) a *canonical cone* if each c.b. of K is a c.c.b.; (ii) a *closing cone* if every c.b. of K is closed and (iii) a *hyper cone* if every c.b. of K is an h.c.b. and K is closed.

First, let us give an example of a c.b. which is not a c.c.b.

<u>Example 15.3.3</u>: Let $X = \phi \oplus sp\{\alpha\}$, $\alpha_n = 1/2^n$, $n \geq 1$ and endow X with the ℓ^1-norm $\|\cdot\|_1$. Then $\{e^n;e^n\}$ is an S.b. for $(X,\|\cdot\|_1)$. Let $K \equiv K\{e^n;e^n\}$ and $B = \text{con}(\{e^n:n \geq 1\} \cup \{\alpha/2\})$. B is a c.b. of K. For, let $y \in K$, then $y = \beta_1 e^{i_1} + \beta_2 e^{i_2} + \ldots + \beta_n e^{i_n} + \beta_{n+1}\alpha$ with all β_i's being positive and, if $r = \sum_{i=1}^{n} \beta_i + 2\beta_{n+1}$, then $y/r \in B$. Here $\alpha \notin B$, $\in F(\{e^n\})$. But $G(B) = \{e^n\}$;

of. the proof of Proposition 15.2.4. Thus $\alpha \in F(G(B))$, giving $F(G(B)) \subsetneq B$.

Different conic bases

Different behaviour of conic bases yields different types of cones. We therefore study sufficient conditions which ensure a c.b. to be either a c.c.b. or an h.c.b. To begin with, let us start from

Lemma 15.3.4: Let B be a c.b. for K and $\{y_n\} = G(B)$. Then, for $z = \sum_{n \geq 1} a_n y_n$ in B and $n \geq 1$, the element $z_n = \sum_{i \geq n+1} a_i y_i (1 - \sum_{i=1}^{n} a_i)$ exists and belongs to B provided that $\sum_{i \geq n+1} a_i y_i \neq 0$.

Proof: Observe that $a_i \geq 0$, $i \geq 1$ and $0 < \sum_{n \geq 1} a_n \leq 1$. Further, $\sum_{i=1}^{n} a_i < 1$, otherwise $a_i = 0$, $i \geq n+1$ and so $\sum_{i \geq n+1} a_i y_i = 0$. Therefore $0 < \sum_{i \geq n+1} a_i$ and $\sum_{i=1}^{n} a_i < 1$.

If $\sum_{i=1}^{n} a_i = 0$, the result trivially follows. Let, therefore, $\alpha = \sum_{i=1}^{n} a_i > 0$. There exists a unique $\lambda > 0$ with $\lambda \sum_{i \geq n+1} a_i y_i \in B$, so that, by the convexity of B, $u = \alpha(\alpha^{-1} \sum_{i=1}^{n} a_i y_i) + (1-\alpha)(\lambda \sum_{i \geq n+1} a_i y_i) \in B$. If $z-u = -\beta k$, where $\beta > 0$ and $k \in K$, then $u-z \in (B-B) \cap K$. Consequently $u = z$ by Theorem 1.5.18 and we get $\lambda = 1/(1-\alpha)$, yielding $z_n \in B$. If $z-u = \beta k$ with $\beta > 0$ and $k \in K$, the required result follows once again in a similar manner. □

Theorem 15.3.5: Let B be a c.b. for K. If $0 \notin \bar{B}$, then B is a c.c.b.

Proof: Put $\{y_n\} = G(B)$ and assume that the required result is false; that is, there exists z in B with $z \notin F(\{y_n\})$. Regarding $\{y_n\}$ to be a t.b. for K and invoking Proposition 15.2.2 and 15.2.4 along with Definition 15.2.5, we find a unique $\{\alpha_n\}$ with $\alpha_n \geq 0$, $n \geq 1$ such that $z = \sum_{n \geq 1} \alpha_n y_n$, where $0 < a \equiv \sum_{n \geq 1} \alpha_n < 1$.

Next, we claim that $z_n = \sum_{i \geq n+1} \alpha_i y_i \neq 0$ for every $n \geq 1$. Indeed, if $z_n = 0$ for some $n \geq 1$, then $z = \sum_{i=1}^{n} \alpha_i y_i$ and $\alpha_i = 0$ for $i \geq n+1$. Observe that $(a^{-1}-1)z \neq 0$ and belongs to K as well as to B-B (observe that

$\sum_{i=1}^{n} \alpha_i y_i / \sum_{i=1}^{n} \alpha_i \in B$ by the convexity of B). However, this gives a contradiction (cf. Theorem 1.5.18(iii)). Hence, by the lemma, $u_n \in B$ for every $n \geq 1$, where $u_n = \sum_{i \geq n+1} \alpha_i y_i / (1 - \sum_{i=1}^{n} \alpha_i)$. But $u_n \to 0$ and so $0 \in \bar{B}$, a contradiction. Thus $F(\{y_n\}) = B$. □

<u>Corollary 15.3.6</u>: A closed c.b. B for K is a c.c.b. and hence it is the closed convex hull of G(B).

For, $0 \notin B = \bar{B}$ and make use of Proposition 15.2.2 and Theorem 15.3.5.

<u>Note</u>: The converse of Theorem 15.3.5 is not true, for consider

<u>Example 15.3.7</u>: Once again, let us consider the c.b. $B = \{\alpha \in K\{e^n;e^n\}: \alpha_n \geq 0, \sum_{n \geq} \alpha_n = 1\}$ for the space $(\ell^1, \sigma(\ell^1, \phi))$ considered in Example 15.2.10. Clearly, $0 \in \bar{B}$. Also B is c.c.b. since K is canonical (cf. proof of Proposition 15.4.2).

<u>Theorem 15.3.8</u>: Let $\{x_n;f_n\}$ be restricted further so that $f_n \in X^*$, $n \geq 1$. Suppose that B is a c.b. for the cone $K \equiv K\{x_n;f_n\}$ such that $0 \notin \bar{B}$. Then B is closed.

<u>Proof</u>: Since K is closed, it is enough to prove that B is closed in K. Put $\{y_n\} = G(B)$; then $\{y_n\}$ is an S.b. for K and let $\{g_n\}$ be the corresponding s.a.c.f. Consider an x_o in $K \setminus B$. Then $x_o = \sum_{n \geq 1} g_n(x_o) y_n$, where $1 \neq \sum_{n \geq 1} g_n(x_o) < \infty$ and $g_n(x_o) \geq 0$ for $n \geq 1$ (since B is a c.c.b.). Put $a_n = g_n(x_o)$ and consider two cases.

<u>Case 1</u>: Let $\sum_{n \geq 1} a_n = 1+\varepsilon$ for some $\varepsilon > 0$. Then there exists N such that $\sum_{i=1}^{N} a_i \geq 1 + \varepsilon/2$. Define a neighbourhood v by

$$v = \bigcap_{i=1}^{N} g_i^{-1} \{[-\frac{\varepsilon}{4N}, \frac{\varepsilon}{4N}]\}.$$

For x in $(x_o + v) \cap K$, we easily find that $\sum_{n \geq 1} g_n(x) \geq 1+\varepsilon/2 > 1$ and so $x \notin B$. Thus $((x_o + v) \cap K) \cap B = \emptyset$ and so $K \setminus B$ is open.

Case 2: Let $\sum_{n \geq 1} a_n = 1-\varepsilon$ for some ε with $0 < \varepsilon < 1$. Choose u in $\mathcal{B}_T$ with $u \cap B = \emptyset$. We can find N in $\mathbb{N}$, $v, w_o, w_1, \ldots, w_N$ in $\mathcal{B}_T$ and positive reals $s_1, \ldots, s_N$ such that

$$v + v \subset \frac{\varepsilon}{2} u, \quad w_o + w_o \subset v, \quad \sum_{i \geq N+1} a_i y_i \in w_o \tag{*}$$

and

$$w_i + w_i \subset w_{i-1}, \quad g_i^{-1} \{[-s_i, s_{i-1}]\} \subset w_i, \tag{**}$$

where $i = 1, \ldots, N$. Put $s = \min \{s_1, \ldots, s_N; \varepsilon/2N\}$ and $w = \cap \{g_i^{-1} \{[-s,s]\}: 1 \leq i \leq N\}$. We show that $B \cap ((w \cap w_o) + x_o) = \emptyset$.

If this is not true, we find $x \in B$ and $x - x_o \in w \cap w_o$ so that by using (**) and putting $b_n = g_n(x)$, $n \geq 1$, we get

$$\sum_{i=1}^{N} (b_i - a_i) y_i \in w_o. \tag{+}$$

Since $|b_i - a_i| \leq s$ for $1 \leq i \leq N$, we easily find that $\sum_{i=1}^{N} b_i \leq 1-\varepsilon/2$. Consequently $\sum_{i \geq N+1} b_i \geq \varepsilon/2$ and so, by Lemma 15.3.4, $\sum_{i \geq N+1} b_i y_i / \sum_{i \geq N+1} b_i \in B$, since b_n's are positive. But u is balanced and $u \cap B = \emptyset$, therefore

$$\sum_{i \geq N+1} b_i y_i \notin \frac{\varepsilon}{2} u. \tag{++}$$

On the other hand, from (*) and (+), we have

$$(x - x_o) - \left(\sum_{i=1}^{N} (b_i - a_i) y_i - \sum_{i \geq N+1} a_i y_i \right) \in v + v$$

$$\Rightarrow \sum_{i \geq N+1} b_i y_i \in \frac{\varepsilon}{2} u,$$

and this contradicts (++). Therefore, we find that $(x_o + w \cap w_o) \cap K \subset K \setminus B$ and so B is again closed in K. □

Theorem 15.3.9: Suppose $f_n \in X^*$, $n \geq 1$ and let B be a c.b. for the cone $K \equiv K\{x_n; f_n\}$ such that $B = K \cap f^{-1}\{1\}$ for some strictly positive linear functional f on X. Then the following statements are equivalent: (i) B is closed; (ii) $0 \notin \bar{B}$; and (iii) $f|K$ is continuous at the origin.

Proof: (i) ⇒ (ii) Immediate, since $0 \notin B$.

(ii) ⇒ (i) Apply Theorem 15.3.8.

(iii) ⇒ (ii) Since $f(B) = \{1\}$, for some u in $\mathcal{B}_T$, $K \cap u \subset f^{-1}[(-1,1)]$; therefore $B \cap u = \emptyset$; that is, $0 \notin \bar{B}$.

(ii) ⇒ (iii) In view of (ii), we find some u in $\mathcal{B}_T$ with $B \cap u = \emptyset$. Let $\varepsilon > 0$ and $x \in K \cap (\varepsilon u)$. There exist a unique $\alpha > 0$ and $b \in B$ with $x = \alpha b$. Hence $b \in \alpha^{-1} \varepsilon u$, where $\varepsilon/\alpha > 1$. Therefore, $|f(x)| = \alpha < \varepsilon$ for all $x \in K \cap (\varepsilon u)$. □

Canonical and closing cones can be recovered in the following results.

Proposition 15.3.10: Let the cone K be metrizable and complete. Then for every positive linear functional f on X, $f|K$ is continuous at zero.

Proof: The topology on K may be defined by a sequence $\{p_n\}$ of 0-continuous pseudonorms on K with $p_1 \prec p_2 \prec \dots \prec p_n \prec \dots$. If $f|K$ is not continuous at zero, we can find $\{y_n\}$ in K with $p_n(y_n) < 1/2^n$ and $f(y_n) > 1$ for $n \geq 1$. Here $\sum_{n \geq 1} y_n$ converges to y in K. Choose N so that $f(y) < N$. As $\sum_{n \geq N+1} y_n \in K$, we find $f(\sum_{n \geq N+1} y_n) < N-N = 0$, a contradiction. □

Proposition 15.3.11: Let the cone K be metrizable and complete, then K is a canonical cone.

Proof: If B is any c.b., then $B = K \cap f^{-1}\{1\}$ for some strictly positive functional f on X by Proposition 1.5.17. Now make use of Proposition 15.3.10, Theorem 15.3.9 (iii) ⇒ (ii) and Theorem 15.3.5. □

Proposition 15.3.12: Let $\{x_n;f_n\}$ be an S.b. for its cone K which is assumed to be metrizable and complete. Then K is closing. In particular, every cone associated with a t.b. $\{x_n;f_n\}$ for an F-space is closing.

Proof: This follows by Propositions 1.5.17 and 15.3.10, Theorems 15.3.9 and 2.2.12. □

To characterize an h.c.b., we first prove

Proposition 15.3.13: Let $f_n \in X^*$ and (X,T) an l.c. TVS. If $K-K$ is barrelled, then every c.c.b. B of K is an h.c.b., and hence a closed c.b.

Proof: As usual, we write the c.s. $\{y_n\}$ for $G(B)$ and denote by $\{g_n\}$ the corresponding s.a.c.f in X^*. Then for each x in K, $\sum_{n\geq 1} g_n(x) < \infty$. For x in $K-K$, let $f(x) = \lim \sum_{i=1}^{n} g_i(x)$; then $f \in (K-K)^*$. Denote by g the extension of f with $g \in X^*$. Since B is a c.c.b., $g(x) = 1$ for every x in B; that is, $B = K \cap g^{-1}\{1\}$. As K is clearly closed, the result follows. □

Theorem 15.3.14: Let (X,T) be barrelled and $\{x_n;f_n\}$ an S.b. for (X,T). Let B be a c.c.b. for $K \equiv K\{x_n;f_n\}$ and $\{y_n;g_n\}$ an S.b. with $\{y_n\} = G(B)$. Then B is an h.c.b. if and only if $\sum_{n\geq 1} g_n(x) < \infty$ for each x in X.

Proof: Let B be an h.c.b. Then $B = K \cap f^{-1}\{1\}$ for some strictly positive $f \in X^*$. Since $f(y_n) = 1$ for $n \geq 1$, $f(x) = \sum_{n\geq 1} g_n(x)$ for x in X. The converse follows on the lines of the proof of Proposition 15.3.13. □

The following example shows that a c.c.b. may not be an h.c.b; it also gives a cone which is canonical and closing but not a hyper cone.

Example 15.3.15: Recall the cone K of Example 11.3.3. Since $\sum_{n\geq 1} \langle\alpha,f^n\rangle < \infty$ for every α in c_0, $\{x_n\}$ is a c.s. of itself. Put $y^{2n-1} = x^{2n-1}/2$ and $y^{2n} = x^{2n}$, $n \geq 1$. Then $\{y^n\}$ is a c.s. of $\{x_n\}$ and the set $B = F(\{y^n\})$ is a closed c.b. of K; the last part follows by considering the s.a.c.f. $\{g^n\}$ corresponding to $\{y^n\}$ and using Proposition 15.2.2 along with the fact that $\{g^n\}$ is equicontinuous on c_0. Here B is not an h.c.b. Indeed, let $\alpha = \{0, -1/2, 0, -1/4, \ldots\}$, then $g^{2n-1}(\alpha) = 1/n$ and $g^{2n}(\alpha) = -1/2n$, $n \geq 1$. Hence $\sum_{n\geq 1} g^n(\alpha)$ diverges and so, from Theorem 15.3.14, B is not an h.c.b. By Propositions 15.3.11 and 15.3.12, K is canonical and closing.

15.4 BOUNDED CONIC BASES

In this last section, we study the impact of a *bounded conic base* (b.c.b.) on the corresponding c.s. and vice versa. Our fundamental assumption on (X,T), $\{x_n;f_n\}$ and K is the same as outlined at the end of Section 15.1. Let us begin with a simple

Proposition 15.4.1: Let B be a c.b. of K. Then B is $\sigma(X,X^*)$-bounded if and only if G(B) is $\sigma(X,X^*)$-bounded; in particular, if (X,T) is an l.c. TVS, then B is a b.c.b., that is, B is T-bounded if and only if G(B) is T-bounded.

Proof: In view of Mackey's theorem, we need to prove only the first part of this proposition. Since $G(B) \subset B$, it is enough to establish the sufficiency of the result. Put $\{y_n\} = G(B)$; then, for x in B and f in X*,

$$|f(x)| \leq \sum_{n \geq 1} \alpha_n |f(y_n)| \leq M,$$

since $x = \sum_{n \geq 1} \alpha_n y_n$ with $0 < \sum_{n \geq 1} \alpha_n \leq 1$ and $|f(y_n)| \leq M$, $n \geq 1$ for some $M > 0$. □

The existence of a b.c.b. is useful in determining the canonical character of the cone in the form of

Proposition 15.4.2: Let (X,T) be an l.c. TVS and K be ω-complete. If K has a b.c.b. B, then K is a canonical cone.

Proof: Let $\{y_n\}$ denote the c.s. of $\{x_n\}$ with $\{y_n\} = G(B)$ and $\{g_n\}$ the corresponding s.a.c.f. Since $\{g_n(u)\} \in \ell^1$ for u in K-K, the map $R: K\text{-}K \to \ell^1$ is clearly an injection, where $Ru = \{g_n(u)\}$. Making use of the boundedness of $\{y_n\}$ (cf. Proposition 15.4.1) and the ω-completeness of K, we find that R is an algebraic isomorphism from K-K onto ℓ^1. Since $Ry_n = e^n$, K-K and ℓ^1 are order isomorphic under R. By Proposition 15.3.11, R[K] is canonical and so is K. □

Exercise 15.4.3: Let the cone K have a b.c.b. Show that $\sum_{n \geq 1} |f_n(x)| p(x) < \infty$ for every x in K-K and p in D_T.

For the next result on the characterization of a b.c.b., it will be convenient to introduce

Definition 15.4.4: Let $\{y_n\}$ be an arbitrary sequence in a real TVS (Y,S). A sequence $\{z_n\}$ in Y is said to be a *normalization* of $\{y_n\}$ if there exists a sequence $\{\alpha_n\}$ with $\alpha_n > 0$, $n \geq 1$ so that $z_n = \alpha_n y_n$ for $n \geq 1$ and $\{z_n\}$ is regular and bounded in (Y,S). Further, $\{y_n\}$ is said to be *normalizable* if there exists a normalization $\{z_n\}$ of $\{y_n\}$.

<u>Theorem 15.4.5</u>: Let (X,T) be an l.c. TVS and $\{x_n\}$ be normalizable. Then the following statements are equivalent:

(i) K has a b.c.b.

(ii) $\sum_{n\geq 1} |f_n(x)|p(x_n) < \infty$ for each x in K-K and p in D_T.

(iii) Each normalization $\{y_n\}$ of $\{x_n\}$ is a c.s. of $\{x_n\}$.

(iv) For each normalization $\{y_n\}$ of $\{x_n\}$, there is a c.b. B of K with $y_n \in B$, $n \geq 1$.

(v) For some normalization $\{y_n\}$ of $\{x_n\}$, there is a c.b. B of K with $y_n \in B$, $n \geq 1$.

<u>Proof</u>: (i) ⇒ (ii) Cf. Exercise 15.4.3.

(ii) ⇒ (iii) Let $\{y_n\}$ be a normalization of $\{x_n\}$ with $y_n = \alpha_n x_n$, $n \geq 1$. Since $\{y_n\}$ is regular, $p_o(y_n) \geq 1$ for $n \geq 1$ and some p_o in D_T. If $\sum_{n\geq 1} \beta_n y_n$ converges in K, then

$$\sum_{n\geq 1} \beta_n p(y_n) = \sum_{n\geq 1} \alpha_n \beta_n \, p(x_n) < \infty \, , \ \forall \, p \in D_T .$$

Hence in particular, $\sum_{n\geq 1} \beta_n < \infty$.

(iii) ⇒ (iv) See Proposition 15.2.2.

(iv) ⇒ (v) Trivial.

(v) ⇒ (i) Here $\{y_n\} = G(B)$ by Proposition 15.2.4 and apply Proposition 15.4.1. □

For another characterization of a b.c.b., we require

<u>Definition 15.4.6</u>: Let $\{x_n;f_n\}$ be a b.s. for a Banach space $(X, \|\cdot\|)$ such that $\{x_n;f_n\}$ is an S.b. for the cone $K \equiv K\{x_n;f_n\}$. Then $\{x_n\}$ is said to be of *type* ℓ_+ if, for each normalization $\{y_n\}$ of $\{x_n\}$, there exists a constant μ depending upon $\{y_n\}$ such that

$$\mu \sum_{i=1}^{n} \alpha_i \leq \left\| \sum_{i=1}^{n} \alpha_i y_i \right\| ,$$

for any $n \geq 1$ and all choice of positive scalars $\alpha_1,\ldots,\alpha_n$.

Theorem 15.4.7: Let $\{x_n;f_n\}$ be a b.s. for a Banach space $(X, \|\cdot\|)$ with $\{x_n;f_n\}$ being an S.b. for the cone $K \equiv K\{x_n;f_n\}$. Then the following statements are equivalent:

(1) $\{x_n\}$ is of type ℓ_+.

(2) For each normalization $\{y_n\}$ of $\{x_n\}$, there exists a constant $L > 0$ such that

$$L \sum_{i=1}^{n} g_i(y) \leq \|y\| \,;\ \forall\, y \in K,$$

where $\{g_n\}$ is the s.a.c.f. corresponding to $\{y_n\}$.

(3) For some normalization $\{y_n\}$ of $\{x_n\}$, there exists $L > 0$ so that

$$L \sum_{i=1}^{n} g_i(y) \leq \|y\| \,,\ \forall\, y \in K.$$

(4) For some normalization $\{y_n\}$ of $\{x_n\}$, there exists a constant $L > 0$ such that

$$L \sum_{i=1}^{n} \alpha_i \leq \| \sum_{i=1}^{n} \alpha_i y_i \| ,$$

for each $n \geq 1$ and every choice of positive scalars $\alpha_1, \ldots, \alpha_n$.

(5) There exists a constant $M > 0$ such that

$$\sum_{n \geq 1} \|f_n(y)x_n\| \leq M \|y\| ,\ \forall\, y \in K.$$

Moreover, (1) through (5) are equivalent to the statements (i) through (v) of Theorem 15.4.5.

Proof: The implications (1) $\Rightarrow$ (2) $\Rightarrow$ (3) $\Rightarrow$ (4) are obvious.

(4) $\Rightarrow$ (5) Put $\beta = \sup \|y_n\|$. Then, for y in K and $n \geq 1$,

$$\sum_{i=1}^{n} \|g_i(y)y_i\| \leq \beta \sum_{i=1}^{n} g_i(y) \leq M \| \sum_{i=1}^{n} g_i(y)y_i \| ,$$

where $M = \beta/L$. Since $g_i(y)y_i = f_i(y)x_i$, $i \geq 1$, the inequality in (5) follows by letting $n \to \infty$.

For the last assertion, we find that (5) implies statement (ii) of Theorem

15.4.5. To complete the proof, it is, therefore, enough to show that (iv) of Theorem 15.4.5 implies (1).

(iv) ⇒ (1) Let $\{y_n\}$ be any normalization of $\{x_n\}$ and let B be the c.b. containing $\{y_n\}$. By Proposition 15.3.12, K is closing. Hence B is closed and so $0 \notin \bar{B}$; consequently, there exists $\mu > 0$ with $\|x\| \geqslant \mu$ for each x in B. Choose any finite set $\alpha_1,\ldots,\alpha_n$ of positive scalars with $\sum_{i=1}^{n} \alpha_i \neq 0$. Then $y = \sum_{i=1}^{n} \alpha_i y_i / \sum_{i=1}^{n} \alpha_i$ is in B and hence

$$\mu \sum_{i=1}^{n} \alpha_i \leqslant \| \sum_{i=1}^{n} \alpha_i y_i \| ,$$

and this proves (1). □

Note: Theorems 15.4.5 and 15.4.7 extend results in [173].

Unbounded conic bases

Having discussed bounded conic bases and some of their applications, we finally touch upon the other extreme of a b.c.b., namely, the existence of an *unbounded conic base* (u.c.b.). Let us begin with

Proposition 15.4.8: Let $\{x_n;f_n\}$ be a shrinking base for an l.c. TVS (X,T). Then the cone $K \equiv K\{x_n;f_n\}$ has no bounded and closed c.b.

Proof: Let K contain a b.c.b. B which is also closed. Put $G(B) = \{y_n\}$ and denote by $\{g_n\}$ the s.a.c.f. corresponding to $\{y_n\}$. Since B is bounded and $\{y_n;g_n\}$ is also a shrinking base, for each f in X* and $\varepsilon > 0$, there exists N in $\mathbb{N}$ such that

$$\sup_{x\in B} |(f - \sum_{i=1}^{n} f(y_i)g_i)(x)| \leqslant \varepsilon, \quad \forall \ n \geqslant N;$$

in particular, $\sup \{|f(y_n)|: n \geqslant N\} \leqslant \varepsilon$. Therefore $y_n \to 0$ in $\sigma(X,X^*)$ and so $0 \in B$. The contradiction arrived at proves the desired result. □

Proposition 15.4.9: Let (X,T) be an l.c. TVS. If $\{\gamma_n\}$ is a sequence of strictly positive scalars such that $\{\gamma_n x_n\}$ is regular in (X,T), then K has a u.c.b.

Proof: Invoking the proof of Proposition 15.2.7, we find that $\{2^n \gamma_n x_n\}$ is a c.s. of $\{x_n\}$, which is clearly unbounded. If B denotes the c.b. corresponding to this c.s., then B is unbounded by Proposition 15.4.1. □

Corollary 15.4.10: If (X,T) is normed, then K has a u.c.b.

Corollary 15.4.11: Let (X,T) be an l.c. TVS and let K contain a c.b. B which is closed. Then K contains a u.c.b.

For, the c.s. generated by B is regular.

Corollary 15.4.12: Let (X,T) be a Fréchet space and let K contain a c.b. B. Then K possesses a u.c.b.

For, let $\{y_n\}$ be the c.s. corresponding to B. If $\{y_n\}$ is not regular, we select a subsequence $\{y_{n_i}\}$ of $\{y_n\}$ such that $\sum_{i \geq 1} 1.y_{n_i}$ converges in K. Since $\sum_{i \geq 1} 1$ diverges, this contradicts the c.s. character of $\{y_n\}$.

Proposition 15.4.13: Let (X,T) be a Fréchet space. If K has a u.c.b., then $\{\gamma_n x_n\}$ is regular for some $\{\gamma_n\}$ with $\gamma_n > 0$, $n \geq 1$.

Proof: Let $\{y_n\}$ be the c.s. corresponding to the u.c.b. B. As above, $\{y_n\}$ is regular. □

Remark: The completeness in the Fréchet character of (X,T) in the foregoing result cannot be overlooked, for we have

Example 15.4.14: Consider ℓ^1 equipped with the l.c. topology T generated by q_i and q, where $q_i(\alpha) = |\alpha_i|$, $i \geq 1$, and

$$q(\alpha) = \sum_{n \geq 1} |\alpha_{2n}|, \; \alpha \in \ell^1.$$

Then $\{e^n;e^n\}$ is a t.b. for (ℓ^1,T). By Corollary 15.4.10, the cone K of $(\ell^1, \|\cdot\|_1)$ has a u.c.b. Observe that K is also the cone for (ℓ^1,T) and so the cone K of (ℓ^1,T) has a u.c.b. But, for any $\{\gamma_n\} \subset (0,\infty)$, $\gamma_{2n+1} e^{2n+1} \to 0$ in T and so $\{\gamma_n e^n\}$ is not regular.

Note: In [99], a cone K is called *well-based* if K has a b.c.b. B with $0 \notin \bar{B}$.

Consequently, Proposition 15.4.8 can be rephrased as "the cone of a shrinking base is not well-based". An alternative proof of this statement is also possible by making use of Proposition 12.2.2 and the result [99]: a cone K in an l.c. TVS (X,T) is well-based if and only if K* has a $\beta(X^*,X)$-interior point.

Proposition 15.4.8 and Corollary 14.4.10 extend similar results in [173].

References

[1] Amemiya, I. and Kōmura, Y., Über nicht-vollständige Montel Räume; Math. Ann. (1968), 273-277.

[2] Arsove, M.G., The Paley-Wiener theorem in metric linear spaces; Pac. J. Math., 10 (1960), 365-379.

[3] Arsove, M.G. and Edwards, R.E., Generalized bases in topological vector spaces; Studia Math., 19 (1960), 95-113.

[4] Bachelis, G.F. and Rosenthal, H.P., On unconditionally converging series and biorthogonal systems in Banach spaces; Pac. J. Math. 37 (1971), 1-5.

[5] Banach, S., Théorie des Operations Linéaires; Chelsea Pub. Co., New York, 1955.

[6] Bennett, G., A new class of sequence spaces with application in summability theory; J. reine ang. Math., 266 (1974), 49-75.

[7] Bennett, G. and Cooper, J.B., Weak bases in (F)- and (LF)-spaces; J. Lond. Math. Soc., 44 (1969), 505-508.

[8] Bennett, G. and Kalton, N.J., FK-spaces containing c_o; Duke Math. J., 39 (1972), 561-582.

[9] Berberian, S.K., Lectures in Functional Analysis and Operator Theory; Springer-Verlag, Berlin, 1974.

[10] Bessaga, C. and Pelczynski, A., On a class of B_o-spaces; Bull. Aca. Pol. Sc., 5 (4) (1957), 375-377.

[11] Bessaga, C. and Pelczynski, A., An extension of Krein-Milman-Rutman theorem concerning bases to the case of B_o-spaces; Bull. Aca. Pol. Sc.5 (4) (1957), 379-383.

[12] Bessaga, C. and Pelczynski, A., On bases and unconditional convergence of series in Banach spaces; Studia Math., 17 (1958), 151-164.

[13] Bessaga, C. and Pelczynski, A., Wlasnosci baz przestrzeniach typu B_o (Polish); Prace Mat., 3 (1959), 123-142.

[14] Bočkarev, S.V., Existence of a basis in the space of functions analytic in the disc and some properties of Franklin's system; Math. USSR Sbornik, 24 (1974), 1-16.

[15] Boland, P.J., Holomorphic Functions on Nuclear Spaces; Ser. B, No. 16, Dept. Anal. Mate.Uni. Santiago Compostela, Spain, 1976.

[16] Boland, P.J. and Dineen, S., Holomorphic functions on fully nuclear spaces; Bull. Soc. Math. (France), 106 (1978), 311-336.

[17] Bourbaki, N., Espaces Vectoriels Topologique, No. 1189; Hermann, Paris, 1953.

[18] Bourbaki, N., Espaces Vectoriels Topologique, No. 1229; Hermann, Paris, 1964.

[19] Carter, L.H., An order topology in ordered topological vector spaces; Trans. Amer. Math. Soc., 216 (1976), 131-144.

[20] Ciesielski, Z., Properties of the orthonormal Franklin systems; Studia Math., 23 (1965), 141-157.

[21] Ciesielski, Z., Properties of the orthonormal Franklin systems II; Studia Math., 27 (1966), 289-323.

[22] Ceǐtlin, J.M. Unconditional bases and semi-orderedness; AMS Translations, 90 (1970), 17-25.

[23] Cook, T.A. Schauder decompositions and semi-reflexive spaces; Math. Ann., 182 (1969), 232-235.

[24] Cook, T.A., Weakly equicontinuous Schauder bases; Proc. Amer. Math. Soc., 23 (1969), 536-537.

[25] Cook, T.A., Some examples of biorthogonal systems in locally convex spaces; Bull. Soc. Roy. Sc. Liège, 41 (1972), 151-154.

[26] Cook, T.A. and Ruckle, W.H., Absolute Schauder bases for C(X) with the compact-open topology; Proc. Amer. Math. Soc., 59 (1976), 111-114.

[27] Courage, W.H. and Davis, W.J., A characterization of M-bases; a preprint.

[28] Das, N.R., Bi-locally Convex Spaces and Schauder Decompositions, Ph.D. Dissertation, Ind. Inst. Tech., Kanpur, 1982.

[29] Davie, A.M., The approximation problem for Banach spaces; Bull. Lond. Math. Soc. 5 (1973), 261-266.

[30] Davis, W.J., Dual generalized bases in linear topological vector spaces; Proc. Amer. Math. Soc., 17 (1966), 1057-1063.

[31] Davis, W.J., Dean, D.W., and Lin, Bor-Luh., Bibasic sequences and norming basic sequences; Trans. Amer. Math. Soc., 176 (1973), 89-102.

[32] Davis, W.J., and Johnson, W.B., On the existence of fundamental and total bounded biorthogonal systems in Banach spaces; Studia Math., 45 (1973), 173-179.

[33] Day, M.M., The space L^p with $0 < p < 1$; Bull. Amer. Math. Soc. 46 (1940), 816-823.

[34] Day, M.M., On the basis problem in normed spaces; Proc. Amer. Math. Soc., 13 (1962), 655-658.

[35] Day, M.M., Normed Linear Spaces; Springer-Verlag, Berlin, 1973.

[36] De Grande-De Kimpe, N., Equicontinuous Schauder bases and compatible locally convex spaces; Indag. Math., 36 (1974), 276-283.

[37] De Grande-De Kimpe, N., On a class of locally convex spaces, with a Schauder basis; Indag. Math., 79 (1976), 307-312.

[38] De Grande-De Kimpe, N., On Λ-bases; J. Math. Anal. Appl., 53 (1976), 508-520.

[39] De Wilde, M., Réseaux dans les Espaces Linéaires à Semi-Normes; Mem. Soc. Roy. Sc., Liège, 18, 1969.

[40] De Wilde, M., On the equivalence of weak and Schauder basis (short summary); Studia Math., 38 (1970), 457.

[41] De Wilde, M., On the weak and Schauder decompositions, Studia Math., 41 (1972), 145-148.

[42] De Wilde, M., and Houet, C., On the increasing sequences of absolutely convex sets in locally convex spaces; Math. Ann.; 192 (1972), 257-261.

[43] Dieudonné, J., On the biorthogonal systems; Michi. Math.J., 2 (1953), 7-20.

[44] Drewnowski, L., The weak basis theorem fails in non-locally convex F-spaces; Can. J. Math., 29 (1977), 1069-1071.

[45] Drewnowski, L., F-spaces with a basis which is shrinking but not hypershrinking; Studia Math., to appear.

[46] Dubinsky, Ed., The Structure of Nuclear Fréchet Spaces, LN 720; Springer-Verlag, Berlin, 1979.

[47] Dubinsky, Ed. and Ramanujan, M.S., λ-Nuclearity, Mem. Amer. Math. Soc., 128, Rhode Island, 1972.

[48] Dubinsky, Ed. and Retherford, J.R., Bases in compatible topologies; Studia Math., 28 (1967), 221-226.

[49] Dubinsky, Ed. and Retherford, J.R., Schauder bases and Köthe sequence spaces; Trans. Amer. Math. Soc., 130 (1968), 265-280.

[50] Dugundji, J., Topology; Allyn and Bacon, Boston, 1966.

[51] Dunford, N., and Schwartz, J.T., Linear Operators, Pt. I; Interscience Pub., New York, 1966.

[52] Duren, P.L., Theory of H^p-spaces; Academic Press, New York, 1970.

[53] Duren, P.L., Romberg, B.W., and Shields, A.L., Linear functionals on H^p-spaces; J. reine ang. Math., 238 (1969), 32-60.

[54] Edwards, R.E., Functional Analysis; Holt, Rinehart and Winston, New York, 1965.

[55] Enflö, P., A counterexample to the approximation problem in Banach spaces; Acta Math., 130 (1973), 309-317.

[56] Fefferman, C., A Radon-Nikodym theorem for finitely additive set functions; Pac. Math. J., 23 (1967), 35-45.

[57] Figiel, T. and Pelczynski, A., On Enflö's method of construction of Banach spaces without the approximation property; Preprint.

[58] Floret, K., Bases in sequentially retractive limit spaces, Colloq. Nuclear Spaces and Ideals in Operator Algebras; Studia Math., 38 (1970), 221-226.

[59] Foguel, S.R., Biorthogonal systems in Banach spaces; Pac. J. Math., 7 (1957), 1065-1072.

[60] Franklin, P., A set of continuous orthogonal functions; Math. Ann., 100 (1928), 522-529.

[61] Frink, O. Jr. Series expansions in linear vector spaces; Amer. J. Math., 63 (1941), 87-100.

[62] Fullerton, R.E., Geometric properties of a basis in a Banach space; Proc. ICM, Amsterdam, 1954; North-Holland, Groningen, 1954, p. 109.

[63] Fullerton, R.E., Geometric structure of absolute basis system in a linear topological space; Pac. J. Math., 12 (1962), 137-147.

[64] Garling, D.J.H., The β-and γ-duality of sequence spaces; Proc. Camb. Phil. Soc., 63 (1967), 963-981.

[65] Garling, D.J.H., On topological sequence spaces; Proc. Camb. Phil. Soc., 63 (1967), 997-1019.

[66] Garling, D.J.H., and Kalton, N.J., Proc. Int. Colloq. Nuclear Spaces and Ideals in Operator Theory, Pt. III; Studia Math., 38 (1970), 474.

[67] Gelbaum, B.R., Expansions in Banach spaces; Duke Math. J., 17 (1950), 187-196.

[68] Gelbaum, B.R., Notes on Banach spaces and bases; An. Aca. Brasil Ci., 30 (1958), 29-36.

[69] Goffman, C., and Pedrick, G., First Course in Functional Analysis; Prentice-Hall, Englewood Cliffs, 1965.

[70] Goldberg, R.R., Methods of Real Analysis; Blaisdell Pub. Co., Waltham, 1964.

[71] Gregory, D.A., and Shapiro, J.H., Nonconvex linear topologies with the Hahn-Banach extension property; Proc. Amer. Math. Soc., 25 (1970), 902-905.

[72] Grothendieck, A., Sur les applications linéaires faiblement compactes d'espaces du type C(K); Can. J. Math., 5 (1953), 129-173.

[73] Grothendieck, A., Produits Tensoriels Topologiques et Espaces Nucléaires; Mem. Amer. Math. Soc. 16, Rhode Island, 1955.

[74] Grothendieck, A., Topological Vector Spaces; Gordon and Breach, New York, 1975.

[75] Gupta, M., and Das, N.R., Bi-locally convex spaces; Portug. Math., to appear.

[76] Gupta, M., and Kamthan, P.K., Quasi-regular orthogonal systems of subspaces; Proc. Roy. Irish Ac., 80A (1980), 79-83.

[77] Gupta, M., and Kamthan, P.K., Dominating sequences and functional equations; to appear in Period. Math. Hungr., 15 (1984), 31-41.

[78] Gupta, M., and Kamthan, P.K., Unconditional bases and the order structure; Preprint.

[79] Gupta, M., Kamthan, P.K., and Das, N.R., Bi-locally convex spaces and Schauder decompositions; Ann. Mat. p.ed. appl. (IV) 33, (1983), 267-284.

[80] Gupta, M., Kamthan, P.K., and Ruckle, W.H., Symmetric sequence spaces, bases and applications; J. Math. Anal. Appl., to appear.

[81] Gupta, M., Kamthan, P.K., and Ruckle, W.H., Schauder Decompositions and Applications; a monograph under preparation.

[82] Gupta, M., Kamthan, P.K., and Sofi, M.A., A note on Nikoĺskiĭ's inequality; to appear in Period. Math. Hungr., 15 (1984), 25-30.

[83] Gurevich, L.A., On unconditional bases (Russian); Uspehi Mate. Nauk (N.S.), 8 (1953), 153-156.

[84] Gurevich, L.A., Conic tests for bases of absolute convergence; Problems of Math. Phy. and Theory of Functions (II) (Russian); Naukova Dumka, Kiev, 1964, p. 12-21.

[85] Halmos, P.R., Measure Theory; Van Nostrand, Princeton, 1950.

[86] Hampson, J.K., and Wilansky, A., Sequences in locally convex spaces; Studia Math., 65 (1973), 221-223.

[87] Hardy, G.H., and Littlewood, J.E., Some properties of fractional integrals II; Math. Zeit., 34 (1932), 403-439.

[88] Heins, M., Complex Function Theory; Academic Press, New York, 1966.

[89] Hewitt, E., and Ross, K.A., Abstract Harmonic Analysis I; Springer-Verlag, Berlin, 1963.

[90] Hewitt, E., and Stromberg, K., Real and Abstract Analysis; Springer-Verlag, Berlin, 1965.

[91] Hofler, J.T., Continuous lattice ordering by Schauder basis cones; Proc. Amer. Math. Soc., 30 (1971), 527-532.

[92] Hoffman, K., Banach spaces of Analytic Functions; Prentice-Hall, Englewood Cliffs, 1965.

[93] Horváth, J., Topological Vector Spaces and Distributions I; Addison-Wesley, Reading, 1966.

[94] Hyers, D.H., Pseudonormed linear spaces and Abelian groups; Duke Math. J., 5 (1939), 628-634.

[95] Issacs, M.I., Character Theory of Finite Groups; Academic Press, New York, 1976.

[96] Iyer, V.G., On the space of integral functions I; J. Ind. Math. Soc., 12 (1948), 13-40.

[97] Iyer, V.G., On the space of integral functions II; Quart. J. Math. Oxford Series, 1 (1950), 86-96.

[98] James, R.C. Bases and reflexivity of Banach spaces I; Ann. Math., 52 (1950), 518-527.

[99] Jameson, G., Ordered Linear Spaces, LN. 141; Springer-Verlag, Berlin, 1970.

[100] Johnson, W.B. Markuschevich bases and duality theory; Trans. Amer. Math. Soc., 149 (1970), 171-177.

[101] Johnson, W.B., On the existence of strongly summable Markuschevich bases in Banach spaces; Trans. Amer. Math. Soc., 157 (1971), 481-486.

[102] Johnson, W.B., and Rosenthal, H.P., On w*-basic sequences and their applications to the study of Banach spaces; Studia Math., 43 (1972), 77-92.

[103] Jones, O.T., Continuity of seminorms and linear mappings on a space with a Schauder basis; Studia Math., 34 (1970), 121-126.

[104] Julia, G., Exemples de structure des systèmes duaux de l'espace hilbertien; C.R. Ac. Sc., Paris, 216 (1943), 465-468.

[105] Kaczmartz, S., and Steinhaus, H., Orthogonalreihen; Pol. Aca. Scs., Warszawa, 1935.

[106] Kadec, M.I., Biorthogonal systems and summation bases (Russian); Izdat. Akad. Nauk Azerbaidzan, Baku, (1961), 106-108.

[107] Kadec, M.I., and Pelczynski, A., Basic sequences, biorthogonal systems and norming sets in Banach and Fréchet spaces (Russian); Studia Math., 25 (1965), 297-323.

[108] Kalton, N.J., A barrelled space without a basis; Proc. Amer. Math. Soc., 26 (1970), 465-466.

[109] Kalton, N.J., Schauder decompositions and completeness; Bull. Lond. Math. Soc., 2 (1970), 34-36.

[110] Kalton, N.J., Unconditional and normalized bases; Studia Math., 38 (1970), 243-253.

[111] Kalton, N.J., Schauder bases and reflexivity; Studia Math., 38 (1970), 255-266.

[112] Kalton, N.J., Normalization properties of Schauder bases; Proc. Lond. Math. Soc., 22 (1971), 91-105.

[113] Kalton, N.J., Some forms of the closed graph theorem; Proc. Camb. Phil. Soc., 70 (1971), 401-408.

[114] Kalton, N.J., Mackey duals and almost shrinking bases; Proc. Camb. Phil. Soc., 74 (1973), 73-81.

[115] Kalton, N.J., Basic sequences in F-spaces and their applications, Proc. Edin. Math. Soc., 19 (1974), 151-167.

[116] Kalton, N.J., Orlicz sequence spaces without local convexity; Proc. Camb. Phil. Soc., 81 (1977), 253-277.

[117] Kalton, N.J., Quotients of F-spaces; Glasgow. Math. J. 19 (1978), 103-108.

[118] Kalton, N.J. and Shapiro, J.H., Bases and basic sequences in F-spaces; Studia Math., 56 (1976), 47-61.

[119] Kamke, E., Theory of Sets; Dover Pub., New York, 1950.

[120] Kamthan, P.K., A note on the space of entire functions; Lab. J. Sc. Tech., 7A (1969), 144-145.

[121] Kamthan, P.K., Various topologies in the space of entire functions, Lab. J. Sc. Tech., 9 (1971), 143-151.

[122] Kamthan, P.K., A study on the space on entire functions of several variables; Yokohama Math. J., 21 (1973), 11-20.

[123] Kamthan, P.K., Regular and bounded bases; Tamkang J. Math., 7 (1976), 203-205.

[124] Kamthan, P.K., Normalized bases in topological vector spaces; Portug. Math., 38 (1979), 55-65.

[125] Kamthan, P.K., A lemma on the convergence of linear maps; Preprint.

[126] Kamthan, P.K. and Gupta, M., Expansion of entire functions of several complex variables having finite growth; Trans. Amer. Math. Soc., 192 (1974), 371-382.

[127] Kamthan, P.K., and Gupta, M., Characterization of bases in topological vector spaces; Tamkang. J. Math., 7 (1976), 51-55.

[128] Kamthan, P.K., and Gupta, M., Schauder bases and sequential duals; Bull. Roy. Soc. Sc. Liège, 46 (1977), 153-155.

[129] Kamthan, P.K., and Gupta, M., Uniform bases in locally convex spaces; J. reine ang. Math., 295 (1977), 208-213.

[130] Kamthan, P.K., and Gupta, M., Weak Schauder bases and completeness; Proc. Roy. Irish Ac. 78A (1978), 51-54.

[131] Kamthan, P.K., and Gupta, M., Schauder decompositions and their applications to continuity of maps; J. reine ang. Math., 298 (1978), 104-107.

[132] Kamthan, P.K., and Gupta, M., Sequence Spaces and Series, LN65; Marcel Dekker, New York, 1981.

[133] Kamthan, P.K., and Gupta, M., Behaviour and Stability of Schauder Bases; a monograph under preparation.

[134] Kamthan, P.K., and Ray, S.K., Decompositions in topological vector spaces; Ind. J. Pure Appl. Math., 6 (1975), 237-246.

[135] Kamthan, P.K., and Ray, S.K., On decompositions in barrelled spaces; Colloq. Math., 34 (1975), 73-79.

[136] Kamthan, P.K., and Sofi, M.A., λ-bases and λ-nuclearity; J. Math. Anal. Appl., 99 (1984), 164-188.

[137] Kantrovič, L.V., Vulih, B.Z., and Pinsker, A.G., Partially ordered groups and partially ordered linear spaces; Uspehi Mat. Nauk, 6 (1951), 31-98; Eng. Transl. Amer. Math. Soc., 27 (1963), 51-124.

[138] Karlin, S., Positive operators; J. Math. Mech., 8 (1959), 907-937.

[139] Kelley, J.L., General Topology, Van Nostrand, Princeton, 1955.

[140] Kelley, J.L., and Namioka, I., Linear Topological Spaces; Van Nostrand, Princeton, 1963.

[141] Khue, N.V., Tests for convergence basis (Russian); Izv. Vysš Učebn. Zaved Mate. 69 (1968), 68-74.

[142] Klee, V.L., On the Borelian and projective types of linear subspaces; Math. Scand., 6 (1958), 189-199.

[143] Knowles, R.J., Schauder Bases, Bounded Finiteness and Summability Bases; Dissertation, Uni. Massachusetts, Amherst, 1972.

[144] Knowles, R.J., The strongest locally convex topology consistent with a Schauder basis; Bull. Roy. Soc. Sc. Liège, 44 (1975), 165-168.

[145] Knowles, R.J., and Cook, T.A., Incomplete reflexive spaces without Schauder bases; Proc. Camb. Phil. Soc., 74 (1973), 83-86.

[146] Köthe, G., Topological Vector Spaces I; Springer-Verlag, Berlin, 1969.

[147] Köthe, G., Topological Vector Spaces II; Springer-Verlag, Berlin, 1979.

[148] Krasnoselskii, M.A., Positive Solutions of Operator Equations; P. Noordoff, N.V. Groningen, 1964.

[149] Krein, M., Milman, D., and Rutman, M., On a property of a basis in a Banach space (Russian); Khark. Zap. Mat. Obsh., 4 (1940), 182.

[150] Kreyszig, E., Advanced Engineering Mathematics, John Wiley and Sons, New York, 1968.

[151] Levin, M. and McArthur, C.W., Order characterizations of unconditional and absolute Schauder bases; Rev. Roum. Math. Pures. Appl., 18 (1973), 53-63.

[152] Levin, M. and Saxon, S., A note on the inheritance properties of locally convex spaces by subspaces of countable codimension; Proc. Amer. Math. Soc., 29 (1971), 97-102.

[153] Lindenstrauss, J., and Pelczynski, A., Contributions to the theory of classical Banach spaces; J. Func. Anal., 8 (1971), 225-248.

[154] Lindenstrauss, J., and Tzafriri, L., Classical Banach Spaces I; Springer-Verlag, Berlin, 1977.

[155] Lohman, R.H. and Stiles, W.J., On separability in linear topological spaces; Proc. Amer. Math. Soc., 42 (1974), 236-237.

[156] Lusternik, L.A., and Sobolev, V.J., Elements of Functional Analysis; Hindustan Pub. Co., Delhi, 1974.

[157] Markushevich, A.I., Sur les bases (au sens) dans les espaces linéaires; Dokl. Adad Nauk, 41 (1943), 227-229.

[158] Markushevich, A.I., Sur les bases dans l'espace des fonctions analytiques (Russian); Mat. Sbornik; 17 (1945), 211-252; Eng. Transl. Amer. Math. Soc., 22 (1962), 1-42.

[159] Markushevich, A.I., Theory of Functions of a Complex Variable, Vol.I; Prentice-Hall, Englewood Cliffs, 1965.

[160] Markushevich, A.I., Theory of Functions of a Complex Variable, Vol. II; Prentice-Hall, Englewood Cliffs, 1965.

[161] Markushevich, A.I., Theory of Functions of a Complex Variable, Vol. III; Englewood Cliffs, 1967.

[162] Marti, J.T., Theory of Bases; Springer-Verlag, Berlin, 1969.

[163] Marti, J.T., On locally convex spaces ordered by a basis; Math. Ann., 195 (1971), 79-86.

[164] McArthur, C.W., The weak basis theorem; Colloq. Math., 17 (1967), 71-76.

[165] McArthur, C.W., A convergence criterion with applications in locally convex spaces; Duke Math. J., 34 (1967), 193-200.

[166] McArthur, C.W., The projective equicontinuous topology, Proc. Conf. Projections, Clemson Uni., 1967; A Preprint.

[167] McArthur, C.W., Convergence of monotone nets in ordered topological vector spaces; Studia Math., 34 (1970), 1-16.

[168] McArthur, C.W., In what spaces is every closed normal cone regular; Proc. Edin. Math. Soc, 17 (1970), 121-125.

[169] McArthur, C.W., Developments in Schauder basis theory, Bull. Amer. Math. Soc. 78 (1972), 877-908.

[170] McArthur, C.W., Advanced Topics in Functional Analysis; Florida St. Uni., 1972/73 (unpublished lecture notes).

[171] McArthur, C.W., and Retherford, J.R., Uniform and equicontinuous Schauder bases of subspaces; Can. J. Math., 17 (1965), 207-212.

[172] McArthur, C.W. and Retherford, J.R., Some applications of an inequality in locally convex spaces; Trans. Amer. Math. Soc., 137 (1969), 115-123.

[173] McArthur, C.W., Singer, I., and Levin, M., On the cones associated with biorthogonal systems; Can. J. Math., 21 (1969), 1206-1217.

[174] Mitiagin, B.S., Approximative dimension and bases in nuclear spaces; Russian Math. Surveys, 16 (1961), 59-127.

[175] Namioka, I., Partially Ordered Linear Topological Spaces, 24; Mem. Amer. Math. Soc., Rhode Island, 1957.

[176] Natanson, I.P., Constructive Function Theory III; Frederick Ungar Pub. Co., New York, 1965.

[177] Newns, F., On the representation of analytic functions by infinite series, Phil. Trans. Roy. Soc., 245 (1953), 429-468.

[178] Nikol'skiĭ, V.N., The best approximation and a basis in Fréchet spaces (Russian); Dokl. Akad. Nauk SSSR, 59 (1948), 639-642.

[179] Pelczynski, A., A note on the paper by I. Singer "Basic sequences and reflexivity of Banach spaces"; Studia Math., 21 (1962), 371-374.

[180] Pelczynski, A., A proof of Eberlein-Šmulian Theorem by an application of basic sequences; Bull. Ac. Polon. Sc., 12 (1964), 543-548.

[181] Pelczynski, A., Some problems on Banach spaces and Fréchet spaces; Israel J. Math., 2 (1964), 132-138.

[182] Pelczynski, A., Some open questions in functional analysis; dittoed lecture notes, Louisiana St. Uni., 1966.

[183] Pelczynski, A., and Rosenthal, H.P., Localization technique in L^p-spaces; Studia Math., 52 (1974/75), 263-289.

[184] Peressini, A.L., Ordered Topological Vector Spaces; Harper and Row, New York, 1967.

[185] Phelps, R.R. Lectures on Choquet's Theorems; Van Nostrand, Math. Studies 7, New York, 1966.

[186] Pietsch, A., Nuclear Locally Convex Spaces; Springer-Verlag, Berlin, 1972.

[187] Potepun, A.V., Generalized regularity of K-lineals and the properties of the order topology; Dokl. Akad. SSSR, 207 (1972), 541-543.

[188] Przeworka-Rolewicz, D., and Rolewicz, S., Equations in Linear Spaces; PWN, Warszawa, 1968.

[189] Ray, S.K., Decompositions of Topological Vector Spaces; Ph.D. Dissertation, Indian Inst. Tech., Kanpur, 1974.

[190] Retherford, J.R., w*-bases and bw*-bases in Banach spaces; Studia Math., 25 (1964), 65-71.

[191] Retherford, J.R., and McArthur, C.W., Some remarks on bases in linear topological spaces; Math. Ann., 164 (1966), 38-41.

[192] Robertson, A.P., and Robertson, W., Topological Vector Spaces; Camb. Uni. Press, Cambridge, 1966.

[193] Robertson, W., Contribution to the General Theory of Linear Topological Spaces; Thesis, Camb. Uni., Cambridge, 1954.

[194] Robinson, W., Extension of basis sequences in Fréchet spaces; Studia Math., 45 (1973), 1-14.

[195] Ruckle, W.H., The infinite sum of closed subspaces of an F-space; Duke Math. J., 31 (1964), 543-554.

[196] Ruckle, W.H., Representation and series summability of complete biorthogonal sequences; Pac. J. Math., 34 (1970), 511-528.

[197] Ruckle, W.H., Topologies on sequence spaces; Pac. J. Math., 42 (1972), 235-249.

[198] Ruckle, W.H., Sequence Spaces, RN. 49; Pitman, London, 1981.

[199] Rudin, W., Real and Complex Analysis; McGraw-Hill, New York, 1966.

[200] Rudin, W., Functional Analysis, Tata McGraw-Hill, New Delhi, 1976.

[201] Russo, J.P., Monotone and e-Schauder Bases of Subspaces, Dissertation; Florida St. Uni., Tallahassee, 1965.

[202] Saxon, S., Basis Cone Base Theory, Dissertation; Florida St. Uni., Tallahassee, 1965.

[203] Schaefer, H.H. Halbgeordnete lokalkonvexe Vektorräume; Math. Ann., 135 (1958), 115-141.

[204] Schaefer, H.H., Halbgeordnete lokalkonvexe Vektorräume II; Math. Ann., 138 (1959), 259-286.

[205] Schaefer, H.H., Topological Vector Spaces; Macmillan, New York, 1966.

[206] Schauder, J., Zur Theorie stetiger Abbildungen in Funktionalräumen; Math. Zeit., 26 (1927), 47-65.

[207] Schauder, J., Eine Eigenschaft des Haarschen Orthogonalsystems; Math. Zeit., 28 (1928), 317-320.

[208] Schmetterer, L., Introduction to Mathematical Statistics; Springer-Verlag, Berlin, 1974.

[209] Shapiro, J.H., Linear Functionals on Non-locally Convex Spaces, Dissertation; Uni. Michigan, Ann Arbor, 1969.

[210] Shapiro, J.H., Examples of proper, closed and weakly dense subspaces in non-locally convex F-spaces; Israel J. Math., 7 (1969), 369-380.

[211] Shapiro, J.H., Extension of linear functionals on F-spaces with bases; Duke Math. J., 37 (1970), 639-649.

[212] Shapiro, J.H., On the weak basis theorem in F-spaces; Can. J. Math., 26 (1974), 1294-1300.

[213] Shields, A.L., and Williams, D.L., Bounded projective duality and multipliers in spaces of analytic functions; Trans. Amer. Math. Soc., 162 (1971), 287-302.

[214] Singer, I., Weak* bases in conjugate Banach spaces; Studia Math., 21 (1961), 75-81.

[215] Singer, I., Basic sequences and reflexivity of Banach spaces; Studia Math., 21 (1962), 351-369.

[216] Singer, I., Weak* bases in conjugate Banach spaces II; Rev. Roum. Math. Pures Appl., 8 (1963), 575-584.

[217] Singer, I., On the basis problem in topological linear spaces; Rev. Roum. Math. Pures Appl., 10 (1965), 453-457.

[218] Singer, I., Bases in Banach Spaces I; Springer-Verlag, Berlin, 1970.

[219] Singer, I., On minimal sequences of type ℓ_+ and bounded biorthogonal systems in Banach spaces; Czech. Math. J., 23 (1973), 11-14.

[220] Singer, I., On the extension of basic sequences to bases; Bull. Amer. Math. Soc., 80 (1974), 771-772.

[221] Smith, S.W., Cone relationships of biorthogonal systems; Pac. J. Math., 35 (1970), 787-794.

[222] Stiles, W.J., On the properties of subspaces of ℓ_p $(0 < p < 1)$; Trans. Amer. Math. Soc., 149 (1970), 405-415.

[223] Stiles, W.J., Some properties of ℓ_p, $0 < p < 1$; Studia Math., 42 (1972), 109-119.

[224] Taylor, A.E., Introduction to Functional Analysis; John Wiley and Sons, New York, 1963.

[225] Taylor, A.E., General Theory of Functions and Integration; Blaisdell, Waltham, 1965.

[226] Terzioğlu, T., Die diametrale Dimension von lokalkonvexen Räumen; Collect. Math., 20 (1969), 49-99.

[227] Titchmarsh, E.C., Theory of Functions; Oxford Uni. Press, Oxford, 1939.

[228] Tumarkin, Ju. B., On locally convex spaces with basis; Soviet Math. Dokl., 11 (1970), 1672-1675.

[229] Vulikh, B.Z., Introduction to the Theory of Partially Ordered Spaces; P. Noordhoff, Groningen, 1967.

[230] Webb, J.H., Sequential convergence in locally convex spaces; Proc. Camb. Phil. Soc., 64 (1968), 341-364.

[231] Webb, J.H., Schauder bases and decompositions in locally convex spaces; Camb. Phil. Soc., 76 (1974), 145-152.

[232] Weill, L.J., Unconditional Bases in Locally Convex Spaces, Dissertation; Florida St. Uni., 1966.

[233] Weill, L.J., Unconditional and shrinking bases in locally convex spaces; Pac. J. Math., 29 (1969), 467-483.

[234] Wilansky, A., Functional Analysis; Blaisdell, New York, 1964.

[235] Wilansky, A., Modern Methods in Topological Vector Spaces; McGraw-Hill, New York, 1978.

[236] Wong, Y.C., and Ng. K-F., Partially Ordered Topological Vector Spaces; Clarendon Press, Oxford, 1973.

[237] Woods, P.C., Banach-Steinhaus Spaces, Decomposition of Operators in Locally Convex Spaces, Dissertation; Florida St. Uni., Tallahassee, 1972.

[238] Zippin, M., A remark on bases and reflexivity in Banach spaces; Israel J. Math., 6 (1968), 74-79.

[239] Zygmund, A., Trigonometric Series, Vol. I and II; Camb. Uni. Press, Cambridge, 1968.

ADDITIONAL REFERENCES

[240] Mil'man, V.D., and Tumarkin, Ju.B., Properties of sequences in locally convex spaces; Soviet Math. Dokl., 10 (1969), 63-66.

[241] Weill, L.J., Stability of bases in complete barrelled spaces; Proc. Amer. Math. Soc., 18 (6) (1967), 1045-1050.

[242] Lorentz, G.G., Approximation of Functions; Holt, Rinehart and Winston, New York, 1966.

Abbreviations

a.p. 55
a.w.S.b. topology 32

b.c.b. 229
bl.s. 139
b.m-convergent 13
b.m-S.b. 117
BP 19
b.s. 103
BSP 8
b.sy. 103

c.b. 20
c.c.b. 224
con(A) 1
c.s. 222

dim X 3
DP 16

e-S.b. or T-e.S.b. 94

HBEP 156
h.c.b. 224

IL-topology 6

l.b. TVS 5
l.c.s.s. 13
l.c. topology 5
l.c. TVL 18
l.c. TVS 5
LF-space 6
l.i. 138
l.s.s. 13
l.topology 4

M-base 108

n.b. 50

(O)-convergence 18
O l.c. TVS 16
OTVS 16
OVS 15

q.r. 114

s.a.c.f. 23
s.a.f. 138
S.b. 25
S.bl.e. 160
S.bl.s. 160
SIL-topology 6
S.p. 25
sp{A} 3
SRILF-space 6
s.s. 11
s-S.b. 117

t.b. 23
TVL 18

TVS 4

u.c.b. 233
u-convergent 13
u.q.r. 114
u-S.b. 117

VL 18

WBT 87
weak a.w.S.b. topology 32

Index

Amemiya-Kōmura, example of, 50, 51

Base (see also conic base)
 M-, 108
 M-dual, 108
 M-generalized, 108
 Markushevich, 108
 normal, 51
 Schauder, 25
  bounded multiplier, 117
  boundedly complete, 117
  conditional, 125
  e-, 56, 94
  γ-complete, 117
  $\mathfrak{S}$-uniform, 128
  shrinking, 117
  stretching, 128
  subseries, 117
  unconditional, 117
 topological, 23
  Schauder property of, 25
Bǒckarev's example, 79, 82

Completion, 5
Coefficient, Bǒckarev-Fourier, 82
Cone
 b-, 16
  strict, 16
 canonical, 224
 closing, 224
 hyper, 224
 minihedral, 16
 normal, 16
 regular, 16
  fully, 16
  ω-, 16
 ω-, 16
 $\mathfrak{S}$-, 16
 strict, 16
 well-based, 234
Conic base, 20
 bounded, 229
 canonical, 224
 hyper, 224
 unbounded, 233
  existence of, 233
Convergence
 (O)-, 18
 (r)-, 18
  diagonal property of, 19
 regulator of, 19
 (*)-, 18
Convergent series
 absolutely, 13
 bounded multiplier, 13
 subseries, 13
 unconditionally, 13
 unordered, 13

Decomposition property, 16
Disc algebra, 72
Dual

algebraic, 3
α-, 11
$\aleph_o$-, 52
bound, 13
finite, system, 51
Köthe, 11
M-, base, 108
M-, system, 114
order, 17
order bound, 17
sequential, 8
topological, 4

Embedding, canonical, 5
Enfló's example, 55, 58
Example of
Amemiya-Kōmura, 50, 51
Bočkarev, 79, 82
Enfló, 55, 58
Expansion operator, 23
Extreme ray, 15
Extreme subset, 15

Functional
coordinate
associated sequence of, 23
Minkowski, 5

Haar system, 75
Hahn-Banach, extension property, 156

Isomorphism, topological, 4

Köthe dual, 11

Lattice operation, 18

Map
coefficient, 106
continuity of, 40

Norm
pseudo, 4
semi, 4
consistent, 144
ω-, 144
F-, 4
lattice, 18, 199
monotone, 16
solid, 18

Operator
adjoint, 10, 24
algebraic, 10
continuity of, 43
expansion, 23, 103
dual, 103
remainder, 24
transpose of, 10
ω-span, 24
Order-
bounded, 15
complete, 16
convergence, 18
isomorphic, 16
Orthogonal complement, 106

Point, quasi-interior, 16
Property
approximation, 55
boundedness, 19
extension, Hahn-Banach, 156
(R), 16

Property of l.c. TVS
BSP-, 8
ω-, 8
S-, 8
Property of TVS
extension, Hahn-Banach, 156

Remainder operator, 24

Schauder-
base, 25
property, 25
system, 73
Schauder base (see Base)
Schauder system, 73
Sequence
associated with functionals, 138
coordinate, 23
basic, 66
Schauder, 66
semi, 132
biorthogonal, 103
absolute, 172
M-base, 108
M-dual base, 108
M-generalized base, 108
M-semibase, 108
quasi-regular, 114
unordered, 114
of type ℓ_+, 231
block, 139
extension of, 160
compansion, 222
complete, 66
independent, ω-linearly, 24
irregular, 5
length of, 11
n-th section of, 11
normal, 5
normalizable, 230
normalization of, 230
normalized, 5
positive, 11
strictly, 11
regular, 5
Sequence space, 11
AK-, 13
K-, 13
monotone, 11
normal, 11
perfect, 11
Series, unordered
bounded, 14
Cauchy, 14
(see also Convergent series)
Set
complete, 106
full, 16
fundamental, 106
norming, 143
order-closed, 19
order-convex, 16
orthogonal complement, 106
sums-limited, 41
total, 106
Space
block, 160
C-, 8
F-, 5
F*-, 5
finite relative to, 51
Fréchet, 5

fully complete, 7
inductive limit of, 6
 sequentially retractive, 6
 strict, 6
locally bounded, 5
locally convex, 5
Mazur, 8
minimal, 153
non-c_o, 131
non-ℓ^1, 129
nuclear, 10
 ℓ^1, 10
ω-barrelled, 8
ω-complete, 5
ω-separable, 49
ordered vector, 15
 Archimedean, 16
 locally convex, 16
 regularly ordered, 17
 topological, 16
  locally full, 16
  locally order-convex, 16
P-, 8
Pták, 7
S-, 8
semibornological, 8
σ-barrelled, 8
σ-infrabarrelled, 8
ultrabornological, 7
W-, 8
webbed, 7
 strictly, 7
System
 biorthogonal, 103
  complete, 106
  equivalent, 112
  extension of, 105
  M-base, 108
  M-dual base, 108
  M-generalized base, 108
  maximal, 105
  minimal, 104
  similar, 112
  total, 106
 finite dual, 51
 Haar, 75
 M-dual, 114
 Schauder, 73
 topologically free, 104

Theorem
 Banach-Mazur, 87
 completion, 36
 continuity, 28
 Eberlein-Smulian, extension of, 158
 Nachbin-Namioka-Schaefer, 20
 Nikol'skii, 68
 weak basis, 87
  failure of, 90
Topology
 associated, 32
  $\hat{\sigma}$-, 32
  $\tilde{\sigma}$-, 32
  $\tilde{T}$-, 32
 compatible, 148
 equivalence of, 6
 inductive limit, 6
  strict, 6
 linear, 4
 locally convex, 5
 normal, 11
 T_1-polar, 148

Unit vector, n-th, 11
Unity, 11

Vector lattice, topological, 18
 locally convex, 18
  T-order complete, 19

Weak order unit, 15
Web, 6
 absolutely closed, 6
 absolutely convex, 6
 strict, 6
 of type ℭ, 6
Wedge
 dual, 17
 generating, 15